Instruments for Physical Environmental Measurements

with special emphasis on atmospheric instruments

VOLUME I

Second Edition

J. Y. Wang, Ph.D.
Professor of Meteorology
San Jose State University
San Jose, California

Catherine M. M. Felton
Associate Professor of Meteorology
San Francisco State University
San Francisco, California

Copyright © 1975 by Milieu Information Service

Copyright © 1983 by Kendall/Hunt Publishing Company

Library of Congress Catalog Card Number: 83-81017

ISBN 0-8403-3098-7

KENDALL/HUNT PUBLISHING COMPANY
Dubuque, Iowa

Printed in the United States of America

B   403098   01

FOREWORD

During the preparation of this book, numerous environmental modeling techniques for the prediction and control of the geosphere were developed. Although not all models are useful, their objectives are rational. Since these models attempt to simulate natural mechanisms with the aid of mathematical and statistical formulations, they serve as a guide for field instrument operations. Today, instrumentation consists of a complete data acquisition system instead of a single device to sense an environmental parameter at a locale. In other words, modern instrumentation may include data collection, storage, processing, and dissemination involving information retrieval and communication. Strictly speaking, instrumentation is developed according to the modeling design. For example, in the recent Global Atlantic Tropical Experiment Program, known as GATE, several new instruments and platforms including electronic communication systems were developed. The GATE field investigation covered a 20 million square mile area of the tropical land and sea. Instruments on 40 ships, more than 60 buoys, 13 aircraft, 6 types of satellites, and at nearly a 1000 land stations observed and recorded weather and ocean phenomena from the top of the atmosphere to about 5000 feet below the sea surface. This gigantic program was possible only because of the well designed modeling system.

Over the past decades, instrumentation has progressed from a piecemeal to an interdisciplinary approach, as can be seen in the literature. And yet, to the best of our knowledge, few books were organized to cover the measurement techniques in the atmosphere, hydrosphere, and lithosphere in a single volume. In this book, the author attempts to stress a total environmental approach by giving an in-depth description of some basic scientific techniques as they apply to the various systems of the geosphere. In writing this, much effort was expended in the collection of up-to-date information on instruments and instrument systems from many sources.

The first volume of this book will focus on the fundamental knowledge pertaining to the measurement of the atmosphere, hydrosphere, and lithosphere of the earth. It covers the requirements and specifications as well as the data acquistion systems and basic principles and mechanisms of instruments, with emphasis on atmospheric measurement. The second volume will discuss environmental monitoring techniques and devices, such as those utilized in pollution measurement, radioactivity and noise monitoring, and earth and space environmental inventory.

This publication is intended to be a textbook in physical environmental instrumentation for the upper division and graduate levels of universities and colleges. It can also be used as a reference work for engineers and scientists who are interested in the practical aspects of geoscience instrumentation. Therefore, the primary intent of the author is to give the reader sufficient knowledge of basic instruments in terms of their principles and mechanisms. The secondary intent is to provide the necessary assumptions and theories, mainly expressed in mathematical and statistical formulations. Recognizing the vast dimensions of instrumentation in geoscience applications, this publication cannot encompass them all. As a remedy, characteristics of each instrument category are summarized, and pertinent references provided.

This book is a result of the combined efforts of many individuals and groups whose help and encouragement provided the author with the incentive to persevere. To express his gratitude, the author has made a list of names appearing in the Appendix (Volume II), in which the contributions of individuals and organizations are listed. Especially, the author wishes to thank Mr. Milt Silverstein of Milieu Information Service for his contributions and participation in various stages of manuscript editing and preparation. Appreciation is also extended to Mr. Arthur Bayce for his contribution related to diffusion monitoring, and a portion of geological measurements, and of environmental systems; and to Mr. Ballard W. George for his assistance in noise measurement. The author is deeply indebted to the staff and students of the Department of Meteorology at San José State University for their valuable assistance and encouragement on the various aspects of this book.

San José State University
San José, California

J.Y. Wang
1975

FOREWORD TO 2ND EDITION

By the end of the third year following its publication, all copies of the first edition had been sold out. Meanwhile, the author had received comments from several scientists and engineers in the United States and abroad, in addition to improvements he made on the text while using it in his classes. Also, additional information became available to the author with respect to innovations and improvements of geoscience instrumentation since the first edition.

To incorporate all of this new information, the publisher requested that the author revise the manuscript and, in so doing, convert it from syllabus to book format.

The original outline of the first edition was preserved, but numerous changes were embodied. This included the addition of new materials on the technologies of instruments and measurements. Also, improvements were made on the text, illustrations, and data source selections.

Last but not the least, the contributions made by Dr. Catherine M.M. Felton as co-author enhanced the quality of this new edition significantly. The competent work of Ms. Nancy E. Burns as editor, Ms. Nobuko Ishigaki as graphic artist and typist, and Ms. Sandy Batey and Ms. C.M. Mancino as typists are gratefully acknowledged.

J.Y. Wang

VOLUME I
TABLE OF CONTENTS

VOLUME II

Chapter

8   Air Quality and Atmospheric Acoustic Instruments

PART C   HYDROSPHERIC INSTRUMENTATION

9   Surface Water Measurements
10   Deep Ocean Measurements
11   Water Quality Measurements
12   Hydrologic Measurements

PART D   LITHOSPHERIC INSTRUMENTATION

13   Pedological Measurements
14   Geological Measurements

PART E   SPACE INSTRUMENTATION

15   Upper Atmospheric Measurements
16   Satellite Applications

PART F   SOME BASIC INFORMATION

17   The World Geophysical Stations
18   Data Acquisition Instrumentation
19   Future Prospectus
20   Epilogue

       Appendices
       Subject Index

INSTRUMENTS FOR PHYSICAL ENVIRONMENTAL MEASUREMENTS
With Special Emphasis on Atmospheric Instruments

CHAPTER 1   INSTRUMENTATION IN THE GEOSPHERE

Air, water and land comprise the physical surroundings of man. Collectively, they may be called the geosphere or the earth's environment, which includes three systems: atmosphere, hydrosphere, and lithosphere. The gaseous envelope of the earth is the atmosphere and the entire watery envelope of the earth is the hydrosphere. The solid crust and the cool, upper mantle of the earth are the lithosphere.

In order to understand the behavior of the geosphere, instruments are required for making observations and measurements of its physical and chemical properties. The dynamics and complexities of these properties are still not fully understood. The demand for more instrumental surveillance is obvious. Fortunately, the tendency toward research and development in new instruments is increasing and a huge inventory of instruments and instrumentation is now available. The reader, therefore, must familiarize himself with both the prototypic and commercial instruments for his own interests.

The first part of the present chapter will provide some fundamental information on instruments and instrumentation in general. The remainder of the chapter will deal with measurements in the atmosphere, hydrosphere and lithosphere, with special emphasis on the scope, characteristics and problems of these three systems.

1.1 Instruments and Instrumentation. An instrument is a device for making observations and measurements[1] and for the control of energy and mass of its immediate or distant surroundings, while instrumentation refers to the use and application of an instrument or instruments. Instrumentation may encompass one or more instruments in addition to the instrumental platform, communications and data acquisition facilities in order to make it practical. We shall examine the general functions of an instrument first and then discuss the general characteristics of instrumentation.

An instrument is a mechanical, electrical, optical, chemical or even biological device for measuring the atom, the world or the universe. It can be a combination of two or more of these devices. It can also be a controlling device, to an appreciable extent, of the human physical environment. Measurements are accomplished by sensing the signal from the environment represented by either energy or mass, or both, arriving at the sensing element (or sensor) of an instrument. The response or interaction of a sensor to a signal defines the

---

[1] While observations are usually made by direct human "vision," measurements are conducted through the use of tools or instruments. However, these terms are used quite loosely and alternatively.

basic principle of measurement. A mercury thermometer, for example, measures the sensible heat of the air (an energy signal) whereas a raingage usually measures the amount of rainfall (a mass signal). While the sensor of the former is the mercurial bulb, that of the latter is the gage receiver. A mercurial barometer, on the other hand, measures the atmospheric pressure exerted by the weight of a column of air extending from the mercury surface to the top of the atmosphere. It is, therefore, a measure of the force of gravity (energy) exerted on a unit of air column (mass) above the mercury surface. Thus, all instruments are designed for measuring mass and/or energy.

Two common mechanisms used in instrument design are classified as mechanical or electrical. They are illustrated schematically in Figures 1-1 and 1-2, respectively. In Figure 1-1, the input signal detected by a sensor, which

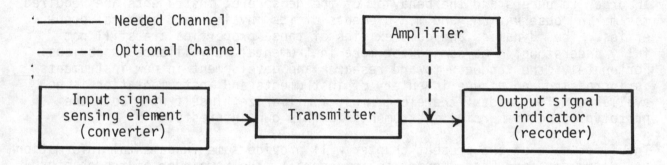

Figure 1-1 A Mechanical System of Instruments

often acts as a converter or transducer,[2] is conducted or transmitted by such mechanical means as fluid expansion or lever movement to an indicator. The transmission may or may not be accompanied by a magnification of the signal (i.e., with an amplifier). Finally the transmitted signal is detected through an indicator or registered by a recorder. While the former is a non-recording device and usually requires visual aid, the latter is commonly registered by a paper strip chart. An ordinary mercury thermometer is an example of a non-recording device; the heat flux from the air is the input signal, the mercury in the bulb is the sensor, the fine column in the thermometer stem is the transmitter, the thick glass wall of the stem itself is the amplifier and the thermometer scale is the measuring device or indicator. Obviously, the output signal of the indicating device has to be read visually. Here the transmission mechanism is the expansion or shrinkage of mercury in the bulb. In

---

[2]A converter is a device for converting energy from one form to another. It is also known as a transducer. For example, a thermocouple transduces heat energy into electric energy.

the thermograph, on the other hand, the bimetallic strip or Bourdon tube[3] is the sensor; a lever mechanism serves as the transmitter and amplifier; and the cylindrical revolving drum and chart together with its recording pen are the recording device. The chart is usually read visually but could be transcribed electronically.

In the electrical mechanism shown in Figure 1-2, the input signal is detected by a sensor, converted to an electrical signal by a transducer, magnified by an amplifier and electrically transmitted to the detector. A feedback device, or a servo system, is often employed so that the input signal will be checked for its exactness with the output signal of the detector before recording. Thus, any noise (the unwanted signal) that may be received from the sensor or produced by the electric or electronic system can be automatically eliminated. No sensor can be designed to be completely free of noise. However, a maximum signal to noise ratio is highly desirable for all instruments.

In addition to the instrument itself, instrumentation may include (a) one or more instrument platforms to expose the sensing devices, (b) telecommunication of data to appropriate locations, and (c) computer controlled data acquisition systems.

Two types of platforms are generally recognized: fixed and movable. Examples of fixed platforms on the earth's surface are ground-based stations, such as masts and towers over land and anchored buoys and ships over water. Fixed platforms aloft include high mountain stations, extremely high towers, tethered balloons and zero-gravity stations such as geostationary satellites. Strictly speaking, neither a geostationary satellite nor an anchored ship is a fixed platform because of its roll, pitch, yaw and elevational changes. Surface movable instrument platforms, on the other hand, are mainly mobile units or mobile stations over land, and ships and free buoys over the sea. Movable platforms aloft are free balloons (including constant-level, tracking, and pilot balloons, radiosondes, rawinsondes and dropsondes), aircraft and spaceships. For underwater platforms, submarines are mobile units whereas undersea laboratories are fixed stations.

With the aid of modern electronic technology, messages can be sent over long distances in a fraction of a second. Almost every point on the earth's surface and on the ocean floor as well as in outerspace is covered. Today the

---

[3] A bimetallic sensor is a curved strip formed by welding together two metals differing in their coefficients of expansion. The changes of curvature undergone by the strip with changes of temperature are used to actuate a recording pen by means of a lever mechanism. When applied to thermometry, the Bourdon tube sensor (a closed curved tube of elliptical cross-section) is filled with liquid. The expansion of liquid due to a temperature rise causes an increase in radius of curvature of the tube which, in turn, activates a recording pen through a lever mechanism. When applied to barometry, the Bourdon tube is evacuated.

4

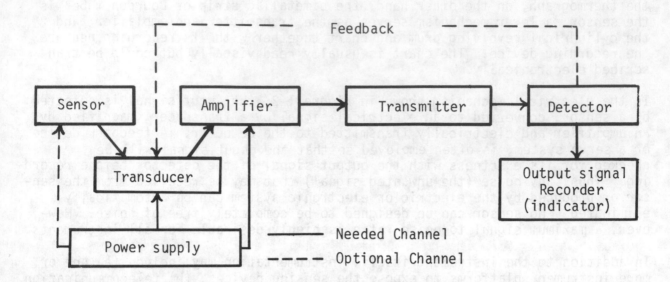

Figure 1-2 An Electrical System of Instruments

availability of electronic computers, radar systems, camera imagery and communication satellites, together with their navigational aids, facilitates
high speed data transmission. Various portions of the entire electromagnetic
spectrum are intensively utilized in these highly complex systems. Generally,
however, two major methods of data transmission are recognized: wire (or
cable) and non-wire (radio and other means). The wire system, which handles
the bulk of data transmission for most developed nations, still relies heavily
upon either the wire teletype or facsimile facilities over the ground. In the
non-wire system, radio transmission is the major tool for upper air and outerspace. In fact, most developed nations have their own national teletype,
radio and facsimile networks as illustrated by several maps shown in Chapter
18. However, data transmission by telephone and cable lines, and even by
mail, is still in use and remains necessary. More information on telecommunication networks will be given when a special type of instrumentation is
discussed, especially in Chapter 18.

As to data acquisition systems, the data obtained from an instrument are
readily recorded on paper strip charts, stored on punched tape, cards, micro-
film, or magnetic tapes, and displayed on a CRT (cathode ray tube) screen.
Computations from data stored on cards or tapes are accomplished directly with
an electronic computer. Automation of sensing, transducing, amplifying, telecommunicating, detecting, recording, processing, and even displaying data without the aid of human attention comprises an ideal instrumentation system, but
it is rather expensive in terms of both initial and maintenance costs. Nevertheless, instrumentation with a built-in computer facility enhances data reduction and statistical analysis, depending on the type of physical environmental measurements. A less expensive system for recording is the digital
printout device and for data acquisition, the micro-processor. More powerful,
compact and sophisticated computer systems are to be expected in the foresee-
able future. For a long run, however, a built-in computer is much more

economical than a strip chart recorder.

1.2 <u>Measurements in the Geosphere</u>. A comparison of the volumes and masses of the atmosphere and hydrosphere with the total volume and mass of the earth is shown in Table 1-1. As a fluid substance, the atmosphere is about 278 times greater by volume than the hydrosphere but the atmosphere is only 1/273 the mass of the hydrosphere. By having a large volume and small mass, the atmosphere is more susceptible to changes in its physical properties, such as temperature, pressure and humidity, than the hydrosphere. In fact, much of the change in the state (liquid, gas and solid) of the hydrosphere is governed by atmospheric conditions. Atmospheric interactions at its lower boundaries (ocean and land) and within itself comprise almost all of the human physical environment. Wind stress waves and soil erosion are examples of its interactions with hydrosphere and lithosphere, respectively. Interactions within the atmosphere through such processes as radiation, turbulence and convection, together with the Coriolis force and gravity, produce almost all the weather; and yet, these interactions are not fully understood. Therefore, measurements of various atmospheric properties and motions in the different spatial and time frameworks were the first assignment for the World Weather Watch of the World Meteorological Organization.[4]

Table 1-1 Estimates of the Volume and Mass of the Geosphere

Geospheric Systems	Volume $(x10^9 \ km^3)$	Mass $(x10^{21} \ kg)$	Bases for the Volume Estimation
Atmosphere	382.00	0.0052	Estimated for the first 60 km above MSL, constituting over 98% of the entire earth's atmosphere.
Hydrosphere	1.37	1.4300	Estimated for the average depth of 4 km for Pacific, Atlantic and Indian Ocean, constituting 97.2% of the world's free water. Most of the remaining 2.8% is in icecaps.
Lithosphere	1100.00	5.983	Estimated on the basis of the average radius of the earth, including mass of the atmosphere and ocean.

[4] The World Weather Watch (WWW) is a cooperative operation among 70 nations for making global weather observations and measurements under the auspices of World Meteorological Organization (WMO).

6

Atmospheric measurements have a greater need for high spatial and temporal resolution in order to understand atmospheric behavior. This is necessary because of its huge volume and high mobility. Some atmospheric constituents such as water vapor and atmospheric ions rapidly fluctuate, thereby causing additional problems. Presently, the world's weather stations cover only some 10% of the earth's surface and are far from being representative. Fortunately, with the advent of weather satellites during the last decade, global coverage is possible. Observations made from space platforms include cloud cover, temperature and moisture profiles of the atmosphere, and surface temperature of cloud tops, ocean, and land in addition to wind estimates. Moreover, with constant-level balloons and dropsondes as navigation aids, atmospheric pressure, wind and humidity signals can be obtained accurately by using satellites as relay stations. The main problem of satellite measurement is the ground truth required for verification and data interpretation.[5] Satellite systems continue to be developed. Information from operational satellites are now routinely used for a variety of applications including weather usage. With the assistance of the Global Atmospheric Research Program (GARP) as a component of the World Weather Watch of WMO, the improvement of our weather observation network is in the making.

As to the hydrosphere, water has a greater thermal capacity than either air or soil. Thus, the time and space variations of water temperature are far smaller than those of the air. While air is predominantly turbulent in flow, water is predominantly laminar in flow. For instance, the phenomenon of a permanent thermocline is a result of laminar flow and is observed in all oceans of the world, with the exception of the shallow mixing layer. The hydrosphere interacts with the lithosphere through two main processes, denudation and deposition, and interacts with the atmosphere through evaporation, precipitation and wind stress waves. In addition, the influences of oceanic currents on the climates of the continent are pronounced. Furthermore, since over 71% of the earth's surface is covered by water, the hydrosphere plays an important role in the overall human physical environment.

Most oceanological and limnological measurements[6] and observations are confined to the surface, particularly those which relate to air-water interactions. In comparison with the frequency, areal coverage, and instrumentation used for routine atmospheric observations, hydrological measurements are far less frequent, cover only a fraction of the oceans and utilize classical instruments. This is particularly true for deep ocean soundings which are made only sporadically and over limited areas or along a few navigation routes. However, sur-

---

[5] The ground truth refers to the calibration of remote sensing data against that of actual ground measurements; data interpretation refers to the logical, rational and statistical explanation of the signals or signatures obtained from a remote sensing device. This process also is called surface truth.

[6] Oceanological measurements sample the biological, physical and chemical properties of the world's oceans and seas; limnological measurements sample the above named properties of freshwater bodies. The properties of brackish and hypersaline water bodies are measured using techniques adapted from both oceanological and limnological fields.

face measurements taken remotely by satellite are providing detailed information about some of the hydrospheric variables. And yet, we know as much about the general distribution of the physical and chemical properties in the oceans as well as their motions, as we know about the atmosphere. This is true as the hydrosphere is much more homogeneous in its horizontal structure and far less mobile than the atmosphere. To gain the same amount of knowlege, perhaps less effort can be expended in oceanic and limnologic surveys than in atmospheric ones.

One of the major difficulties appearing in hydrospheric measurements is the choice of a platform. Marine observations, for example, are complicated by the rolling, pitching and yawing of ship and buoy. Accuracies of meteorological measurements such as those obtained by the use of mercury barometers, raingages and anemometers are greatly affected by motion on shipboard. This also applies to the measurements of waves and swells from these platforms. In addition, marine instruments are always subject to corrosion which further limits their accuracy. As there are few inhabitants over the ocean, the costs of installation and maintenance of marine stations are tremendous. On the contrary, sonar probes have been utilized in recent decades, of which the principle application has been the measurement of the topography of the ocean basins. The sonar system is less expensive than other available methods for this purpose.

Lithospheric measurements employ two distinct methods of approach: the <u>direct</u> and the <u>indirect</u>. The former is used primarily for the survey of soils and minerals, whereas the latter is used in geological surveys. Pedologists determine the physical, chemical and biological properties as well as the behavior of soil, whereas mineralogists are interested in the structure and the physical and chemical properties of rocks and minerals. Examples of the physical measurements of the soil profile are soil compaction and aeration, thermal characteristics, field capacity and moisture content; chemical measurements include the chemical composition and processes within organic and inorganic soils; and biological measurements encompass the amount and distribution of microbes and soil dwellers including small animals. Geologists, on the other hand, study the earth's crust, mantle and core, indirectly measuring its density, temperature, pressure and chemical composition. Indirect measurements are necessary because subsurface layers are inaccessible to direct measurement. In fact, the deepest bore holes in search of petroleum now penetrate only a little more than 6 km, which is less than one thousandth of the distance of the earth's center.

The bulk of measurements in the lithosphere have been conducted so far in inhabited areas, leaving vast areas of the continents and ocean basins untouched. Measurements in inhabited areas are further restricted to soil heterogeneity. With the exception of soil type identification and surface temperature measurements, most of the important physical variables have not been thoroughly investigated. Thus, the distributions of soil moisture, soil temperature and soil compaction or aeration are not fully understood in many parts of the continents. This is particularly true for soil air composition, including water vapor. For example, the concentration of carbon dioxide in the root zone, determining the photosynthesis/respiration process, has not been measured. Much remains to be done in soil environmental measurements. Geological surveys exist in nearly every nation of the world, yet the direct exploration of the earth's interior has not been fully achieved. Earthquake forecasts, for example, still remain in the experimental stage. Lithospheric measurements are far behind those of the atmosphere and hydrosphere. The surfaces of the continents are extremely heterogeneous in horizontal structure, while those of the ocean basins are more or less homogeneous. The interior of the earth is perhaps the most homogeneous due to its extremely high pressure and temperature; but its characteristics are not fully understood.

Despite the present shortcomings as described above, modern instrumentation has provided information about our environment totally unknown in the past. Electron microscopes, for example, allow observation of living cells at a magnification of 2,000,000 times their original size. Hence, many viruses and bacteria which cause diseases can be detected. Gamma ray mass spectrometers measure radiation from the nuclei of radioactive atoms. Missile systems, including controls for the rocket, have explored some of the mysteries of outerspace. And yet, much of man's physical environment on earth has not been fully explored.

The purpose of this book is then to introduce the reader to classical and modern instrumentation for the observation, measurement and control of the physical environment. Our mission throughout the book is to explore all aspects of man's physical environment in order to gain a better understanding of its ultimate role in shaping our future human environment.

Although biological measurements are not emphasized in this book, the impacts of living organisms on the three systems of the geosphere cannot be ignored. A typical example of this is the contribution by green plants and soils to evapotranspiration as measured by a lysimeter. Therefore, whenever necessary, biological measurements are included.

Finally, the overall presentation of this book is in a concise form in order to preserve the uniqueness of each chapter. Thus, when nomenclature in instrumentation is introduced, it is defined concisely without much elaboration. Since brief and comprehensive definitions for most terminology are impossible, the beginner may not learn the entire meaning of each term at once. But, he will grasp it gradually through its use in the remainder of this book. One learns through personal experience and not from a glossary or dictionary. Stated another way, one obtains the true meaning of each term through a working definition. It is therefore recommended that the reader refer back to either the definitions presented in the text or in the glossary (Appendix 2) whenever necessary. Likewise, the diagrams, charts and tables provided have minimal explanation. They are elaborated upon later and serve as frames of reference for the remainder of this book. For convenience, unit conversions and symbols are available for the reader in Appendix 1.

PART A

FUNDAMENTAL CONSIDERATIONS

Prior to the discussion of instruments and instrumentation, some basic knowledge of the dimensions and variables of the physical environment of the earth is required. Then, topics on the requirements and specifications of the instruments for measuring these variables are presented. Finally, the various types of instrument systems or instrumentation are introduced. All of these topics, which constitute the foundation of the book, are included in the next four chapters of Part A.

Chapter 2 describes the dimensions of the physical environment, specifying its spatial and time scales, called the moving scales for short. The physical environment is here identified by three spheres, namely the atmosphere, hydrosphere and lithosphere. These spheres constitute the three basic systems of the geosphere. Thus the moving scales of the three systems are the central theme of Chapter 2.

Chapter 3 discusses the various signaling phenomena of the three systems. These phenomena are identified by such characteristics as radiation, temperature, pressure, moisture and velocity of fluids and solids. Collectively, they identify the electromagnetic field, thermofield, pressure field, hydrofield and dynamic field, respectively. In Chapter 3, the coordination of these characteristics with their respective moving scales in each of the three systems is the main topic of discussion.

Chapters 4 and 5 deal with the basic characteristics of instrumentation. Chapter 4 covers such specifications as the range, accuracy, representativeness, sensitivity and reliability of an instrument. It follows that Chapter 5 provides the basic knowledge of instrumentation for environmental engineering systems including static versus dynamic systems, Eulerian versus Lagrangian systems, and in situ versus remote sensing systems, designed specifically for physical environmental measurements as opposed to laboratory experiments. Some laboratory instruments are not suitable for field experiments because they undergo drastic changes when exposed in the field. Therefore, the design criteria and requirements for field instruments differ appreciably from those for laboratory instruments. The basic characteristics of field instrumentation are described in these two chapters in terms of the respective physical properties of the three systems and their moving scales.

CHAPTER 2 DIMENSIONS OF THE GEOSPHERE

Of the three systems within the geosphere, normally the atmosphere has the greatest mobility in its moving scale for instrumentation. The hydrosphere ranks second and the lithosphere, third.[1] Therefore, atmospheric instruments should detect the various forms of eddies over a wide range of scales in a changing environment. Such atmospheric properties as radiation, temperature, pressure, wind, humidity and cloudiness should be monitored in a small increment of time and space. Figure 2-1 illustrates the rapid fluctuations of air

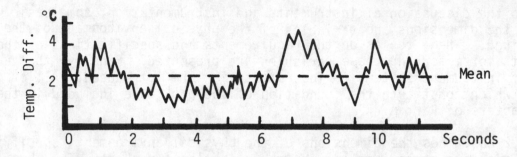

Figure 2-1 Illustration of Temperature Fluctuation

temperature at a fixed location within seconds. As a rule, the spatial variations in air temperature over a heterogeneous surface can be several degrees within a few meters of distance.

2.1 Dimensions of the Atmosphere. Our discussion in this section deals with the statics (spatial dimensions) and then the dynamics (moving scale dimensions) of the atmosphere.

2.1.1 Vertical Scales. Since the invention of space vehicles, the study of atmospheric sciences has been distinctly divided into the lower atmosphere (from the ground to about 60 km) and the upper atmosphere (60 km upwards).[2] The former is known as meteorology whereas the latter is aeronomy. Although the top of the atmosphere is ill-defined, it may reach out to several times the earth's radius (6371 km) where it gradually merges into the interplanetary medium. Due to the enormous changes in atmospheric constituents and density from the lower to the upper atmosphere, techniques of measurements differ

---

[1] With the exception of soil erosion and volcanic eruptions, by and large soil profiles and, sometimes, geological structures are much more stabilized than the ocean or atmosphere.

[2] The generally accepted height is 60 km above the earth's surface, but sometimes the boundary layer has been taken as low as 20 to 30 km and as high as 90 km.

drastically between the two. Ionospheric air measurements,[3] for example, depend upon rocket, satellite and ground-based radio-echo techniques[4] for measuring the type and concentration of ions and free electrons. These measurements cover a range of about 60 km in altitude upwards to perhaps 1000 km where dissociation and ionization become fundamental properties. In the lower atmosphere neither of these properties are of significance in the high air density conditions. While the upper atmospheric density may be measured directly, the lower atmospheric density is determined only indirectly from temperature and pressure observations.[5]

When the vertical atmospheric temperature distribution is taken into consideration, the atmosphere is divided into the regions of troposphere, stratosphere, mesosphere, and thermosphere. Where the atmospheric chemistry is concerned,[6] two distinct layers can be identified: the homosphere and the heterosphere. Based upon the static ionization of the daytime or the nighttime atmosphere, the ionosphere may also be classified into the D-Region, E-Region, F_1-Region and F_2 Region. When dynamic ionization is considered, a distinct layer termed the magnetosphere is clearly distinguished. There are two approaches for the identification of the atmospheric structure: (1) the physical and chemical properties and (2) the ionization of the atmosphere. In considering the atmospheric physical properties, the vertical temperature variations are commonly used to differentiate such atmospheric layers as the troposphere, stratosphere, mesosphere and thermosphere. When the atmospheric chemical properties are taken into consideration, the distinction is made between the heterosphere and homosphere, with its subdivision called the ozonosphere. As mentioned above, the ionization of the upper atmosphere is distinguished as dynamic and static ionization. Based upon the above two properties, the atmospheric structure is elucidated in Figure 2-2. The dimensions, characteristics and boundary layers of the troposphere, stratosphere, mesosphere, thermosphere and exosphere, however, are presented in Table 2-1.

A discussion of upper atmospheric instruments and instrumentation is given in Part E of Volume II. Although such atmospheric probes as satellites, missiles and rockets are available, routine atmospheric soundings are confined mainly

---

[3] The ionosphere consists of a weakly ionized gas or plasma controlled by both solar and cosmic radiation. Enveloping the entire earth, the ionosphere is one of the important layers of the upper atmosphere.

[4] While ground-based techniques provide a convenient method of studying ionization at heights of 90 km and above, sounding rockets reaching to 800 km and satellites measure the entire vertical profile of the ionosphere.

[5] Air density, ρ, of the lower atmosphere can be obtained by the Equation of State for a perfect gas: $P = \rho RT$, where R is the universal gas constant, P, the atmospheric pressure, and T, the atmospheric temperature in degrees Kelvin. Methods of direct measurement for the upper atmosphere are given in Chapter 15.

[6] The heterosphere starts from 85 or 90 km altitude upward and consists of four gaseous layers: the molecular nitrogen layer (85 to 200 km); the atomic oxygen layer (200 to 1100 km); the helium layer (1100 to 3500 km), and the atomic hydrogen layer (3500 to 10,000 km).

to the first 50 km of the lower atmosphere. This is partially due to the fact that all weather activities take place in the first 10 km (about 250 mb) where most of the atmospheric mass is concentrated. The other reason is the high cost factor of upper atmospheric instrumentation. The emphasis has thus been placed on the lower atmosphere instead of the upper atmosphere. We are concerned with the human living environment, concentrating our attention here on the structure and dimensions of the lower atmosphere.

Within the first 50 km, four atmospheric layers can be distinguished. They are the micro-, the surface, the meso- and the upper-air layers. The dimensions and characteristics of the four layers are tabulated in Table 2-2. This table lists those dimensions which are generally accepted. A detailed examination of the four atmospheric layers is given below.

The frictional layer, which encompasses the micro-, surface, and mesolayers is the region wherein air motion is appreciably influenced by surface friction. It extends from the earth's surface to about 600 m (2000 ft or about 950 mb) above the ground. Some atmospheric scientists, however, have taken approximately the first 1000 m (900 mb) as the frictional layer, based upon the theoretical Ekman Layer determination. Above this layer the influence of surface friction is assumed to be negligible. This is designated as the free atmosphere where airflow is essentially geostrophic, due to a balance between the Coriolis force and the pressure gradient force. In fact, only the upper-air layer of Table 2-2 is considered as the free atmosphere, and all the others are in the region of the frictional layer.

The microlayer is found in the lowest portion of the frictional layer. It ranges from a few millimeters to as high as 300 m above the earth's surface, including the first 10 m described as the ideal anemometer height. In practice, however, the anemometer height is 2 m above the ground. The vertical extent of the microlayer, however, has to be determined by such variables as the surface characteristics (topography and types of surface coverage) and the meteorological variables to be measured. While the upper boundary of the microlayer is flexible and determined by the user, its lower boundary can readily be specified by such surfaces as grassland, forest crown, bare land, ice-covered land and oceanic surfaces. Its lateral boundary can also be specified by such boundaries as a forest edge, mountain bluff, and even a small hill-dome. Instruments employed for measuring the micro-environment, known as microclimatic or micrometeorologic instruments, are specially designed for high spatial and time resolution.

The surface layer superimposed upon the microlayer may be designated as the layer between the instrument floor of a weather shelter (1.5 m) and the anemometer height (10 m). Instruments used for measuring meteorological variables in this layer are commonly called surface instruments. The surface instruments are operated by the various national weather services as a routine practice throughout the world. They are mainly employed on land stations but some of them are located on board ships and/or buoys over the ocean.

The mesolayer may extend from the anemometer height up to some 600 m to 1500 m (850 mb). The mesoscale measurements include such local weather phenomena as thunderstorms, squall lines, tornadoes and dust devils. These phenomena, which

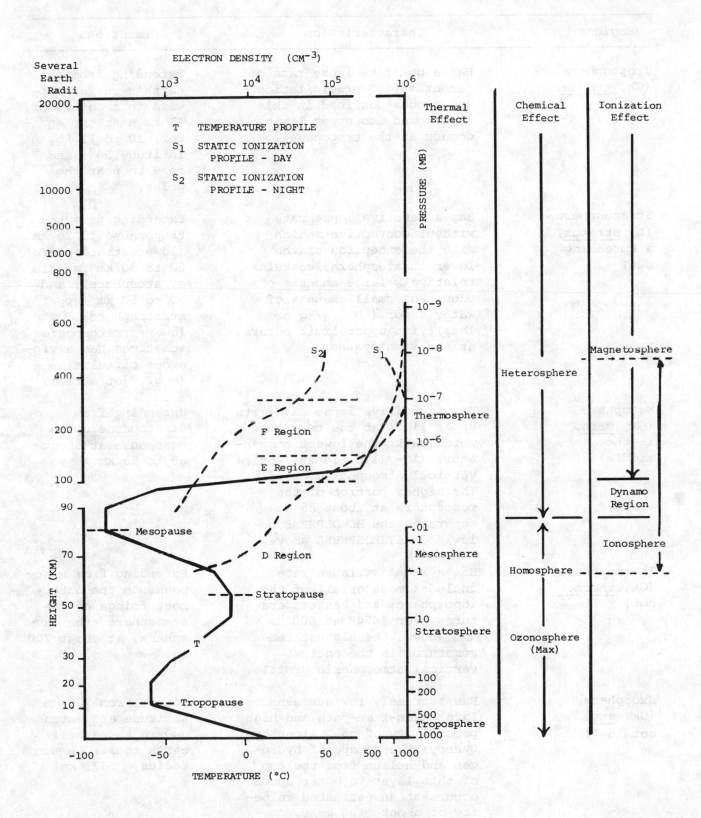

Figure 2-2 Atmospheric Structure

Table 2-1 The Vertical Dimensions of the Atmosphere[+]

Regions	Characteristics	Dimensions
Troposphere (Gk, _tropos_, turn)	Has a positive lapse rate of temperature;[*] precipitation and clouds confined in this layer; and its upper limit occurs at the tropopause.	Extending from the earth's surface upward to about 16 or 17 km near the equator; 10 to 11 km in latitude 50°; and 7 to 9 km near the poles.
Stratosphere (L, _stratus_, a spreading out)	Has a negative lapse rate without convective motion with the exception of the lower stratosphere; contains relatively large amounts of ozone but small amounts of water vapor (10^{-2} g/kg or less); its upper limit occurs at the stratopause.	Extending from the tropopause to 20 km (lower stratosphere); 20 to 30 km (middle stratosphere); and 30 to 50 km (upper stratosphere). The lower stratosphere sometimes has a vigorous circulation (e.g., jet stream).
Mesosphere (Gk, _meso_, in the middle)	Has a positive lapse rate; its upper limit at the mesopause which marks the lowest temperature ($\sim$ - 119C) in the entire vertical atmospheric profile. The higher portion of the mesosphere at about 85 km separates the HOMOSPHERE below and HETEROSPHERE above.	Extending from stratopause up to mesopause at about 80 to 85 km.
Thermosphere (Gk, _thermos_, hot)	Has a negative lapse rate; includes most or all of the ionosphere; and has temperatures over 1474K at 500 km and above, the highest temperatures in the entire vertical atmospheric profile.	Extending from mesopause to the outermost fringe of the atmosphere, the exosphere, at about 700 km.
Exosphere (Gk, _exo_, outside)	Has extremely low gas density, large mean-free-path and high probability of collision frequency; the escape of hydrogen and helium from the top of this layer to outer space occurs at an estimated velocity of about 11.2 km/s.	Starts from 700 km altitude and extends upward to several earth radii; (1 earth radius = 6371 km).

[+] Based upon the thermal structure of the neutral atmosphere.

[*] The positive rate refers to a decrease of temperature with height and, conversely, a negative rate refers to an increase of temperature with height. In the troposphere it is about 6.5C/km.

earth's surface up to about
1 or 2 m; depending upon the
sphere of influences of the
underlying surface, the
upper limit can be as high
as 300 m. | Characterized by strong
boundary effects in-
cluding frictional
forces; represented by
the fine structure of
physical processes oc-
curring in the atmo-
sphere such as diffusion
and turbulence. |
| Surface
Scale* | Confined generally to the
layer between 1.5 m (the
weather shelter height) to
10 m (the anemometer height)
above the earth's surface. | Dominated by the in-
fluences of local en-
vironment, particularly
its surface frictional
forces; however, the
boundary effects are
comparatively smaller
than those of the micro-
scale. |
| Mesoscale | Usually confined to the
layer between 150 m to 1 km
above the earth's surface
but may extend up to about
1.5 km. | Characterized by such
meso weather phenomena
as thunderstorms, squall
lines, frontal zones and
sea breezes but surface
frictional effect is
still appreciable in the
lower mesolayer. |
| Upper-Air
Scale* | Extends from about 600 m
to 1 km upward to about
35 to 50 km; includes
tropospheric and strato-
spheric air. | Characterized by the
physical processes oc-
curring in the _free_
atmosphere where the
surface frictional force
assumed to be nil. |

*
The asterisk refers to the layer that is measured through routine weather
observations such as those conducted by the various national weather services.
Layers without asterisk refer to those measured by research scientists.

are often missed on the normal synoptic network, are beyond the microscale but
within the synoptic scale (see Table 2-3).

As mentioned, the upper-air layer is the free atmosphere. The probes used for the measurement of this layer are known as upper-air instruments, and include radio-theodolite, tracking radar and pilot balloon. Most of these measurements are performed by using the balloon as a platform or tracer of air movement. Others depend upon the use of aircraft or ground-based platforms.

It must be noted here that instruments used for upper-air layer measurements are not the same as those for the upper atmosphere. The latter rely upon such high altitude platforms as missiles, rockets and satellites. There is, of course, an overlap in the measurement of the upper-air layer and the upper atmosphere. Satellite instrumentation, for example, covers the entire atmosphere, while missiles and rockets are specifically designed for measuring parameters of the upper atmosphere. Of course, differentiation must be made between the measurements involving mesospheric and mesometeorological instrumentation. It must also be noted that the definitions of the four atmospheric layers described above are generally accepted but not rigidly designated. Their dimensions may be altered by the individual user but they are all below stratopause.

2.1.2 Horizontal Scales. The horizontal scales of the atmosphere may vary from a few millimeters to tens of thousands of kilometers, geared mainly to the surface characteristics of the various geographical regions. These scales may be represented by such atmospheric phenomena as illustrated below.

PHENOMENA	LINEAR RANGE
Small scale turbulence	5 mm to 5 cm (up to 100 m)
Tertiary circulation	10 m to 500 m
Cumulus clouds	100 m to 5 km
Thunderstorms	1 km to 10 km
Frontal zones	100 km to 1000 km
Secondary circulation	300 km to 4,000 km
Planetary waves (primary circulation)	2,000 km to 20,000 km

The above may be classified into four main horizontal scales, namely microscale, mesoscale, synoptic scale, and macroscale. The latter may be better termed the planetary scale. Although the delineation of the four scale classification is rather arbitrary, each may be specified by a range of geometric dimensions. Table 2-3 shows the four horizontal scales together with their three combinations (i.e., micro-mesoscale, meso-synoptic scale, and synoptic-macroscale).

With the exception of weather satellites, most of the present routine weather observations are confined to surface and upper air measurements. Their scales represent the planetary or regional circulations in general. Microscale and mesoscale investigations of the atmosphere have been done only as part of research efforts. Obviously, since they are conducted only sporadically, their observations are rather limited in time and space distribution.

Micrometeorological and surface measurements, on the other hand, are affected by the presence of green vegetation in terms of evaporation, evapotranspiration, surface roughness over land, and the effects of waves over the sea. Thus, an abrupt change in surface characteristics such as bare soil surface as opposed to water surface and grassland as opposed to cement pavement creates a noticeable transitional zone. Sometimes, a transitional zone which usually covers a narrow band of the horizontal scale may serve as the lateral boundary of a microclimate.

Table 2-3 The Dimensions of the Horizontal Scale of the Atmosphere

Scale	Length, km	Area, km^2	Illustrations*
MICRO	< 3	< 9	Dock, airport, railroad yard, stadium, wheat field, orchard, small farm, pond and small lake.
MICRO-MESO	2 to 15	4 to 225	Large farm, town, urban center, suburb, harbor and lake.
MESO	10 to 100	100 to 10^4	Metropolitan area, county, ranch, large lake, and hill.
MESO-SYNOPTIC	65 to 500	4,225 to 25×10^4	Large county, region, state, stream basin, bay, large lake, forest and small mountain.
SYNOPTIC	300 to 1,500	9×10^4 to 225×10^4	Large state, large region, river, basin, sea, gulf, and mountain.
SYNOPTIC-MACRO	1,000 to 8,000	100×10^4 to $6,400 \times 10^4$	Large sea, continent and small ocean and mountain range.
MACRO	> 6,500	> $4,225 \times 10^4$	Large continent, hemisphere, ocean, and perhaps a planet or solar system.

*The boundary effects are significant to most of the scaling above, hence statics instead of dynamics are illustrated here. The dynamic dimensions of atmospheric phenomena are given in the text.

18

2.1.3 <u>Moving Scales</u>. The lower atmosphere is characterized by a fluid motion scale or simply a moving scale which is composed of numerous small and large eddies. A small eddy can be seen as a tiny dust whirl over a street corner with a diameter of less than a meter and a height of about two meters. It dissipates within minutes, if not seconds. A large eddy may be represented by a tropical cyclone with a diameter on the order of 50 km and a height of about 8 to 12 km. It may last for a day or more.

A wide range of eddies, which may co-exist simultaneously, are observed to be constantly in motion in the atmosphere. The interactions among all these eddies are of great significance to atmospheric scientists. It is obvious that we are concerned with the speed and direction in which these eddies are moving. Thus we are dealing with their time and spatial scales (four-dimensional dynamic system). Illustrations of the moving scale for several atmospheric events are given in Table 2-4.

Table 2-4 Time and Space Dimensions of Atmospheric Events

Atmospheric Events	Size, km Horizontal	Vertical	Duration
Small Cloud	about 0.15	1.5 to 3.0	10 to 30 minutes
Tornado and Waterspout	.01 to 0.5	about 0.8	10 min. to 1 hr.
Large Thunderstorm	about 15	6.5 to 9.0	1 to 6 hrs.
Cyclone and Anti-cyclone	300 to 4,000	3.5 to 8.0	1 to 5 days
Hurricane	50 to 80	8.0 to 12.8	1 to 6 days
General Circulation	2,000 to 20,000	8.0 to 9.6	season to 1 year

In order to understand these phenomena, the following criteria are established:

1. The size or volume of an eddy is specified through its physical or, perhaps, chemical properties or processes;
2. The residence time of the eddy is designated as the total time from its formation to its dissipation;
3. The requirements of instrumentation performance are specified for obtaining information regarding the moving scale of such an eddy respective to its time and spatial resolutions.

As a glob of fluid mass, an atmospheric eddy retains its identity or integrity over a limited <u>time</u> and <u>space</u> while moving within the surrounding fluid. Some eddies may live for a fraction of a second while others may last for months.

Also, small-scale eddies (approximately 1 cm or less for microscale turbulence) are co-existent with or embedded in the large-scale eddies (some thousands of kilometers for cyclone and anticyclone). While the large-scale eddies are responsible for much of the meridional transport implicit in the general circulation, the small-scale eddies reinforce the mixing within a large eddy. For example, the vertical stratification of heat and water vapor in a large-scale eddy is achieved through a vertical mixing process.

In view of the techniques for measuring the various sizes and life-spans of atmospheric eddies, an examination of their sizes (three dimensional) and residence time together with a consideration of instrumentation must first be made.

(1) Space. In the analysis of the field of airflow (i.e., kinematics), simple geometric configurations are always imposed upon the descriptions of the various natural flow patterns (i.e., the three-dimensional winds). In streamline analysis, such linear or curvilinear configurations as translation, vorticity (or rotation), divergence, convergence, and deformation are the bases for the analysis. In the three-dimensional model such configurations as cubical, rectangular, cylindrical, spherical, and dome-shaped have been implemented. However, all of these do not adequately describe the shape and size of eddies occuring in the atmosphere. All eddies are essentially irregular, therefore, a fine scale is required.

In order to determine the irregular structure of a sizeable eddy, physical variables such as pressure, temperature, humidity, clouds, visibility, and sometimes, atmospheric chemical composition in addition to winds must be measured. The three-dimensional structure of an eddy is thus determined. But the presently available synoptic network cannot resolve and identify microscale and mesoscale eddies such as tornadoes, dust devils, and clear air turbulences. As a remedy, either a several-fold increase in synoptic stations over a restricted region, or an improvement of the currently available remote-sensing devices equipped with space platforms, would be needed.[7] In the recent decade, the use of weather satellites as space-based platforms together with the use of radar stations as ground-based platforms has been successful. And yet, an adequate description of the three dimensional boundary of an eddy still poses a major problem. Thus, the detailed characteristics of a depression or vortex as well as those of a transitional zone remain unknown. This is particularly true for a transitional zone between two extremely diversified micro-environments mentioned earlier. How many synoptic stations do we need then for the task? Can we afford to install and maintain an ideal global network? What is an ideal network, after all? Many more questions have been asked, discussed and debated for years, but the problem remains unsolved.

The density of a global synoptic network depends upon the types of sensor, the locations (three dimensional) for exposing the sensor, and the roughness of the natural surface over which the physical variates are measured. Current

---

[7] To be sure, remote-sensing techniques potentially are more economic and offer high spatial and time resolutions. But this does not rule out the need of point-source measurements.

synoptic stations are equipped with only surface and upper-air instruments. It is assumed that the former measure physical variates in the frictional layer while the latter operate in the free atmosphere. But neither the measurements in the free atmosphere nor those in the frictional layer are adequate. This is attributed to the lack of micrometeorological and mesometeorological instrumentation in conventional synoptic stations. This does not pose a major problem for some variates but it does for others. Barometric measurements, for example, do not require as many stations as the measurement of rainfall or wind. The latter are subject to local effects whereas the former are not. For surface measurements, surface roughness plays the most important role among the many environmental conditions. In fact, the roughness of terrain is one of the important factors determining the density of the surface network. For upper-air measurements, the roughness parameter is insignificant above the Ekman layer (about 1 km). Finally, it must be noted that all the routine surface observations are made on one level only, which further limits their application.

Figure 2-3 schematically shows the horizontal and vertical resolution of instrument platforms for homogeneous and heterogeneous surfaces. A homogeneous surface refers to a smooth (or wavy) water or ice surface as well as flat terrain or desert land. A heterogeneous surface refers to such surfaces as urban areas, rough terrain, forest regions, and rugged mountains. The actual station network for physical environmental measurements including weather stations are given in Chapter 17 of Volume II.

(2) _Time_. The residence time of eddies also varies greatly, similar to their spatial dimensions. The time for making an observation, the frequency of observation, and the time resolution of an eddy in a given locality are geared to the diurnal and seasonal changes of atmospheric variates. During the course of a day, high resolution is required for the transitional periods, and low resolution is sufficient at nightfall with medium resolution for daytime measurements. This is shown in Figure 2-4, in which the two transitional periods are clearly shown. The resolution range depends also upon the types of variates to be measured. In the ideal situation, for turbulence studies during the transitional period, the sampling rate for wind speed can be as high as 12 measurements per minute, whereas for temperature, one or two measurements per minute are required. Under fair weather conditions during the midday hours, a temperature sampling rate of 1 or 2 per hour would be sufficient. Similarly, at nightfall a sampling rate of 1 per 2 hours will be adequate. However, during a windy day the appropriate frequency of temperature sampling would be 2 to 3 measurements per hour. In reality, however, the frequency of measurement depends upon the objectives of the user.

As shown in Figure 2-4, it is essential to know that there are two distinct regimes during the course of a day: the daytime and the nighttime. After sunrise, the incoming solar radiation is the controlling factor. The atmosphere is characterized by high temperature, low relative humidity, high turbulence, gusts, evaporation and evapotranspiration. After sunset, outgoing infrared radiation dominates. The reverse conditions then exist in the atmosphere. The transitional periods occur after sunrise and before sunset, with a time span of about 1 or 2 hours for each period depending upon the characteristics of the locality.

Figure 2-3 Vertical and Horizontal Resolution

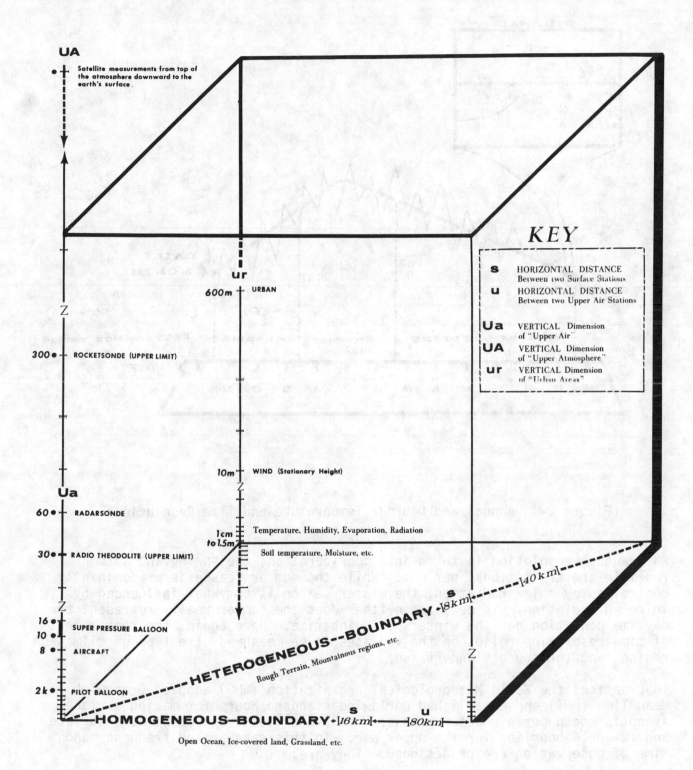

UA

Satellite measurements from top of
the atmosphere downward to the
earth's surface.

ur

600m — URBAN

Z

300 ● — ROCKETSONDE (UPPER LIMIT)

10m — WIND (Stationary Height)

Z

Ua

60 ● — RADARSONDE

1cm
to 1.5m — Temperature, Humidity, Evaporation, Radiation

30 ● ● — RADIO THEODOLITE (UPPER LIMIT)   Soil temperature, Moisture, etc.

Z

16 ●
10 ● ● — SUPER PRESSURE BALLOON
8 ● — AIRCRAFT

Z

2k ● — PILOT BALLOON   Rough Terrain, Mountainous regions, etc.

HETEROGENEOUS—BOUNDARY →|8 km|←   →|40 km|← u

s

HOMOGENEOUS—BOUNDARY →|16 km|←   →|80km|← u

s

Open Ocean, Ice-covered land, Grassland, etc.

KEY

s   HORIZONTAL DISTANCE
Between two Surface Stations

u   HORIZONTAL DISTANCE
Between two Upper Air Stations

Ua VERTICAL Dimension
of "Upper Air"

UA VERTICAL Dimension
of "Upper Atmosphere"

ur VERTICAL Dimension
of "Urban Areas"

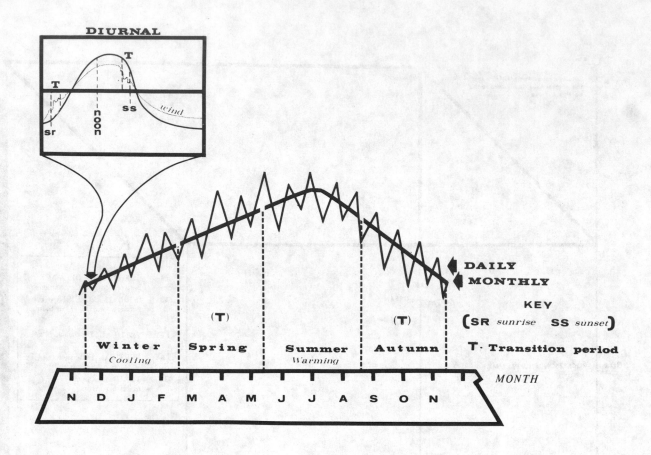

Figure 2-4   Annual and Diurnal Temperature and Time Resolutions

When annual resolution is taken into consideration, the spring and autumn
resemble the transitional periods. While the summer season is predominantly
controlled by solar radiation, the winter season is somewhat influenced by
infrared radiation. To be sure, neither does the summer really represent the
daytime condition nor the winter, the nighttime. Here again, the requirement
of time resolution relies on the variates to be measured, the type of climatic
region, and the project undertaken.

In practice, the World Meteorological Organization (WMO) adopted Greenwich
Mean Time (GMT) and established worldwide _standard_ hours for making surface
synoptic and upper air observations. The 3-hourlies are used for the surface
and 12- or 6-hourlies for the upper air. In this case, fixed frequency and
time of observation are predesigned. They are:

SURFACE:	0000	0300	0600	0900		
	1200	1500	1800	2100		GMT
UPPER AIR:	12 hourlies		0000		1200	GMT
	6 hourlies		0000	0600	1200	1800 GMT

Aside from the above routine practice, the performance of the instruments used also affects the choice of time scale. These are discussed in conjunction with spatial and time variations, or simply resolution criteria.

(3) Resolution Criteria. The requirements of time and spatial resolution (four-dimensional) of an instrument for measuring one or more eddies depend upon (a) the physical or chemical variates and size of the eddy to be measured, (b) the climatic regime where the eddy is located, (c) the exposure of the sensor with respect to an eddy, and (d) the type of instrument sensor and platform. Of these four factors the first, having been described briefly, is elaborated upon in the next chapter on signaling phenomena. The last three factors, which are interrelated and are closely associated with the performance of instruments, are discussed here conjointly.

As mentioned earlier in Chapter 1, the sensors of an instrument are provided for monitoring either the energy and/or the mass of its physical environment. Cold or warm advection as associated with a frontal passage in the middle latitudes, for example, is an energy and/or mass transfer from polar or tropical regions, respectively. This is the energetic mechanism of the meridional flow. High resolutions in time and space are required for measuring such physical characteristics as winds, temperature, humidity and atmospheric pressure along the frontal zone. Obviously, low resolution is sufficient within the respective warm and cold airmasses themselves. It must be noted that a front represents a dynamic and lateral boundary. A physical boundary such as the wall of a large manmade structure or the bluff of a mountain range acts as a permanent lateral boundary which affects the characteristics of physical variates in the vicinity. The physical and dynamical boundaries are considered in the classification of types of boundary. An illustration of boundary types in terms of the various media, namely gaseous, liquid, and solid, is given in Table 2-5.

In the polar regions, on the other hand, a high spatial resolution for incoming, and reflected, and infrared radiation measurements is important under stagnant weather conditions and during the summer months of the year. This is simply because under such weather conditions the radiative flux, both incoming and outgoing, becomes the single most significant variable among the many atmospheric variables. Similarly, high resolution for heat and moisture is necessary in rain forests where temperature and humidity are the dominating factors.

The choice of instruments and especially their sensors varies with the differences of each climatic regime. In order to meet the climatic requirements, a certain sensor used in one region may not be useful in another. The freezing temperatures of a cold region reduce the sensitivity and the proper functioning of a hair hygrometer; at the other extreme, the high humidity in the rain forest renders a number of instruments useless. In the latter case, water-proof instruments must be designed. Therefore, the time and spatial scales which an instrument detects depend on the climatic conditions of a locality.

Table 2-5 Illustration of the Types of Atmospheric Boundary Conditions

| | Physical Boundary | | Dynamic Boundary | |
	Solid	Liquid	Liquid/Solid	Gaseous
Lower	Surfaces of the land and vegetation; ice and snow.	Water surface.	Shallow fog (ground fog); dew deposition.	Ground layers of cold air drainage or super-heated air.
Lateral	Walls of a fence, structure, bluff, mountain slope, and forest edge.	Waterfall; a high tidal wave.	Fog bank, cloud wall;* waterspout.	Frontal surface and other surfaces of discontinuity, such as foehn wall and a squall line.
Upper	Ceiling of a cave, a structure or a crown of vegetation.	Beneath a waterfall.	Ceiling of clouds, haze, dust and smoke.	Temperature inversion, tropopause, stratopause, mesopause, ozone layer.

* The lateral surface of the sideview of a cloud bank.

When the accuracy of the time scale is taken into consideration, a movable instrument platform, in general, requires much greater accuracy than a fixed one. In routine practice, surface measurements are generally made through human vision over a fixed platform. It may take 5 minutes or more for an experienced observer to complete his observation beginning at each of the WMO standard hours for surface synoptic observations. In radiosonde practice for upper soundings, for example, the time elapsed between the release and the termination of a balloon flight may require over 2 hours. Therefore it is concluded that most routine weather observations are not kept to an exact time. Usually, the measuring time is much longer than the response time of a sensor. The discrepancy in time scale is not very serious in steady weather conditions but it may be serious for severe storm detection. On the other hand precise timing is necessary for a fast moving platform. A weather satellite, for example, may travel as fast as 7.6 km s^{-1}. In either the tracking of a satellite from the ground-based station or the observation of weather phenomena from the satellite (the space-based station), the time scale is on the order of one millisecond. Without such a high speed, measurements from satellites would be rendered impossible. This applies to a number of space vehicles and some other atmospheric instruments.

In conclusion, the discussions presented so far are concerned with the dimensions and boundary conditions of the lower atmosphere. Emphases are placed upon the ideal situation of "What is needed?" This paves the way for "What has been done?" to be discussed in the remainder of the book.

Dimensional analyses of atmospheric eddies encompass: (1) specifications of the lower, lateral and upper boundaries, (2) identification of the moving scales in time and space, and (3) determination of the requirements of temporal and spatial resolution of measurements.

2.2 <u>Dimensions of the Hydrosphere</u>. As can be seen from Table 2-6, over 97% of the world's water is contained in the oceans; most of the remainder is locked up in icecaps and glaciers. Strange as it may seem, less than seven-tenths of a percent of the world's water is available as <u>fresh water</u>. And yet, this trivial volume of fresh water is the main resource of life supporting systems for most of the world's terrestrial animals and plants. These oceans and seas are integral parts of the hydrologic cycle, acting as water and heat sources and sinks, particularly as moderators of the earth's climate.

In this section, our attention is concentrated on the structure of the oceans and lakes in connection with their physical variates. However, emphasis is placed on the <u>vertical</u> structure. Finally, the section closes with specifications of the temporal and spatial resolution of the hydrosphere.

2.2.1 <u>The Vertical Structure</u>. The ocean covers some 3.6×10^8 km^2 (1.4×10^8 mi^2), or about 71% of the earth's surface, with an average depth of about 4 km (2.5 mi). It has a maximum depth of over 11 km (>7 mi). The first 100 m of the surface layer, known as the <u>mixed layer</u>, is warm in temperature, usually characterized by an isothermal layer, and is uniform in composition. The mixing depth varies in different regions of the same ocean and is proportional to the drag force of the local wind. The higher the wind velocity, the deeper the mixing depth.[8] This layer is also the <u>photosynthetic</u> or <u>photic zone</u> which is illuminated by sunlight with intensity decreasing logarithmically to 1 per cent of its surface value. The layer below the warm mixed layer is characterized by a rapid decrease of temperature with depth, and is known as the <u>thermocline</u>. Its average thickness is about 1500 m. Below the thermocline layer is the <u>deep water layer</u> in which the temperature decreases slightly with depth. Figure 2-5 and Table 2-11 show the vertical temperature variations of the Atlantic. It must be noted in Figure 2-5 that the term epilimnion refers to the mixed layer and hypolimnion to the deep water layer. Also note the direction of the underwater currents as indicated by arrows. These subsurface currents are superimposed on several oceanic surface currents (not shown) in which the Gulf Stream predominates.

Two distinct layers of the photofield are identified: the photic layer (the solar penetration layer), and the aphotic layer (the dark layer). The former

---

[8] In blowing over the water surface, the wind drags the surface layer along to make eddies, or sets up waves in which the water circulates in small orbits, producing mixing.

Table 2-6$_a$ Spatial Dimensions of Global Water Distribution[1]

	Surface Area (x 10^5 km^2)	Volume (x 10^5 km^3)	Percent of Total	Remarks
1. Atmosphere	5,100	0.13	0.001	Mostly gaseous form.
2. Ocean	3,600	13,220.00	97.200	Mostly liquid.
3. Land[2]	--	86.37	0.649	Mostly liquid.
(a) surface water	--	2.30	0.020	Rivers, streams and lakes, mainly freshwater with some saline.
(b) sub-surface water	--	84.07	0.629	Upper and deep-lying waters.
4. Icecaps and Glaciers	180	292.00	2.150	Solid form only.
Total:	--	13,598.50	100.000	

[1]Data based on estimates of R.L. Nace, U.S. Geological Survey (1964) with minor changes.

[2]No estimates available for surface or subsurface waters in the continents. The former is subject to change seasonally or annually while the latter has no reliable method of estimation at present.

Table 2-6$_b$ Meridional Distributions of Surface Areas Covered by Oceans[3]

Latitude Zone, N	80-90	70-80	60-70	50-60	40-50	30-40	20-30	10-20	0-10	0-90°N
Percent	93.4	71.3	29.4	42.8	47.5	57.2	62.4	73.6	77.2	60.6%

Latitude Zone, S	0-10	10-20	20-30	30-40	40-50	50-60	60-70	70-80	80-90	0-90°S
Percent	76.4	78.0	76.9	88.8	97.0	99.2	89.6	24.6	0.0	80.9%

[3]The total surface area of the oceans is estimated to be about 70.8% of the global surface of the earth.

can be further divided into two zones according to the amount of light penetration. The upper zone (0 to 100 m) is the photosynthetic zone in which light intensity decreases to 5 per cent of its surface value, and the lower (100 to 200 m) or twilight zone, in which the intensity decreases from 5 to 0 per cent of its surface value. The energy falling on the water surface is absorbed within the first 200 m of the water column, with approximately 95 per cent of the incident radiation absorbed in the upper half of the layer. The absorption, however, is wavelength dependent. The shortest wavelengths (between 0.2 and 0.6μm) penetrate beyond 100 m. Infrared radiation at wavelengths greater than 1.2μm is absorbed in the first 10 cm of the water column. For turbid waters the penetration depth for all wavelengths is much less. For clear freshwater lakes, the absorption of the various solar spectra is shown in Table 2-7.

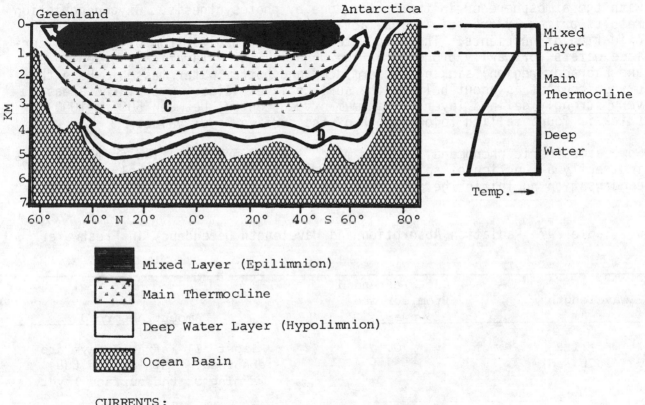

CURRENTS:

A. Atlantic Neutral Water (Surface)
B. Atlantic Intermediate Water
C. N. Atlantic Deep Water
D. Antarctic Bottom Water

Figure 2-5 Vertical Structure of the Atlantic Ocean

About 57 elements are present in seawater, the major constituents being sodium, calcium, potassium, and magnesium salts in the forms of chloride, bromide, sulfate, and carbonate. Sodium chloride (common salt) comprises approximately 90 per cent by mass of the constituents. Salinity is defined as the total amount of solid material in grams contained in 1 kg of seawater when all the carbonate has been converted to oxide, the bromide and iodide replaced by chloride, and all organic materials completely oxidized. Salinity ranges from 33 to 36 parts per thousand (°/oo) in the open ocean, rising to 55 °/oo in land-locked extensions of subtropical seas and close to zero in parts of the ocean heavily diluted by freshwater inflow.

The salinity of ocean waters, which influences the type and distribution of marine organisms, is controlled mainly by vertical changes of water temperature and the rate of surface evaporation. For instance, in the Atlantic Ocean the temperature decrease with depth in the thermocline is correlated positively with the salinity change.

Oxygen enrichment of ocean waters occurs only at the surface due to exchange with the atmosphere or in the photic zone by photosynthesis. Oxygen depletion results primarily from biological activity, degradation of organic debris and oxidation of nutrients. The deep water increase results from oxygen-rich surface waters (primarily produced in the Weddell Sea, Antarctica, Norwegian Sea and Labrador region) sinking to form deep and bottom water masses. Here the oxygen content is about half of the surface layer value. Over large areas a very strongly defined layer of minimum oxygen content between 500 to 1000 m thick is found falling to one-tenth of the surface layer value.

Some atmospheric phenomena, such as mist, fog and precipitation formation, are indirectly a function of salinity, since seawater particles supply the major condensation nuclei in the atmosphere.

Table 2-7 Radiation Absorption and Wavelength Dependency in Freshwater

Wavelengths, μm	Depth extended from surface downward, m	Amount Absorbed
> 1.2 (Infrared)	0.1	Almost all infrared rays are absorbed. This causes the warming of the surface layer.
0.9 to 1.2 (Near infrared)	1	All near infrared rays are absorbed. This further reinforces the warming.
0.6 to 0.9 (Mostly visible light)	10	About 97% of this wave band is absorbed. The remaining 3% unabsorbed portion consists mainly of the blue and green light causing the appearance of blue/green color of the oceans and lakes.
0.2 to 0.6 (Near ultraviolet rays and visible light)	100	About 94% of the total energy of sun that reaches the water surface is absorbed within this wave band and in this layer.

The hydrostatic pressure in a body of water (or water pressure) is another variate. In oceanography, water at the surface is given a zero value, even though the atmosphere exerts a pressure on the water surface. The unit of water pressure is the decibar (db), which is defined as 10^5 dynes cm^{-2}. The water pressure of one db is approximately equivalent to the pressure exerted by one meter of seawater. The total water pressure exerted by an ocean layer extending to 7 km depth is approximately 7000 dbs. This is equivalent to about 700 kg cm^{-2} or 1,480,000 lbs ft^{-2}. Sudden changes of water pressure do not occur above the ocean floor in contrast to the large pressure fluctuations that occur in the atmosphere. Instead, water pressure in the oceans increases steadily and linearly with depth and varies slowly with the march of the seasons. Thus, laminar instead of turbulent flow predominates below the surface turbulence zone.[9] Nevertheless, a complex relationship exists among such variables as temperature, pressure, density, salinity and dissolved oxygen of seawater. Water temperature, for example, may increase downward at a rate of about 1.25C km^{-1} due to adiabatic compression, or heating through sinking. However, adiabatic heating is too small to offset the vertical temperature profile of the ocean. The density of seawater also increases to some extent with the increase of pressure. Salinity, however, varies directly with the increase of temperature and evaporation, as does the solubility of dissolved oxygen.

2.2.2 Horizontal Structure. The discussion, so far, has covered the vertical structure of the hydrosphere, specifically that of the ocean. Unlike the vertical scales, however, the horizontal scales of the hydrosphere are extremely complex due to ocean dynamics and the irregularity of oceanic boundaries. While the dynamics of the ocean are discussed in the next section, the boundary conditions are described here. The lateral and lower boundaries of an ocean are marked by such materials as rocks and sediments forming the physical boundary. The dynamic or upper boundary is the overlying air. In addition, the lateral boundaries of warm and cold water currents, the interfaces between saline and fresh water, or high and low density water, may be taken as dynamic boundaries.

The lateral physical boundaries are continental shelves, submarine canyons and coral reefs of the continental margins. Continental shelves are gently sloping (3 to 6 per cent), shallow water surfaces extending outward from continental margins for distances ranging between approximately 2 to 320 km. The shelf is found at an average distance of 180 m below the surface. Submarine canyons are steepwalled depressions cut below the surface of the ocean basins (in some cases between 1.5 to 3.5 km deep). Coral reefs are massive structures of biogenic origin occurring in waters with a temperature range of 25 to 30C. Reefs are either fringing (shelf attached to land), barrier (offshore coral embayments and lagoons) or atoll (ringlike reefs enclosing a lagoon).

The lower boundaries of the ocean have a variety of topographical forms. These features include the abyssal plains and hills, oceanic rises, and seamounts. The abyssal plains, averaging approximately 0.1 per cent slope, are large extensive basins that are, on the average, approximately 3 km below sea level. For example, the abyssal plains of the eastern Pacific Ocean resemble moats with a width of approximately 50 km and length ranging from 2400 km to 4800 km. Abyssal hills, individually or clustered, protrude 10 m to 100 m above the ocean floor, while oceanic rises reach heights of a few hundred meters above the sea floor. The seamount is the largest of these projections, dome-shaped and having

---

[9] This assumption accounts for circulations due to mean motion. However, recent events in oceanology suggest that mesoscale eddies in ocean circulations can be quite vigorous (McWilliams, 1977).

heights between 3 and 5 km. Their large size, both horizontally and vertically, makes them conspicuous features above the relatively low profile of the abyssal plain. Some of these features are shown on the profile of the Atlantic Ocean, Figure 2-6.

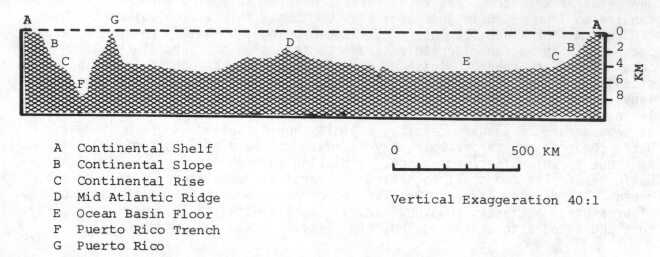

A   Continental Shelf
B   Continental Slope
C   Continental Rise
D   Mid Atlantic Ridge
E   Ocean Basin Floor
F   Puerto Rico Trench
G   Puerto Rico

Vertical Exaggeration 40:1

Figure 2-6 Profile of the Atlantic Ocean - Florida to West Africa

The mid-oceanic ridge system is a great submarine mountain chain, in some areas appearing as single belts and in other areas assuming complex branching structures. The total length of the mid-oceanic ridge system is approximately 6.4 x 10^4 km, its width is between 1500 km and 2500 km, and its average height is about 3.7 km. The network of this submarine mountain chain starts in the Arctic Ocean, whence it intersects the North and South Atlantic Oceans; a section of the ridge branches and extends into the Indian Ocean, and thence through the South and North Pacific Oceans. In the latter two oceans it resembles more of an escarpment and it is displaced away from a mid-oceanic position.

The ocean's crust, when depressed well below the average depth of the ocean floor, forms the structure known as an ocean trench. The trenches have depths between 7 km and 9 km below the ocean surface, with widths of 40 km to 120 km and lengths of 500 km to over 4500 km. Most commonly these trenches occur on the seaward side of long narrow submarine ridges or island arcs. The statistics of Table 2-8 compare the size and depth of the world's oceans to known trench depths.

The lateral boundaries of the ocean basins are highly irregular, a fact evident upon examining topographic maps of continental margins and bathymetric maps of coastal margins and ocean basins. The various dimensions of both lateral and lower boundary features of the oceans are summarized in Table 2-9. Some of these features are identified in a vertical cross-section of the Atlantic Ocean, Figure 2-6.

Boundaries of surface and subsurface waters on the land are highly complex. Rivers, streams, lakes and ponds are subject to fluctuations in their channels or sizes over time periods of months and years. Their boundaries are continually

Table 2-8 World Ocean Trenches

Ocean	Area (to nearest 10^5 km^2)	Average Depth m	Trench Depth m
Pacific	1,600	4,257	11,022 (Marianas)*
Atlantic	800	3,903	9,145 (Puerto Rico)
Indian	725	3,940	7,411 (Sunda)
Arctic	130	1,198	5,418 (Fram Basin)

*It is possible that the deepest part of the ocean is not yet known. For example, the Philippine trench was claimed to be 37,782 ft (∿11,449 m) deep but this figure has not been verified.

being refined. Subsurface waters, soil and ground waters within 800 m or so of the surface, have highly complex boundaries. They fluctuate in size and position both diurnally and monthly. Ground water lying below 800 m accounts for an estimated one-half of the volume of all ground waters. Its boundaries and cycles are unclear.

Fresh water retained on land and frozen in icebergs at sea constitutes a small fraction of the global water supply. The total amount of water estimated to be in these latter two categories is compared in Table 2-6a.

Table 2-9 Dimensions and Slopes (Mean Values) of
Ocean Lateral and Lower Boundaries

Boundary Feature	Mean Dimensions km	Mean Slope per cent
Continental Shelf	2 - 320 (width)	3 - 6
Submarine Canyon	1.5 - 3.5 (depth)	---
Coral Reef	0.5 - 300 (width)	---
Abyssal Plain	50 - 200 (width)	0.1
Abyssal Hill	0.01 - 0.1 (height)	---
Oceanic Rise	0.1 - 0.3 (height)	---
Sea Mount	3 - 5 (height)	---

2.2.3 Moving Scales. The moving or dynamic scales of the hydrosphere are traced through horizontal and vertical transport of watermasses in the atmosphere, ocean, and land. The transport of water occurs as gaseous (water vapor), liquid (precipitation and clouds) and solid phases (icebergs, glaciers and solid precipitation). The hydrologic cycle describes the transport and stage changes of water between atmosphere, hydrosphere and lithosphere. Water is carried by atmospheric and oceanic currents, ground infiltration, percolation, seepage, overland runoff, etc.

An estimated 3.99×10^5 km^3 of water is evaporated annually from free water surfaces. Transpiration from plants contributes smaller amounts. The bulk of this water vapor remains confined to the troposphere. A considerably smaller amount of water vapor is found in the pore spaces of soils and plant tissues. The water vapor content of tropospheric air varies between 7.6×10^{-4} g kg^{-1} and 50 g kg^{-1}. This variability ranges over five orders of magnitude in air at temperatures and pressures of meteorological interest.

Hygrometers developed to date are incapable of measuring such widely fluctuating amounts of water vapor. Furthermore, present measurement methods only sample approximately 9 per cent of the water vapor found in the total atmospheric volume. Thus, the spatial and temporal distribution of water vapor in the atmosphere is far from being thoroughly mapped. However, using data from upper-air soundings, the water vapor channel on weather satellites, and surface measurements, estimates are possible of the vertical and horizontal extent of water vapor globally.

(1) Flux Estimates. Estimates of water vapor flux have varied over time. At present the following findings appear to describe generally the vertical and horizontal fluxes of water vapor.

VERTICAL FLUX

1. Most of the measurable flux is concentrated in the first 7 km (to approximately the 400 mb level).

2. A small amount of the flux may extend upward to 16 km (about 100 mb) over the equator, to 10 km (about 250 mb) in latitutde 50°, and to 8 km (about 350 mb) near the poles.

HORIZONTAL FLUX

1. Zonal Flux Transfer

    a. The zonal flux is complicated by the influences of the topography, as well as the size and configuration of the land mass.

    b. The zonal flux is less important than the meridional transfer because the global evaporation maximum (over 135 cm/year) occurs in the low latitudes and the minimum (less than 5 cm/year) over the poles.

2. Meridional Flux Transfer

    a. The flux moves from evaporation maximum (20° latitude, north and south) poleward.

b. The maximum flux transfer, however, is concentrated between 35 and 40°S and 35 and 45°N. The former is about 50% greater in magnitude than the latter.

c. There is a net flux converging toward the Intertropical Convergence Zone from the north and the south.

d. The order of magnitude of meridional flux is about 10^{15} kg per year.

e. There are counterflows of the flux in the upper troposphere and lower stratosphere. These flows are insignificant due to the very small amounts of water vapor present.

The annual water balance of the world's oceans is given in Table 2-10$_a$. The difference between evaporation, E, and precipitation, P, is balanced by the inflow (or runoff) to the surrounding lands, L, and by the inflow or outflow between oceans, O. Thus

$$E - P = N \dots \dots \dots \dots \dots \dots \dots \dots \dots (1)$$

where N is the net gain or net loss of the ocean. For example, water deficits occur in both the Indian and Atlantic Oceans. The deficit is compensated by runoff from the surrounding lands and by inflow from Pacific and Arctic Oceans, hence

$$E = P + L + O \dots \dots \dots \dots \dots \dots \dots \dots (2)$$

Oceans having a water surplus redistribute the excess as outflow in order to maintain an equilibrium of water level. A negative sign is assigned to the last term of Eq. 2.

Table 2-10$_a$ Annual Water Balance of the Oceans[1]
(cm per year)

Ocean	Evaporation (E)	Precipitation (P)	Inflow from Land (L)[2]	Inflow from Ocean (O)[3]	Gain or Loss[4]
Pacific	114	121	- 6	13	7
Atlantic	104	78	-20	- 6	-26
Indian	138	101	- 7	-30	-37
Arctic	12	24	-23	35	12

[1] Modified from W.D. Sellers (1965).

[2] The negative sign indicates that water is supplied by surrounding lands.

[3] The negative sign indicates that water is supplied by neighboring oceans; the positive sign indicates the outflow of surplus water from the ocean.

[4] The positive sign (or gain) indicates that the precipitation is greater than evaporation; the negative sign (or loss) indicates that evaporation is greater than precipitation.

On the continents, the difference between precipitation and evaporation is the net runoff, R, assuming that the annual variation in water storage over the land is negligible

$$P - E = R \ldots\ldots\ldots\ldots\ldots\ldots\ldots\ldots\ldots\ldots \quad (3)$$

The annual water balance of the continents, illustrated in Table 2-10$_b$, shows that precipitation always exceeds evaporation over the continents.

Table 2-10$_b$ Annual Water Balance of the Continents[1]/

(cm per year)

Continents	Evaporation	Precipitation	Runoff
Europe	36	60	24
Asia	39	61	24
North America	40	67	27
South America	86	135	49
Africa	51	67	16
Australia	41	47	6
Antarctica	∿ 0	3	3

[1]Modified from W.D. Sellers (1965).

Of the 3.99×10^5 km^3 of water evaporated from free water surfaces, only 0.63×10^5 km^3 comes from the continents while the remainder is from the oceans. There is 1.01×10^5 km^3 of precipitation which falls on the continents and 2.98×10^5 km^3 over the oceans. Thus, the total outflow from the continents is 3.8×10^4 km^3. Taking 1.0×10^3 km^3 as the unit and applying the above figures to Eq. 3, then the water budget for continents as a whole is estimated as

$$\begin{array}{cccc} P & - & E & = & R \quad \text{(outflow)} \\ (101) & & (63) & & (38) \end{array}$$

Estimating the water budget of all oceans yields

$$\begin{array}{cccc} P & - & E & = & R \quad \text{(inflow)} \\ (298) & & (336) & & (-38) \end{array}$$

The moving scales, or dynamics, of liquid water occur both vertically and horizontally in ocean circulations. The vertical circulation is controlled mainly by density differences while the horizontal circulation of surface water is controlled by the wind. The bottom water circulation, however, is controlled by high density and salinity and initiated by extremely low temperatures.

As shown in Figure 2-5, the three-layer thermal characteristics of the Atlantic Ocean are superimposed schematically on the four-level meridional water movements. As mentioned earlier, the three layers are the mixed layer, the main thermocline and the deep water layer. The four levels of meridional flow are described in Table 2-11.

Table 2-11 Characteristics of Meridional Flow of
Water Masses in the North Atlantic Ocean

Water Mass	Level of Water Mass Approx. Depth	Characteristics
Upper	0 m to 100-150 m (Ekman layer or mixed layer)	Almost uniform in the distributions of water temperature, salinity and dissolved oxygen; the mixed layer includes photosynthetic and twilight zones; various oceanic currents in the first 100 meters move rapidly.
Intermediate	150 m to 1,000 m (thermocline and halocline layer)	Rapid decreases in temperature (thermocline), salinity (halocline), and dissolved oxygen in the low and middle latitudes; the water flows northward; Aphotic Zone prevails.
Deep	1 km to 3 km	Thermocline continues in this layer; the water flows slowly with a speed of a few centimeters per second, moving southward.
Bottom	3 km to 5 km	Uniformity in temperature, salinity and dissolved oxygen with depth; water moves northward extremely slowly.

The depths shown in Column 2 of Table 2-11 are rough estimates. Depths can vary with different locations in the Atlantic and with the seasons of the year. Greater variability is to be expected for different oceans.

In practice, a routine oceanic survey usually is made on board ship, known as an ocean station, but may also include some island stations. The survey is restricted solely to surface observations. Deep ocean soundings, however, have been conducted to date as part of sponsored research studies. Routine surface

36

observations include state of sea, physical properties of seawater and weather conditions over the ocean. The spatial and temporal dimensions of some of these surface observations as adopted by the WMO are illustrated in Table 2-12. These are the Surface Synoptic Requirements for SHIP codes.

Table 2-12   Dimensions of Ocean Surface Observation

Observations	Range	Dimensions Unit	Average (Approx.)
Sea Waves (state of sea)			
1. Height (meters)	0 to 14 or more	0.456 (1.5 ft)	1.5 to 2.5*
2. Period (seconds)	5 to 14 or more	1	4 to 5*
Sea Swell			
1. Height (meters)	0 to 10	0.456 (1.5 ft)	0.5 to 1*
2. Period (seconds)	1 to as much as 20*	1	6 to 7*
3. Direction (degrees)	0 to 360	10	--
Surface Temp.	About -2 to 30C*	0.1C	--

* Estimated by authors.

Besides the observations listed in Table 2-12, the WMO SHIP code requires observations of the kinds of ice and ice effects on navigation, including the bearing and orientation of the ice edge and the rate of ice accretion on the ship. Also required is the tendency of wave height changes since the last observation. At some island and offshore stations, additional routine measurements are made on the height and period of tidal waves, the wind fetch,[10] the salinity of seawater and the penetration of light.[11] Additional discussion of

[10] The wind fetch is the area over which the wind blows in generating wind waves. It is measured in the direction of the wind.

[11] The light penetration is estimated from water transparency; the conventional method uses a secchi disk to estimate transparency (point at which black and white sections of the disk are no longer distinguishable to the eye).

hydrospheric dimensions is found in the next chapter, in which special attention is given to the various sizes of eddies comprising the oceanic circulation.

2.3 <u>Dimensions of the Lithosphere</u>. The solid earth includes the crust (continents and ocean basins) and the earth's interior (the mantle, the outer core and the inner core). The lithosphere, as defined on p. 1, encompasses the solid crust and the cool upper mantle. However, in this section the discussion is limited to the upper portion of the earth's crust over the continents. Special attention is given to the soil layer of the continents, called the regolith, and its dimensions and dynamics. This is followed by a brief discussion of the upper crust (exclusive of soils) for the earth as a whole.

2.3.1 <u>The Structure of the Pedosphere</u>. The scientific study of soils is called pedology, from which is derived the term pedosphere or soil layers of the earth's crust. Our attention is directed to properties and scales of the soil layer, the upper most portion of the regolith. The regolith is inclusive of both the soil and weathered material derived from the underlying bedrock.

In this context soil is taken to be a collection of natural bodies on the earth's surface, containing living matter and supporting or capable of supporting plants. Soil includes all horizons differing from the underlying rock material as a result of interactions between climate, living organisms, parent materials and relief. The lower limit normally is the greatest depth of the common rooting of native perennial plants. At its lateral margins soil grades to deep water or barren areas of rock, ice, salt or shifting sand dunes.

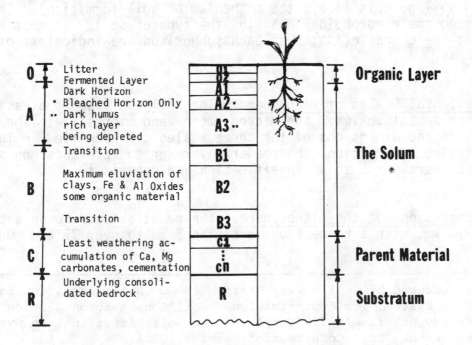

Figure 2-7 Typical Mature Soil Profile

38

Soil horizons are subdivisions recognized in a vertical soil profile. The generalized horizons are illustrated in Figure 2-7; all these layers may not be developed in a particular soil type, however. Related groups of continguous pedons (the smallest volume identified as a soil, ranging in size from 1 to 10 m^2) are collectively called a soil individual or soil series. The vertical and horizontal structure of the pedosphere is modified by composition (organic and inorganic materials, soil air, water and microbes), changes of a physical, chemical and biological nature induced by interactions between the pedosphere and the atmosphere, hydrosphere, and parent materials. Well over 7000 soil individuals have been classified within the United States.[12]

(1) The Vertical Scales of the Pedosphere. As shown in Figure 2-7, the undecomposed and partially decomposed organic matter (including animal bodies), the O1 horizons, and the parent material rocks of the lowest layers, the C and R horizons, do not constitute soil. Organic debris and substratum are the parent materials and bedrock, respectively. The average depth of a typical soil profile ranges from about 0.5 m to 2 m or more.

A true soil or solum consists of two layers. The A-horizon or topsoil lies beneath the organic layer. The A-horizon is the zone of depletion; it is generally rich in organic and mineral materials. The next layer below is the subsoil or the zone of accumulation, the B-horizon, in which most of the leached materials are accumulated. Beneath the solum is the C-horizon and the substratum. Usually, most of the parent materials are located in the C-horizon of the least weathered materials. Each horizon can be further divided into subhorizons with subscripts, both numerical (1,2, ...) and alphabetic (a,b, ...) as subdivision notation for particular characteristics (i.e., B1, accumulation layer for clay from layer A). The total number as well as the depth of each subhorizon depend upon the soil material and soil genesis (i.e., the mechanism of soil formation). The pattern of horizons and their morphology describe the type of soil. It must be emphasized that the characteristics of each subhorizon are indicators of the soil-forming process.

(2) The Horizontal Scales of Pedosphere. The soil can also be classified into three main horizontal scales: the macro-, meso- and microscale. The physical characteristics and dimensions of the three scales are specified in Table 2-13. For more detailed information, the reader may refer to references on soil sciences which pertain to soil classification and genesis.[13]

2.3.2 The Dimensions of the Lithosphere. The crust extends to an average depth of 10 km to 16 km, with a range from less than 5 km to over 75 km. Under the

---

[12] Definitions adopted by Soil Survey Staff. 1960. Soil Classification, A Comprehensive System, 7th Approximation. Soil Conservation Service, USDA; and Soil Survey Staff. 1967. Supplement to Soil Classification System, 7th Approximation. Soil Conservation Service, USDA.

[13] Typical references are (1) USDA. 1957. Soil, The Yearbook of Agriculture 1957. US Gov't. Printing Office; (2) Overdal, A.C. 1960. In: Goode's World Atlas, Ed. Rand McNally & Co., Chicago; (3) Reference 12, ibid.

continents, the average depth of the crust is 35 km, while under the oceans the average depth is 5 km.

It has been theorized that there are 6 primary lithosphere plates floating over the asthenosphere.[14] These 6 plates are separated by deep ocean trenches and mid-oceanic ridges. Each primary plate can be subdivided into several smaller plates which are moving toward deep oceanic trenches at an average speed of 1 mm per year.

The dimensions of the topographic features of the lithosphere (ocean basins and continents) are very important in the surveillance of the earth's surface. The per cent of the earth's surface area in each altitude zone is given in Table 2-14$_a$; the elevation of each latitude zone is shown in Table 2-14$_b$. Zones with a low elevation, of course, are associated to a great degree with zones with a large percentage of ocean surface. The latter was given previously in Table 2-6$_b$.

Tables 2-14$_a$ and 2-14$_b$ illustrate average conditions. In actuality, high mountain peaks which exceed 6 km are not uncommon (see Table 2-15), and some deep ocean trenches are below 7 km (see Table 2-8). Also, the term "mean sea level" does not refer to the boundary between the continent and the ocean. Some basins on the continental areas are below sea level and affect the calculation of the areas shown in Table 2-14$_a$.

The horizontal dimensions of the land areas may range from those of small islands to large continents. Greenland in the North Atlantic, the world's largest island, has an area of about 2,175,600 km² (840,000 mi²), while Australia, the world's smallest continent, has about 7,694,890 km² (2,971,000 mi²). Australia is about 3.5 times larger than Greenland. There are numerous small islands having diameters of a few square kilometers. The horizontal scales are further subdivided into types of land such as barren land, desert, tundra, grassland, forests, wetland and ice-covered land. In addition, specification also is made according to great soil groups (see Table 2-13), land use (urban, crop land, pastures and mining regions), topography, landforms, and watersheds. The dimensions of the horizontal scale are determined by choosing a desirable classification scheme. The spatial dimensions of the horizontal scale as shown previously in Table 2-3 are applicable here. In Table 2-3 the seven spatial dimensions ranging from microscale to macroscale have been identified.

2.3.3 The Moving Scales. In this section, the moving scales of the lithosphere and pedosphere will be described separately. The former applies to such macroscale motions of the lithosphere as glaciation and denudation, while the latter usually refers to the microscale and/or mesoscale motions of the pedosphere including local erosion and soil formation. The time scale of the two range from

---

[14] Below the lithosphere there is a transition zone (about 60 to 250 km below the earth's surface) consisting of molten rock, known as the plastic zone. Below the plastic zone is the asthenosphere which extends downward to the earth's core (about 400 km). While the lithosphere is made of strong and rigid plates (or brittle rock), the asthenosphere consists of fluid-like materials.

40

Table 2-13 The Horizontal Scales of the Pedosphere

Scale	Taxonomic Unit	Characteristics
Macroscale	Old System: Great Soil Orders: (a) Zonal; (b) Intrazonal; and (c) Azonal	Zonal Soil covers an area as large as several states and includes: Tundra, Podzol, Laterite, Prairie, Chernozem, Sierozem, Red Desert Soil, etc. Intrazonal Soil covers an area the size of a typical county or slightly larger and includes: Forest, Rendzina, Bog, Half Bog, Solonchak, Solonetz, Soloth, etc. Azonal Soil includes Alluvial, Lithosols, and some dry sand.
	New System: Soil Orders described by formative elements: Entisol, Vertisol, Inceptisol, Aridisol, Mollisol, Spodosol, Alfisol, Ultisol, Oxisol, Histisol	The new orders, with the exception of Entisols which are approximately equivalent to the Azonal soils, have little in common with the old system. See Footnote 13, Reference 3, on p. 38 for detailed descriptions.
Mesoscale	Soil Series*	Soil Series covers an area the size of a county or smaller, having similar soil profiles except for the texture of the surface soil. They are developed from the same type of parent material. Defined in the new system, series are a group of pedons closely associated and similar in properties in the field.
Microscale	Soil Individual or Type and Soil Phases	Soil Types cover an area the size of a township or smaller, and are the textural subdivision of the Soil Series. Soil Phase is a subdivision of Soil Type.
	Pedon	The pedon is a sampling volume with surface area between 1 m^2 and 10 m^2 in which the nature of its horizons can be studied and range of properties established.

*To differentiate series, types and phases of soil, for example the Miami Series (a soil series), it is subdivided into Miami silt loam (a soil type) and into Miami silt loam, steep phase, Miami silt loam, stony phase, and so on. The new system carries an alphic character from the order in the nomenclature describing the suborder, great group, and subgroup, such as Mollisol (order), Aquoll (suborder), Argiaquoll (great group), and Typic Argiaquoll (suborder).

Table 2-14$_a$ Surface Areas for Altitude Zones

Altitude Zone	Per cent Over Continents	Per cent Under Oceans	Remarks
Above 5 km	0.1		High mountain
4 to 5	0.4		peaks
3 to 4	1.1		
2 to 3	2.2		
1 to 2	4.5		Average height
0 to 1	20.8		
0			Mean sea level (MSL)
-1 to 0		8.5	Continental margins
-2 to -1		3.0	Ocean basins
-3 to -2		4.8	
-4 to -3		13.9	
-5 to -4		23.3	Average depth
-6 to -5		16.4	
Below 6 km		1.0	Deep trenches
Subtotal	29.1	Subtotal 70.9	Total 100.0

Table 2-14$_b$ Mean Elevation of Latitude Zones*
(Includes Continental and Ocean Areas)

Northern Hemisphere		Southern Hemisphere	
Latitude Zone	Elevation, meters	Latitude Zone	Elevation, meters
0 - 10	158	0 - 10	154
10 - 20	146	10 - 20	121
20 - 30	366	20 - 30	156
30 - 40	496	30 - 40	106
40 - 50	382	40 - 50	5
50 - 60	296	50 - 60	5
60 - 70	202	60 - 70	388
70 - 80	220	70 - 80	1,420
80 - 90	137	80 - 90	2,272
0 - 90	284	0 - 90	216

* The global average is 250 meters above MSL.

one year to one thousand years on the geological time. Glaciation, perhaps, may be an exception to the time scale as will be explained shortly.

Table 2-15 Some Principal Mountain Peaks

Continental Peaks	Height, meters	Continental Peaks	Height, meters
Asia		North America	
Everest, China-Nepal	8,796.4	McKinley, Alaska	6,157.6
Godwin Austen (K2), Pakistan	8,560.9	Logan, Yukon	6,015.2
Kanchenjunga, India	8,529.1	Orizaba (Citlaltipetl), Mex.	5,666.7
Etc.		St. Elias, Alaska-Yukon	5,457.0
South America		Africa	
Aconcagua, Argentina	6,980.0	Kilimanjaro, Tanzania	5,860.6
Ancohuma, Bolivia	6,973.3	Europe	
Etc.		El'brus, USSR	5,600.3

(1) Moving Scales of Lithosphere. On a macroscale in an area the size of a region or continent, the moving scales of the lithosphere are identified by such processes as fluvial denudation and continental glaciation accompanied by abrasive corrosion, solution, and deposition.

It has been estimated that fluvial denudation, assuming a maximum flow rate of sediments as both dissolved and solid, ranges from 10 to 15 cm per 100 years. For the eastern and central regions of the United States, however, the average denudation rates are about 0.5 cm per 100 years. High denudation rates are increased, however, on steep, elevated mountains at high altitude, on rocks of high solubility, and with deep forest penetration. Table 2-16 illustrates the denudation rates of some of the world's major rivers. The rate, of course, is highest for a rapidly moving stream, such that some sections of the Nile River of Africa (6656 km) and the Yellow River of China (4592 km) have greater denudation rates than the ones listed in Table 2-16.

Unlike denudation, the normal flow for a glacier proceeds at a rate between several centimeters and one meter per day. The fastest rate occurs at the center and closest to the bottom of the glacier. Through glaciation various landforms have been created in a shorter time interval than those produced by fluvial denudation. Exceptions to this have occurred, such as the massive erosion of the Columbia gorge and adjoining areas attributed to the breaching of a massive glacial lake and rapid draining of its millions of acre-feet of water during the last glacial period in North America. On the other hand, continental drift is the result of seafloor spreading and the movement of tectonic plates. They are very slow processes (about 1 mm/year) requiring one million years or so for a noticeable change.

(2) The Moving Scales of the Pedosphere. Soil is a dynamic body, ever changing

Table 2-16 Estimates of Fluvial Denudation Rates Of Some Principal Rivers

Rivers	Length km	Rate cm/100 yrs.	Remarks
Amazon (S. America)	6240	0.47	Data obtained from the following sources:
Mississippi (N. America)	6016	0.51	1. Judson, S., and D. F. Ritter (1964), J. Geophys. Res., 69.
Congo (Africa)	4352	0.20	2. Livingston, D.A. (1963). U.S. Geol.
Colorado (N. America)	2320	1.65	Survey, Prof. paper 440-G.
Columbia (N. America)	1936	0.38	3. Judson, S. (1968), Amer. Sci. Vol. 56.
			4. Gibbs, R.J. (1967), Geol. Soc. Amer. Bull., 78.

in its response to the environment and at the same time influencing its environment. Two types of mechanisms describe the responses of soil to its environment: long-term interaction and immediate interaction. The former pertains to soil genesis which requires a few hundred to thousands of years to produce a natural soil profile. Natural agents that govern soil formation are weathering, chemical reactions, physical disintegrations and biological activities. A soil is thus recognized in genetically related layers or horizons, developed from parent materials under the influence of climate, organisms, topography and time. Immediate interaction includes both water and wind erosion. Sometimes, the top soil may be washed away by torrential rain in a few days or weeks. At other times, it may take years for the wind to transport the top soil from one region to another. The transportation of loess is a good example, being carried from the semi-arid regions of the west to the midwest of the United States.[15]

---

[15] The deposit of loess is a relatively uniform, fine material, consisting mostly of silt. A large portion of the deposit is to the west of the Mississippi River and most heavily concentrated in the state of Iowa.

CHAPTER 3   ENVIRONMENTAL SIGNALING PHENOMENA

Extraterrestrial signals in the form of electromagnetic and gravitational waves interact with our terrestrial environment and account for many natural phenomena observed in the atmosphere, oceans, and crust of the earth. The auroral lights seen close to the north and south geomagnetic poles of the earth are triggered by charged particles released by solar flares. These particles interact with the earth's magnetic field and gases of the thermosphere to produce a panorama of colors and shapes in the sky. The reflection, refraction, and diffraction of sunlight or moonlight by water droplets and ice crystals in the atmosphere produce **the** spectacular optical phenomena of rainbows, haloes, and coronas. The ocean and solid earth tides are attributed to the gravitational pull of sun and moon, while the atmospheric tides are thought to be caused by the heating of the ozonosphere as it absorbs solar ultraviolet rays. Of greater importance are the circulations induced in the planetary atmosphere by the absorption of solar radiation: water via the hydrological cycle and heat via the diurnal and seasonal cycles of warming and cooling of the earth's surface. Thus, terrestrial signaling phenomena in the atmosphere, hydrosphere, and lithosphere are dominated to a large extent by energy originating extraterrestrially.

3.1 <u>Extraterrestrial Signals</u>. A multitude of extraterrestrial signals cascades into our terrestrial habitat. Solar radiation, lunar, stellar, auroral, and zodiacal light, and cosmic rays are signals identified and measured in the environment. Solar radiation provides an estimated 99.97 per cent of the energy in the earth-atmosphere system. The estimated 3.67×10^{18} kcal (1.54×10^{22} J) reaching the top of the atmosphere produces an average radiant intensity of approximately 2 cal cm^{-2} min^{-1} (1396 W m^{-2}), known as the solar constant. In contrast, the radiation from lunar and stellar sources would require inputs extending over 90 yr and 70,000 yr, respectively, to equal the amount of solar radiation received at the top of the atmosphere in one day. Since the prime source of terrestrial energy is the sun, we shall begin our discussion with the characteristics of solar radiation.

3.1.1 <u>Solar Radiation</u>. In the upper atmosphere from 60 km and above, the entire solar spectrum from cosmic rays to radiowaves is observed. In the lower atmosphere, however, wavelengths below 0.3 μm ordinarily do not reach the earth's surface because of absorption by atmospheric gases, chiefly oxygen and ozone. Only a trace of ultraviolet rays in the range of 0.26 to 0.29 μm has been detected on occasion. In this book, we shall restrict our study to the 0.3 to 3.0 μm spectral range for surface measurements. However, for upper atmospheric studies, the radiation region below 0.3 μm is highly significant to photochemical reactions. The characteristics of solar radiation are specified by its intensity, quality, and duration. A brief review of the three is in order.

In studying the physical nature of radiation, it is necessary to have a clear concept of radiation intensity emitted by a radiating surface. The amount of radiant energy, E, received per unit time, t, is the radiant flux, expressed as dE/dt (cal min^{-1} or kJ s^{-1}). The radiant flux density, F, is the radiant flux per unit area, A, or

$$F = d^2E/dA^2 \cdot dt \quad (ly~min^{-1}~or~mW~cm^{-2}). \ldots \ldots \ldots \quad (1)$$

In Eq. (1) the unit ly min^{-1} or mW cm^{-2} is langley per minute (cal. cm^{-2} min^{-1}) or milliwatt per square centimeter, respectively. The former is a widely used cgs unit of irradiance (radiant flux density), whereas the latter is from the International System of Units (SI). The relation of the two is

$$1~ly~min^{-1} = 698~W~m^{-2} = 69.8~mW~cm^{-2}$$

Since many early studies adopted cgs units, these units and constants are retained in this book. With the aid of conversion tables in Appendix I, the reader can readily convert from cgs into SI units.

Radiation intensity, I, is designated as the portion of flux density normal to the unit area, A. As shown in Figure 3-1, dA is the differential area; ω, the solid angle; θ, the zenith angle; ε, the elevation angle; and r, the radius. The radiation intensity and flux density are related by

$$I = dF/d\omega \cdot \cos \theta \quad (ly~min^{-1}) \ldots \ldots \ldots \ldots \ldots \ldots \ldots \quad (2)$$

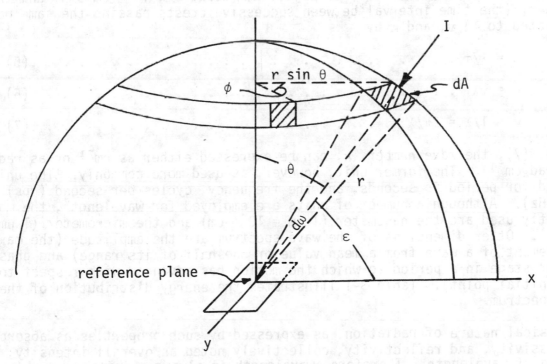

Figure 3-1 Illustration of the Solid Angle of
a Cone of a Beam of Radiation

---

[1]The solid angle, ω, is defined as the ratio of the area, A, of the sphere intercepted by the cone to the square of the radius, r, or $\omega \equiv A/r^2$. Obviously, the area cut out of a sphere by a unit solid angle (i.e., a steradian) is equal to the square of the radius. Also, the whole sphere subtends a solid angle of 4π steradians at the center of the sphere.

46

The integral form of Eq. (2) is therefore

$$F = \int_0^{2\pi} I \cos \theta \, d\omega \; (ly \; min^{-1}) \quad \ldots \ldots \ldots \ldots \ldots \quad (3)$$

When isotropic radiation (i.e., I is independent of direction) is taken into consideration, the relationship $d\omega = 2\pi \sin \theta \, d\theta$ is substituted into Eq. (3) and integrated, yielding

$$F = \pi I \; (ly \; min^{-1} \; or \; mW \; cm^{-2}) \quad \ldots \ldots \ldots \ldots \ldots \quad (4)$$

The annual mean solar intensity received at the top of the atmosphere, where there is no absorption or reflection, is approximately 1.940 ± 0.03 ly min^{-1} or 135.3 ± 2.1 mW cm^{-2} (Coulson, 1975).[2] This extraterrestrial solar intensity is called the solar constant. In applications where preciseness is not required, the solar constant is rounded to 2.00 ly min^{-1} or 139.6 mW cm^{-2}.

The quality of radiation may be expressed by the wavelength, frequency, or wave number. Wavelength, λ, is defined as the distance between successive maxima of a wave. Frequency, ν, is the total number of waves measured in a unit time interval, whereas the wave number, κ, is the total number of waves in a unit distance. The functional relationships of the three can be expressed by the following equations. The velocity of light, c, (about 3×10^{10} cm s^{-1}) and the period, τ, (the time interval between successive crests passing the same point) are related to λ, ν, and κ by

$$c = \lambda/\tau \quad \ldots \ldots \ldots \ldots \ldots \ldots \ldots \ldots \ldots \quad (5)$$

$$c = \nu\lambda \quad \ldots \ldots \ldots \ldots \ldots \ldots \ldots \ldots \ldots \quad (6)$$

$$\kappa = 1/\lambda = 2\pi/\lambda \quad \ldots \ldots \ldots \ldots \ldots \ldots \ldots \ldots \quad (7)$$

From Eq. (7), the wave number, κ, can be expressed either as cm^{-1} or as radians cm^{-1} (rad cm^{-1}). The former unit, however, is used more commonly. The unit employed for period is seconds, and the frequency, cycles per second (cps) or Hertz (Hz). Although a number of units are employed for wavelength, the most frequently used are the nanometer (1 nm = 10^{-7} cm) and the micrometer (1 μm = 10^{-4} cm). Other dimensions of the wave spectrum are the amplitude (the maximum displacement of a wave from a mean value or one-half of its range) and phase (a point or stage in a period to which the motion has advanced with respect to a given initial point). Table 3-1 illustrates the energy distribution of the entire solar spectrum.

The physical nature of radiation, as expressed by such properties as absorptivity, transmissivity, and reflectivity, collectively noted as overall intensity, is a function of wavelength. Therefore, wavelength is always specified when designing a radiometer (an instrument for measuring radiant energy). For a radiometer, the common bandwidth, which is the minimum detectable signal of an instrument usually expressed in a frequency band or wavelength, is 0.1 micron.

---

[2] NASA convened a special committee in 1971 that used eight sets of solar radiation data taken at high altitudes. After each set was weighted according to its reliability, final values of spectral irradiance and the solar constant were adopted.

Table 3-1 Energy Distribution of the Solar Electromagnetic Wave Spectrum +

Radiation Region (spectral band)	Wavelength (approximate)	Radiant Energy	
		Amount (ly min^{-1})	Percent of Total
Cosmic rays	below 1.0 nm	1×10^{-7}*	
X-rays	to 1.0 nm	1×10^{-9}*	
Extreme ultraviolet	to 200 nm	1×10^{-6}*	
Ultraviolet (UV)	0.20 to 0.28 μm	8.0×10^{-3}	0.4
Far UV (C Band)	0.28 to 0.32 μm	2.5×10^{-2}	1.2
Near UV (A Band)	0.32 to 0.40 μm	1.07×10^{-1}	5.4
Visible light A Band (violet, indigo & blue)	0.40 to 0.52 μm	3.5×10^{-1}	18.0
B Band (green & yellow)	0.52 to 0.62 μm	3.0×10^{-1}	15.0
C Band (orange & red)	0.62 to 0.75 μm	2.6×10^{-1}	13.0
Infrared (IR) Near IR (A Band)	0.75 to 1.40 μm	6.4×10^{-1}	32.0
Middle IR (B Band)	1.40 to 3.00 μm	$2.5 \text{ to } 10^{-1}$	13.0
Far IR (C Band)	3.00 to 100.00 μm	4.0×10^{-2}	2.0
Radiowaves GM Band	0.1 to 10 cm	1×10^{-11}*	
DM Band	10 to 100 cm	1×10^{-13}*	
Meter Band	100 cm to 20 m	1×10^{-11}*	
Total		1.98 ly min^{-1}	100%**

A single asterisk under the radiant energy column indicates an approximate value. The double asterisk refers to the value 1.98 ly min^{-1} or $\sim$ 138.2 mW cm^{-2} as 100%; then the energy distribution of the three major solar radiation bands becomes: ultraviolet radiation 7%, visible radiation 47%, and infrared radiation 27%. The total radiant energy of cosmic, x, and extreme ultraviolet rays is too small to be included in the above percentage calculation. This applies to the radiowave region as well. The blue-green band has the highest intensity in the visible radiation region.

+After R.M. Marchgraber (1970) with modifications.

The duration of solar radiation usually is expressed by the time unit of minutes. Occasionally, seconds or a fraction of an hour are employed. Disregarding cloudiness, the period between sunrise and sunset of a day is the possible sunshine hours and the period between the beginning of the civil twilight time to the end of the evening civil twilight time is the total hours of daylight.

3.1.2 Other Heat Sources. The two major sources of heat at the earth's surface are solar energy (extraterrestrial) and the heat flux from the earth's interior (terrestrial). The latter is energy derived from heat released by decay of radionuclides in the earth's interior. Additional heat sources are extraterrestrial energy from the moon, stars, aurora, cosmic rays, meteoritic contrails, geomagnetic storms, and zodiacal light. Of these the full moon contributes the greatest amount of energy, about 3.9×10^{-6} ly min^{-1} (2.72×10^{-4} mW cm^{-2}), whereas zodiacal light contributes about 1.3×10^{-9} ly min^{-1} (9.1×10^{-8} mW cm^{-2}). The intensity of moonlight varies at different phases and its brightness depends on the amount of surface area reflecting sunlight towards the earth.[3]

3.2 Atmospheric Signals. Solar radiation is discussed first, since atmospheric phenomena are controlled mainly by solar radiation and, secondarily, by the gravitational, Coriolis, and centrifugal forces. The receipt of solar radiation results in a cascade of energy transformations in the atmosphere. The differential heating of the earth-atmosphere system produces sensible and latent heat fluxes. Therefore, because of differential heating, differences in atmospheric pressure and density are produced over the earth's surface. This determines the distribution of temperature, wind, humidity, and a number of other atmospheric signals. For the sake of convenience, these signals are grouped under the headings of electromagnetic field, thermofield, hydrofield, dynamic field, and other atmospheric signaling fields. The latter includes visual range, acoustics, atmospherics, and constituents of the atmosphere.

3.2.1 Electromagnetic Field. As shown in Table 3-1, there are three major electromagnetic wave regions in the solar energy spectrum. They are the ultraviolet (UV), visible radiation (VR) and infrared (IR) regions. In nature, the solar energy output below the UV and beyond the IR regions is too weak to have a direct, significant physical effect on the earth's atmosphere. Most of the radiation below the near-UV region does not reach the earth's surface. Therefore, we shall restrict our discussion here to the three major regions. The other regions are mentioned whenever appropriate to the application of an instrument.

---

[3] The albedo of the moon is about 10 per cent. The phases of the moon result from a series of changes in appearance of this satellite viewed from the earth through the synodic month. At new moon, the moon's disk facing the earth is completely dark except for a faint glow due to light originating from the earth and reflected back from the moon. At this point in the moon's orbit, the illuminated side of the moon is facing towards the sun and the unilluminated lunar surface is facing the earth. As the moon orbits the earth from new to first quarter (90°), it is seen on earth as a gradually enlarging crescent to first quarter. Continuing its orbit towards 180°, it is seen on earth as gibbous then full, thus completing its waxing in size and luminosity. The waning phases, as the lunar position changes from 180° through 270° and back to 0°, appear as full, gibbous, third quarter, crescent, and new moon.

The atmosphere is heated principally by energy exchange with the underlying surface. Thus, the absorption, transmission, and reflection of a surface determine the exchange. Of the three, reflection dominates these processes. The fraction of the total incident radiation reflected for a specific waveband is the reflectivity of a surface and for a specific broad spectral band is referred to as its albedo.

The albedo of surfaces such as the ocean, clouds, ice, vegetation, and so on is determined by measurement or estimation of the quantity of reflected to incident solar radiation. There are large differences in the albedo of these common covers of the earth's surface, making them of great importance to radiation balance studies. Different values for albedo over different types of surfaces are dependent on the angle of incident rays and the waveband of the radiation. For example, a freshly fallen snow surface has an average albedo of about 80% in visible light and less than 18% in the IR region. Similarly, the albedo of green leaves lies between 8 and 20% for visible light, less than 10% for ultraviolet and approximately 44% for infrared. In fact, even within the same radiation region, the albedo can differ from one wave band to another for a reflecting surface. Thus, blue light often has a different albedo from red light. In addition, a wet surface has a smaller albedo than a dry surface. An illustration of the albedo differences of wet and dry sand in different wavelengths is shown in Table 3-2.

Table 3-2   Albedos of Wet and Dry Sand*

Surface	IR	V	G	O	R
Wet Sand	19	10	12	15	16
Dry Sand	30	20	23	29	30

*Infrared (IR) centered at 0.8μm; violet (V), 0.4μm; green (G), 0.5μm; orange (O), 0.6μm; and red (R), 0.7μm.

Table 3-3 illustrates the effect of different surfaces on albedo measurements, whereas Table 3-4 shows the effect of the differences in angle of incident rays with respect to albedo for a water surface. Both tables refer to albedo measured in the visible light portion of the spectrum.

Temporal and spatial variabilities of the earth-atmosphere system affect the measurement of radiation. During the day, the system receives much more incoming shortwave (UV and VR) and longwave (IR) radiation than is re-radiated and reflected.

At night the IR outgoing radiation is dominant even though its intensity is much lower than the daytime IR. Complications arise from daily and seasonal variations of the reflecting surface and also from the hourly and daily changes of the constituents of the atmosphere. Constituents of particular concern are the absorbers,

Table 3-3 Albedos of Some Natural Surfaces Under Visible Light

Surface	Albedo (%)	Type of Observation	Observer and Date
VEGETATION	8 to 20*		
Short grass	28.0	Funk-type pyranometer	Arnfield (1975)
Corn field	20.2	Same as above	Same as above
Irrigated potatoes	18.6	Kipp and Zonen solarimeter	Nkemdirim (1972)
Unirrigated potatoes	20.6	Same as above	Same as above
Peas	24.9	Same as above	Same as above
Oats	16.0	Inverted pyranometer	Impens and Lemeur (1969)
Wheat	8 to 29	Not specified	Deamead (1976) Paltridge et al. (1972) Stanhill et al. (1972)
Sugarcane	6 to 18	Pyrheliometer	Chang (1961)
Rice	9 to 30	Not specified	Uchijima (1976)
Cotton stand	25	Not specified	Stanhill (1976)
Pine forest	8.9 to 11.7	Kipp solarimeters	Stewart (1971)
	12	Measured from satellite	Conover (1965)
Salt marsh grass	18	Inverted star pyranometer	Felton (1978)
Salt marsh succulent	14	Same as above	Same as above
SNOW AND ICE	80 to 90*		
Pack ice	80	Not specified	Kondratyev (1969)
Pack ice dirty melting to ice covered with new snow	40 to 90	Not specified	Kraus (1972)
Snow, Antarctica	98	Not specified	Kondratyev (1969)
Snow cover in coniferous forest	35	Not specified	Same as above
Fresh snow, 3-7 days old	59	Measured from satellite	Conover (1965)
CLOUDS	65 to 75*		
Cumuliform			
80°/o cover	29** to 69	Measured from satellite	Conover (1965)
individually	86 to 92	Same as above	Same as above
Cirriform			
thick, with lower clouds	74	Same as above	Same as above
individually	32 to 36	Same as above	Same as above
Stratiform			
80°/o cover	68 to 69	Same as above	Same as above
individually	42 to 64	Same as above	Same as above
OCEAN			
Calm surface	5 to 10	Not specified	Kondratyev (1969)
Moving surface			
clear skies, solar elevation of 70°	3.5	Eppley pyranometers	Payne (1972)
	5.0	Not specified	Kondratyev (1969)
	3.0	Same as above	Hollman (1968)
solar elevation of 10°	28	Eppley pyranometers	Payne (1972)

*
Figures with one asterisk refer to most frequent range of values.

**
The low value was reported for cumulus of fair weather over land, with the underlying surface viewed by the satellite camera contributing to the total albedo.

Table 3-4 Albedo of Water Surface as Measured
at Different Solar Zenith Angles

Zenith Angle (degrees)	40	50	60	70	80	90
Undisturbed water-surface*	2.5	3.4	6.0	13.4	34.8	100
Disturbed water-surface (moving water)	7.0	10.0	16.0	26.0	47.0	100

*
The values may be computed from Fresnel's formula for albedo of sunlight, A, or
$A = 1/2[\sin^2(Z-r)/\sin^2(Z+r) + \tan^2(Z-r)/\tan^2(Z+r)]$, where Z is the zenith angle
and r is the angle of refraction. The albedo for zenith angles below 40° is
about 2%.

namely H_2O, O_3, CO_2, and N_2O in the tropospheric air. Thus, a number of radiometers have been designed to measure the directional fluxes of radiation and
various wavebands of radiation.

In 1971, the WMO published the Guide to Meteorological Instrument and Observing
Practices[4] in which the nomenclature for several types of radiometers was recommended (Table 3-5). The net radiometer measures the difference between incoming
and outgoing radiation or net radiation. Nomenclature for the instrument that
measures the net shortwave radiation is net pyranometer; for that of longwave
radiation, pyrgeometer; and that for both short- and longwave, net pyrradiometer.
The latter, however, is also commonly known as a net radiometer. It must also be
noted that global radiation refers to the measurement of both diffuse sky radiation (skylight) and direct solar radiation (sunlight).

In addition to the quality and intensity of radiation, the total hours of either
sunshine or starshine are measured. A sunshine recorder measures the total hours
of visible sunshine directly and cloudiness indirectly during the day. At night,
a starshine recorder is employed to measure the total hours of starlight (with
reference to the star Polaris) when the sun is 10° or more below the horizon.
The time scale accuracy of a sunshine or starshine recorder is 0.1 to 0.2 hr,
respectively.

Although astronomers have calculated the duration of twilight, its obscuration
by clouds, fog, and haze is not commonly measured. The duration of civil twi-

[4]See WMO - No. 8, 4th Edition (in English and French).

Table 3-5 Nomenclature of Radiometer

Nomenclature	Radiation Flux Measured	Example of Radiometer
PYRHELIOMETER pyre (L. pyra = hearth) helio (Gk. = sun) meter (Gk. metron = measure)	Direct solar beam (or components) on a plane surface at normal incidence, or simply solar radiation.	Abbot Silver Disk Pyrheliometer and Ångström Compensation Pyrheliometer
PYRANOMETER ano (Gk. ano = upward)	Total shortwave radiation (or components) on a horizontal plane surface from a hemispheric sky (or 2π solid angle).*	Eppley Pyranometer; Moll-Gorczynski Pyranometer (solarimeter); Robitzsch Bimetal Pyranometer (actinograph)
PYRRADIOMETER (radiometer = an instrument for measuring radiant energy)	Global radiation on a horizontal plane surface from a hemispheric sky for both short and long wave (or 2π solid angle).	Albedometer and perhaps Eppley Pyranometer
PYRGEOMETER geo (Gk. = the earth)	Net atmospheric IR radiation on an upward facing black plane surface sustained at ambient temperature.	Ångström Pyrgeometer
NET PYRANOMETER	Net shortwave radiation (or components).	Net radiometer equipped sky shield
NET PYRRADIOMETER	Net short- and longwave radiation (or components).	Gier and Dunkle Net Exchange Radiometer, etc.

* Sometimes it has referred to the measurement of shortwave global radiation.

light time[5] for latitudes 0° to 25° is about 21 to 26 minutes for either morning

[5]Since daylight is defined as the interval between sunrise and sunset, the civil twilight time is designated as the interval between sunrise or sunset and the time when the true position of the center of the sun is 6° below the horizon, at which time stars and planets of the first magnitude are just visible and darkness forces the suspension of normal outdoor activities. The nautical and astronomical twilight times depend on the angle subtended between the sea level horizon and the position of the sun. The angles are 12° and 18° for nautical and astronomical twilight time, respectively. The true center of the sun is considered to be 50' below the horizon when the sun's upper limb contacts the apparent horizon.

or evening twilight during different seasons of the year. It lasts approximately 30 to 120 minutes for latitudes 35° to 65°, and 1.5 to 24 hours for latitudes 75° to 90°.

3.2.2 Thermofield. Measurements taken of the thermofield in the atmosphere usually measure air temperature rather than heat. To accurately measure the true air temperature two major problems must be solved: the time response and the radiation error of the temperature sensor. The former is the time required for the sensor to respond to a change in temperature; the latter refers to radiational heating of the sensor, principally by solar radiation. The problem of time response becomes serious if a high speed vehicle is employed as the platform. Radiation error is most severe on sunny, windless days when the thermometer lacks a sufficient radiation shield. It is even more serious for high altitude observations.

Differentiation of day versus night temperatures is necessary in upper-air soundings as well as in biometeorological measurement near the ground. In upper-air measurements, the problem is essentially one of instrumentation. With biometeorological measurements, it is a matter of application. The latter may refer to a variety of engineering applications, for example, the control of heating and refrigeration as well as icing problems on aircraft and highways. This is particularly true in the growth of plants.

3.2.3 Hydrofield. Atmospheric moisture in gaseous, liquid, and solid forms gives rise to spectacular and fascinating signaling phenomena. In fact, nearly all optical and weather phenomena in tropospheric air are associated with certain amounts and forms of atmospheric moisture. However, the variations in moisture content with space and time are so large that there is great difficulty in making measurements and predictions with reasonable accuracy. For this reason there have been many studies on instrumentation for atmospheric moisture determinations.

Different ways of expressing the amount of water vapor in the atmosphere have been introduced and the choice is up to the user. Nomenclature is given in Table 3-6, along with a listing of some common instruments. Expressions for liquid and solid water are presented under Hydrospheric Signals in Section 3.3

3.2.4 Dynamic Field. Wind and atmospheric pressure are two major dynamic fields of great concern to meteorologists. Wind is designated as the horizontal motion of the air, and pressure is the weight of the entire column of air exerted on a horizontal surface. The velocity of wind is essentially a result of variations in the vertical and horizontal pressure distribution. The latter is caused by differential surface heating as well as the rotation and gravitational field of the earth. Of all these variables, wind velocity is the most important variate, as it defines the rate at which heat, foreign matter, and moisture, including clouds, are transported from one region to another. Wind plays an important role in the hydrologic cycle. It influences the oceanic waves; it dominates the weather. Wind as a signaling phenomenon is discussed first.

Table 3-6 Specifications of Atmospheric Moisture

Nomenclature	Symbols and cgs unit	Definition	Common Instrument
Vapor density (absolute humidity)	ρ (g cm^{-3})	Mass of vapor per unit volume of air, or simply vapor concentration.	gravimetric hygrometer
Vapor pressure	e (mb, mm Hg, in. Hg)	Partial pressure of the total atmospheric pressure exerted by water vapor.	diffusion hygrometer
Relative humidity	r (%)	The ratio of vapor pressure of the air to the saturation vapor pressure with respect to water at the same temperature.	hair hydrometer; goldbeater skin hygrometer
Dewpoint temperature	T_d (°C or °F)	The temperature to which the air must be cooled until saturated with respect to water at its existing pressure and moisture.	dewpoint hygrometer
Specific humidity	q (g kg^{-1})	The mass of vapor per unit mass of moist air.	gravimetric hygrometer
Mixing ratio	ω (g kg^{-1})	The mass of vapor per unit mass of dry air.	mixing ratio indicator
Wet bulb temperature	T_w (°C or °F)	The temperature reading due to evaporation from a water saturated sensor.	psychrometer

Since air motion is always oblique and never truly horizontal, the wind direction may be described by its three components along the three axes of a rectangular coordinate system as shown in Figure 3-2. Here, u, v, and w are the wind velocity components due east, north and upward, directions considered positive. The opposite directions (west, south, and downward) are negative. The vertical wind component is about 1 to 10 percent of its horizontal magnitude, with the exception of strong vertical convection found in hurricanes and tornadoes. Even with such small speeds the vertical wind is far more important than its horizontal counterpart. This is simply because all weather phenomena are essentially a product of its vertical instead of its horizontal component. Due to economic reasons, the national weather services the world over have not yet adopted routine vertical wind measurements even though several types of instruments are available for this purpose.

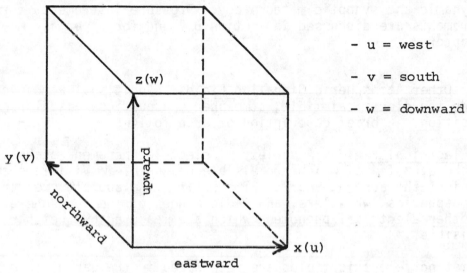

- u = west

- v = south

- w = downward

Figure 3-2 Cartesian Coordinates
and Wind Components

The wind direction is defined as the direction from which the wind is blowing. The direction, which is specified from the true geographical north, is identified by either compass bearings in degrees clockwise or 8, 16, or 32 points of the compass according to the accuracy required. The wind speed is expressed by the unit knot by international agreement (since 1956). Conversions from knots to several alternative speed units are given in Appendix 1.

A scale for estimating wind speed by visual observation was devised by Admiral Beaufort of the British Navy early in the 19th Century. His wind scale, known as the Beaufort Scale (B.S.),[6] was originally designed for use

---

[6] For details, see List, R.J., ed. 1958. Smithsonian Meteorological Tables.

over the ocean. It has since been modified for use over both land and sea.
The relations between the estimated values and the anemometer readings are:
$v = 1.87\sqrt{B^3}$, where v is in mph and B is the corresponding Beaufort Scale number. The pressure equivalents are $P = 0.003 \, v^2$, where P is the pressure in
lbs ft^{-2} and v is in mph.

The atmospheric pressure unit, called the millibar (mb), has been adopted as
the international pressure unit in meteorology. The Système International
d'Unités (S.I.), a refinement of the metric system, defines the standard
pressure unit as a pascal (Pa), equal to 10^{-2} mb (see Appendix I). It is
interesting to note that the Meteorological Office of Britain adopted the
millibar system as early as May 1, 1914.

Vertical and horizontal distributions of pressure worldwide are the bases
for dynamic and synoptic meteorology. Pressure instruments for surface
measurements are discussed in Section 6.3 and for upper air measurements,
in Chapter 7.

3.2.5 <u>Other Atmospheric Signaling Fields</u>. Additional atmospheric signals
include the electrical field, atmospheric acoustics, visual range, and air
composition. A brief description of each follows.

(1) <u>Electrical Field</u>. A variety of electric phenomena occur in the earth's
atmosphere, a few of which are visible or audible. Visible phenomena include
lightning, the aurora, and St. Elmo's fire; the audible ones include sferics
or atmospherics, whistlers, and radio echoes of the ionosphere. There are
many other electrical phenomena which are measured but neither are audible
nor visible.

Two distinct electric fields are recognized in the earth's atmosphere: surface and upper atmosphere. Surface fields are measured between the ground
and cloud layers, whereas the upper atmospheric field is measured in the
ionospheric region from 60 km and upward.

Most of the studies in the past were restricted to surface electric fields.
These studies distinguished two types of electric fields in the lower atmosphere: the undisturbed field (or fair weather electricity) and the disturbed
field (or stormy weather electricity). Measurements have shown that electric
charges, conduction currents, and potential gradients change abruptly from
undisturbed to disturbed states in the electric field. The occurrence of
haze, smoke, fog, low clouds, or rain is sufficient to produce minor electrical
disturbances. Thunderclouds (cumulonimbi) or lightning discharges definitely
generate major disturbances. The vertical electric potential gradient, for
example, under fair weather conditions is on the order of 1.3 volts cm^{-1}; in
thunderclouds accompanied by lightning discharges it can reach as much as
4000 volts cm^{-1}. These large potential gradients occur when the electric
potential between the cloud and ground has reached a critical value on the
order of 10^8 volts.

Surface electric fields are generated in several ways. Surface fields are
produced by ion production resulting from the presence and radioactive decay

by such substances as uranium, radium, and thorium at the earth's surface, and by cosmic radiation ionizing air molecules. The dissipation of atmospheric ions over the earth's surface occurs primarily by ion interaction with liquid aerosols, dust, and other foreign matter.

The dissipation rate of ions is faster over land, where the suspended particles are very large, than it is over the open ocean.

Space electricity, on the other hand, is produced by the ionization induced by x-rays, and extreme ultraviolet ray and cosmic ray interaction with atmospheric gases. The dissipation factor is geared primarily to the presence or absence of sunlight. Spatial and temporal variations are observed in the magnitude of the electric field. For example, great variations are observed between disturbed and undisturbed weather conditions near the earth's surface, whereas there are only small variations in the upper atmosphere electrical field.

Table 3-7 Electrostatic Measurements of the Atmosphere

Measurements	Symbols and Dimensions	Instruments
Conduction current	i, esu $cm^{-2}s^{-1}$	Electrometer
Total conductivity	λ, esu s^{-1}	Double Aspirator, Gerdien
Total ion concentration	n, cm^{-3}	Ebert Ion Counter (for small ions)
Mobility of ions	k, $cm^2 v^{-1}s^{-1}$	Ion Spectrometer

Several types of instruments have been designed for measuring the surface electric field. Besides the instruments for measuring electric field strength, devices are available that measure suspended particles (e.g., aerosols and condensation nuclei), radioactivity, ion concentration and mobility, and air-earth currents. Instruments that measure space electricity detect such things as the concentration of ions and numbers of free electrons in the upper atmosphere. Ionograms are produced regularly on a worldwide basis from these upper air measurements. Special sensors attached to missile, rocket and satellite platforms also monitor the upper atmosphere electric field.

(2) Acoustical Field. The origins of acoustical waves in the atmosphere are associated with two distinct sources: sounds originating from (a) meteorological phenomena, and (b) artifically produced sounds. The former pertain to the eolian sounds of the wind, thunder, impact noises of precipitation, and many others. The latter refers to noises generated as shock waves from nuclear detonations, super-sonic aircraft, and mechanical vibrations from urban activities.

The propogation and attenuation of sounds in the atmosphere, with the exception of shock waves, are recognized by pitch (frequency of the impulses), loudness (energy or intensity of the vibrating source), and quality (complexity of vibrations or numbers of overtones). The shock wave, by contrast, is identified by the abrupt changes of atmospheric pressure and temperature. There have been documented reports of shock waves shattering windows, and heating associated with the extreme high pressures and temperatures of the waves producing in infants and elderly persons skin burns and cardiac arrest. Acoustic waves are classified into three major types as a function of the frequency of vibration and the speed of propagation.

Wave Type	Range of Frequency (Hz)
Infrasonic	< 20 Hz
Audible	20 to 20,000 Hz approx.
Ultrasonic	> 20,000 Hz

The speed of sound, c_s, in the atmosphere is directly proportional to the square root of the absolute temperature, T. for dry air, it is

$$c_s = 20.08\sqrt{T} \ (m \ s^{-1}) \ \ldots \ldots \ldots \ldots \ldots \ldots \ldots \ldots (8)$$

The speed c_s is modified by the presence of water vapor, droplets of fog and cloud, temperature inversions, and wind speed and direction. The amount, size and density of suspended droplets produce complicated relationships in the speed and direction of sound propagation. The attenuation of acoustic intensity depends upon the processes of absorption and scattering similar to those observed with electromagnetic radiation. Attenuation is a function of air and moisture density in general, concentration of fog and cloud droplets, and possibly intensity of air turbulence.

The intensity of sound, I, is expressed in decibel units (db), as follows

$$I \ (db) = 10 \ \log_{10} \ (I_2/I_1) \ \ldots \ldots \ldots \ldots \ldots \ldots \ldots (9)$$

where I_1 is the reference intensity with a value of 10^{-10} mW cm^{-2}. Thus, Eq. (9) gives a measure of relative values of two sound intensities, I_1 and I_2. It is interesting to note that when the intensity of sound reaches the critical level of 130 db, it becomes injurious to the human ear, especially if it continues over a period of time.

(3) Visual Range. Visibility is another atmospheric signaling phenomenon. In practice, visibility or visual range is defined as the greatest distance (in kilometers or miles) in a given direction to which an observer can see and identify visually. Both the vertical and oblique visibility are important for aviation. Unfortunately, this aspect of visual range is still in the research stage. No practical methods for its determination have been established thus far by the National Weather Service.

Although various prototype instruments for visual measurements are available, the daytime visual range can be observed by eye just as accurately as by instrument measurement. However, transmissometers have been used along airport runways for an automatic and continuous observing of daytime and nighttime visual range. A backscattering visibility monitor is useful in areas where a target is not available, such as over the open ocean, along a valley, or on a rapidly moving vehicle.

Factors that determine visual range are the size and color of the target, the background conditions including the azimuth and elevation angle of the sun, and the limitation and difference of each observer's vision. Visual contrast between the target and the background is the most important factor. Thus, it is necessary to have (a) a prominent dark object against the sky at the horizon in the daytime, or (b) a known, preferably unfocused, moderately intense light source at night. After visual ranges have been determined around the entire horizon circle, they are resolved into a single value describing the average prevailing visibility, unless the visual ranges of predesigned directions have to be reported. For details, the reader may refer to the surface synoptic codes of the WMO for visibility determination.

(4) <u>Air Quality</u>. Another atmospheric signal is the chemical composition of the atmosphere. In tropospheric and stratospheric air the permanent constituents, namely N_2, O_2, Ar and the other rare gases, remain constant. But many gases such as CO, CO_2, NO_x, SO_2, O_3, CH_4, NH_3, and so on are present in trace amounts on the order of parts per million (ppm) or parts per billion (ppb). In addition, many submicron size metallic and non-metallic substances remain suspended in the atmosphere. These are known as particulates. Most particulates and some inert gases, such as CO, are good tracers of air movement. In other words, the trajectories of the air can be determined from time and space variations in the concentrations of tracers. Other trace gases, such as O_3, vary greatly with time and space.

3.3 <u>Hydrospheric Signals</u>.[7] The various forms of precipitation, such as showers and drizzles of liquid water, and hailstones and snow of solid water, are magnificent to behold. Even more fascinating are the crystalline forms of snowflakes and the microscopic structures of frost and rime. The various forms of water waves, waterfalls, and waterspouts are other attractions of hydrospheric phenomena. Some of the measurable hydrospheric signals are discussed below.

3.3.1 <u>Snowfall</u>. The rate at which snow falls is called snowfall, and the accumulation of snow is called snow depth. The words <u>snow cover</u> sometimes refer to the areal extent of snow covered ground, and is usually expressed as per cent of total area in a given region. When snow only covers the ridges and peaks of mountains but is not found at lower elevations, it is referred to as a snow cap.

The amount of annual accumulation of snow at higher elevations in the western

---

[7] Water in vapor form is discussed in Section 6.6.

60

United States usually is expressed in terms of the average water-equivalent of the snow pack. In the USA the snow-water equivalent is computed by setting a snow depth of ten inches equal to one inch of rainwater. There is some evidence to suggest that the mountainous areas of the West and High Plains have snow to water conversion ratios less than 0.1, with values on the order of 0.07 at least. This is attributed to the cooler temperatures of formation and, consequently, reduced water vapor content of the air masses producing the snow, yielding a drier snow than that found in midwestern and eastern states (Auer, Jr., 1975). In Britain 30 cm of freshly fallen snow is converted to 2.5 cm of equivalent water. It is obvious that the British method applies the 1/12 ratio or 0.08 for water-equivalent computations whereas the U.S. uses the 1/10 or 0.1 ratio. The specific gravity of snow (or snow density) varies with the type of snow storm and the duration of snow cover. A freshly fallen snow has a density between 0.07 to 0.15 g cm^{-3}, with values as low as 0.004 g cm^{-3} measured. Glacial ice formed from compacted and transformed old snow (called firn) has a maximum density of approximately 0.91 g cm^{-3}. Therefore, the most accurate and simplest method is to melt the snow sample for each determination. The identification of a snow day in British climatology is based upon the occurrence of snowfall within a 24-hr period. This time interval is regarded statistically as a day of snow.

It is important to note the role of snow measurements in the context of water conservation. For the engineer, the construction of a dam or reservoir, the control of floods, the design of irrigation facilities, and even the construction of buildings rely to a great extent upon a knowledge of the amount of snow accumulation (both depth and coverage) in a given region. Meteorologists, on the other hand, are interested in the structure of snowflakes as they relate to the origins and physical processes within a storm system. In this case, atmospheric temperature and, to a lesser extent, atmospheric pressure determine the shapes and forms of snowflakes.

3.3.2 Rain and Hail. Rainstorms and hailstorms are somewhat related phenomena. Precipitation in the form of rain is far more frequent and greater in quantity throughout the tropics and mid-latitudes than precipitation as hail. Therefore, rain is stressed in this discussion.

(1) Rainfall. In general, measurement of rainfall is specified by its intensity, amount, and duration. The intensity of precipitation is the rate of fall per unit time, usually expressed in inches or millimeters of rainwater per hour. For accurate measurements, however, it is measured in mm min^{-1}. In practice, the mean rate of rainfall has been specified as follows:

Intensity of Rain	Rate of Fall
slight	< 0.5 mm hr^{-1}
moderate	0.5 to 4 mm hr^{-1}
heavy	> 4 mm hr^{-1}

The amount of rainfall is expressed in millimeters (mm), which is the WMO standard to be used in preferences to inches (in.). In the United States and a few other countries, however, inches are still employed as the unit of measurement. A depth of 1 mm of rainwater is equivalent to 1 kg m^{-2}, whereas 1 in. of rain is approximately 100 tons per acre. The depth of all solid precipitation, such as snow, sleet, rime, and hail, is converted into millimeters equivalent of rainwater for statistical purposes of climatological records. The total amount of rain falling in one day is specified in mm day^{-1}. Synoptic reports give rainfall amounts for six hour intervals at 0000Z, 0600Z, 1200Z, and 1800Z (GMT).

The duration of rainfall is measured to an accuracy of 0.1 hr intervals. Accurate rate-of-rainfall recorders, such as the Sil intensity gage, record rates for time intervals as brief as 5 s. A rainfall intensity gage utilizing electronic sensors and records is capable of accurately measuring the fall over a few seconds.

A day with rainfall has been defined for statistical purposes in climatological studies. Both in Britain and the United States a rain day is defined as a day with measurable rainfall of 0.01 in. or 0.2 mm or more. Such a definition often is not practical. Should 0.2 mm (0.01 in.) of rain fall on a warm and windy day, it would soon evaporate. Such small amounts of rain have negligible effects on water conservation and human activities. Such a definition of day with rainfall can be meaningless in terms of its significance in growth and development of crops. Instead, a crop rainy day has been established wherein the effects of evaporation are considered.[8] The drop size distribution in a rainstorm is important in the estimation of the intensity and amount of precipitation using information from the return echoes on a radar scope. The typical diameter of a raindrop is of the order of 1 to 2 mm. The probable limiting diameter of a drop, however, is ∿0.2 mm if it is to survive evaporative dissipation for a distance of several hundred meters. The exact survival distance is, of course, a function of relative humidity.

(2) Hailstones. Another manifestation of both atmospheric and hydrospheric signaling phenomena is hail. An individual unit of hail is called a hailstone, usually having a spherical or conical shape with a diameter ranging from 0.25 inch to more than 5 inches (5 mm to over 100 mm). The density of hail can vary between 0.1 and 0.9 g cm^{-3} with a mean of about 0.8 g cm^{-3}. Varieties of hail are classified by structural differences. Some common types are: (a) soft hail, which is crisp, opaque and easily compressible pellets; (b) small hail, which has a soft nucleus and an outer coating of clear ice, also pellet-shaped; (c) hail, which is concentric layers of alternately clear and opaque ice; and (d) small hail, which has a soft hail nucleus and outer coating of clear ice.

All the characteristics of rainfall in terms of its intensity, duration, amount, and size distribution are applicable to hail. Visual observations, direct measurements, and radar detection are employed in hail measurements.

3.3.3 Hydrometeors. Any produce of condensation, sorption, sublimation, or

---

[8]Wang, J.Y. 1961. Moisture: Normals and Hazards, Part A, Rainfall, Phyto-Climate of Wisconsin. Agricultural Experiment Station, University of Wisconsin, 64 pp.

precipitation of atmospheric water vapor, whether in the free atmosphere or at the earth's surface, is considered a hydrometeor. Hail, rain, and snow, mentioned previously, either reach the earth's surface as precipitation or evaporate during their fall as virga. Hydrometeors suspended in the air form mist, fog, or clouds, those lifted from the earth's surface upward form spray or drifting and blowing snow, and those deposited on the surface of the earth form dew, hoarfrost, or rime.

All of the above hydrometeors result from the processes of condensation, precipitation, or sublimation. Sublimation occurring underground produces frost and/or ice. The depth of frost penetration depends upon moisture, temperature, and aeration of the soil. Should a surface be hygroscopic in nature and if its temperature is higher than the dewpoint of the surrounding air, sorption of water vapor occurs on the surface rather than condensation. Sorption processes include: liquid sorption, dissolving of vapor in water soluble solids; adsorption, adhesion of vapor to nonsoluble solid surface; absorption, intake of vapor by organic substances without being dissolved; and chemical sorption, reaction of vapor with chemical fertilizers to form new compounds. A significant amount of moisture gained in farm areas is through the sorption process.

3.3.4 <u>Watermass Transport</u>. Water in its three states is constantly in motion. In the atmosphere, water vapor presumably moves at the same speed and direction as the wind or air currents. A large amount of the world's water vapor is carried in the atmosphere, particularly via meridional transport. In fact, over the years the total transport of water in the atmosphere is far more efficient, in terms of quantity and geographical distribution, than it is in the hydrosphere and lithosphere. On the other hand, water storage is least in the atmosphere (for details see Table 2-6$_a$).

(1) <u>The Continent</u>. Once a mass of water reaches the surface of a layer of soil, as shown in Figure 3-3, it has four possible means of dissipation: (a) it is stored in the soil as soil water or hygroscopic water; (b) it infiltrates or percolates through the layer as gravity water; (c) it flows away along the surface and/or the subsurface as drainage water or runoff; and (d) it evaporates from the soil surface or transpires through vegetation as gaseous water or water vapor. Processes of water loss over the continents, arranged in descending magnitude, are <u>evapotranspiration</u>, <u>runoff</u>, <u>percolation</u>, and <u>storage</u>. Moisture levels change very slowly in soils. Most often soil is unsaturated with respect to liquid water, with moisture levels between wilting point (tension of 15 atmospheres) and field capacity (tension of approximately 0.33 atmosphere). The soil, if saturated with water and allowed to freely drain for 2 or 3 days until free drainage has essentially ceased, leaves water in the micropores of the soil. This moisture state is described as the field capacity. The soil field capacity of most soils, except for heavy clay and peat soils, is less than 45 per cent (expressed as the ratio of water mass to soil dry weight). Excess water in a saturated soil infiltrates through soil layers and percolates as ground water, then seeps to open bodies of water to ultimately enter the ocean. Exceptions are drainages into ephemeral lakes whose waters do not flow out of the basin. Another method of dissipating excess water, of course, is through surface and subsurface runoff. Runoff is more of a problem on sloping land than flat land. To reduce soil erosion and surface runoff, downward percolation has to be promoted in the soil.

The net storage in the soil layer over one year is usually very small. Net global runoff should be zero. The net runoff over the continents exceeds storage but is always smaller than evaporation. An exception is Antarctica as given in Table 2-10b.

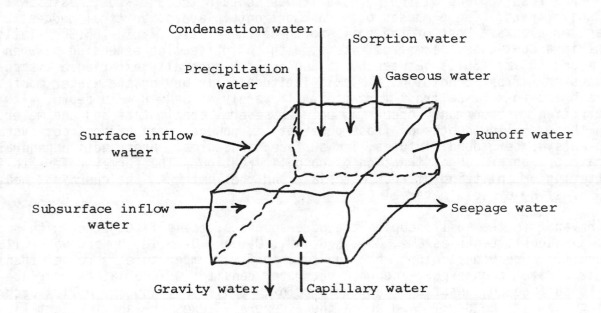

Figure 3-3 Watermass Transport with Respect
to a Layer of Soil

Among all the sources supplying water to a layer of soil, as shown in Figure 3-3, precipitation is the principal source. Modifications to Figure 3-3 must be made for different geographical regions and landscape types. For example, in an arid region dew resulting from condensation and frost resulting from sublimation can be major water sources. Dewfall is the only daily water source in some deserts and marginal lands. Farmlands have sorption as an additional water process because of the abundance of organic and chemical compounds in the soil. Outflow from the uplands becomes a source of water inflow for the adjoining lowlands. Aerable soil has capillary or small pore water available as a water source. Coastal shrub and forest communities collect fog particles on leaves and stems, thereby supply fog drip or fog precipitation to the underlying soil. The plateau of Table Mountain, Cape Town, South Africa, has an annual total of fog drip twice that of rainfall. Trees and other vegetation intercept cloud and drifting fog droplets. The coastal redwood forests (Sequoia sempervirens) of California and Oregon collect as much as 0.13 mm (0.05 in.) of fog drip, equivalent to a moderate shower in a single night.

It can be concluded that the principal gain of water in a soil layer is from precipitation and the major loss is through evaporation or runoff. Generally, the annual water budget of the continents show the total gain from precipitation is greater than the total loss from evaporation. The reverse occurs over most of the oceans. Estimates of the global water balance for the oceans and the continents are given in Tables 2-10$_a$ and 2-10$_b$ and discussed under Section 2.2.3. The transfer of watermass in solid and gaseous forms also occurs over the continents.

The permafrost regions of the high latitudes contain ice and frost masses within the soil layers. These masses occur as horizontal layers, vertical wedges, and irregular blocks. Most of the soils north of the Arctic Circle (66.5° N latitude) exist in a continuous permafrost state, with solidification extending between 300 m and 600 m. Soils between 60° N and 66.5° N generally experience discontinuous permafrost conditions. Middle latitude soils during the winter months are subject to surface frost formation. It rarely exceeds 8 m in depth, with penetration most commonly occurring between several centimeters and one meter. Even these shallow depths of frost penetration, however, can affect root systems of plants, underground fixtures, and surface structures. Frost action generally is categorized as congelifraction or congeliturbation. The former refers to the shattering or splitting of rock material, and the latter to the churning, heaving, and thrusting of soil material.

Water vapor in the soil occupies the pore spaces of the soil together with such soil air constituents as CH_4, CO_2, N_2O, NH_3, O_2, and O_3. The upward vapor flux is usually observed at night when the soil surface temperature is colder than its substratum counterpart and when the water density is lower at the surface due to surface condensation than it is below. Conversely, downward flux occurs during most of the day. Imposed on these general fluxes are shorter term flux reversals within portions of the soil profile in response to localized thermal and moisture gradients. In addition to the diurnal cycle of soil vapor flux, other weather phenomena impose changes in the movement of water vapor in the soil profile. Heavy rainfall, intense radiation, high winds, and freezing conditions affect moisture movement. A heavy rainfall almost completely exchanges the air in the soil pores. Saturated soil conditions produced by prolonged rain infiltration result in pore space being occupied by liquid rather than gaseous water. High radiation levels at the soil surface set up large vapor density gradients, fostering rapid soil drying. Strong winds reduce surface vapor pressure, resulting in moist soil air moving towards the surface and its eventual replacement with drier air.

Water found in the rocks of the earth's crust and mantle has geologic and hydrologic importance. Porous rocks, such as sandstone, are permeable to water. Water is retained by and transmitted as gravity water through permeable rock layers. Some igneous and tightly cemented sedimentary rocks have only minute pore spaces and are relatively impervious to water.

Water locked within the rock matrix over geologic ages is referred to as meteoric water. The connate form of meteoric water is water originally present in the pore spaces of the original material composing the rock; this water is retained in the pores. Magmatic water is expelled from cooling igneous magmas, although it is not exposed at the surface or the air above it.

(2) <u>The Ocean</u>. Energy and mass transfer dominate the dynamic processes of the open ocean and large bodies of water. Small bodies of water, such as ponds, wells, and creeks, are dominated by boundary effects which overly and complicate the mass and energy processes occurring in these small bodies. The primary forces that govern mass transport and, in turn, determine various oceanic signaling phenomena are solar radiation, winds, solar and lunar gravitational attraction, and submarine earthquakes. All of these forces propagate through the boundaries of an ocean. Secondary forces are induced by vertical temperature and salinity gradients. The interactions of these forces are associated with various oceanic eddies.

Various sizes of eddies are observed in the ocean. The most prominent ones are wind waves, tides, and ocean swells. Less noticeable or less frequent are eddies due to upwelling, tsunamis (seismic sea waves), subsurface currents, and surface currents. Until Benjamin Franklin's inspired deduction that the Gulf Stream was a "river of the sea," discrete oceanic currents were not identified as coherent entities. Since then, many submarine currents and tsunamis have been identified. The dimensions and characteristics of these eddies are summarized in Table 3-8.

The clockwise surface oceanic currents in the Northern and counterclockwise in the Southern Hemisphere are established as a result of the three driving forces: excessive heating of equatorial water, the Coriolis effect, and prevailing trade winds. Examples of these giant eddies (macroscale) are the Gulf Stream of the Northern Atlantic and the Kuroshio Current of the Northern Pacific. These eddies and others are constantly interrupted by tidal waves (twice daily), swells over the open sea, and upwellings (mesoscale) along the coastal waters. Also, they interact frequently with wind waves (microscale) and occasionally with the tsunamis. Even subsurface currents and deep ocean currents, which take part in the general oceanic circulation, substantially affect the physical dimension of these eddies.

The various eddies, aside from their interactions and interdependence in the oceans, have tremendous impacts on the atmosphere and lithosphere. These occur through sea-air interaction and land-sea interaction, respectively.

The ocean is the single most important supplier of moisture to the atmosphere. It is estimated that over 84 per cent of the total atmospheric moisture comes from evaporation of ocean waters. As a result of this evaporation, oceans help stabilize the world's climates. In addition, most of the world's weather is geared to the surface temperature. Surface temperature has an effect on the formation of tropical cyclones, monsoons, land-sea breezes, sea fog, and, to some extent, planetary and secondary circulations. In short, the various sizes of atmospheric eddies are controlled to a great extent by the state of the ocean.

Land-sea interactions (i.e., ocean effects on land and vice versa) occur on various scales. Waves generated on the ocean surface often are transformed into surface when frictional interaction between the water wave and nearshore bottom topography steepen the wave while shortening its length. In the process material on the bottom is put into suspension and transported by the wave action. One example of the energy associated with breakers occurred along the Scottish coast. It was estimated that the breakers produced a force of 6000 lbs ft^{-2}, sufficient to carry away a 2600 ton concrete slab. The configurations of pebble beaches,

Table 3-8 Physical Characteristics of Oceanic Eddies

Scale and Types	Dimensions	Characteristics
Microscale		
Wind Wave	Height: 0.1 to 15 m or more Period: 6 to 10 s	Wind stress waves produce an orbital motion with a mass transport velocity from less than 0.1 up to a few knots. The energy generated by such waves drives the movements of ocean swells and currents outside the generation areas. Intense tropical cyclones can generate wave heights as high as 34 meters.
Mesoscale		
Upwelling	Depth: Varies	Wind stress and the Coriolis force shift warm surface water away from the shoreline and cold subsurface water upwells to the surface.
Macroscale		
Ocean swell, Tsunami, Surface current, Undercurrent, Deep ocean current	Depth: Varies Length: 100 to 6,500 km Width: 0.5 to 250 m	The average swell has a height between 5 cm and 10 m; period, 15 to 20 s; speed to 35 mph. The average tsunami has a height of 0.5 m; period, 10 to 30 minutes; speed, 300-400 mph; and length, 100 to 200 km. The Gulf Stream (a surface current) has a depth of 1 to 1.6 km; width, 80 to 240 km; speed, less than 5 mph; and water transport of 75 to 90 million m^3 s^{-1}.

bluffs, sand bars, and reefs are thought to be products of wave-land interaction. Tidal deltas and salt marshes are another feature produced by water-land interaction. Slow moving, deep water ocean currents flow over the benthic zone with speeds on the order of 5 to 50 km yr^{-1}, carrying sediments over the benthic plains and hills.

Diurnal tides can generate tidal bores that extend upstream into estuaries and adjoining rivers. An extreme example is the tidal bore of the Tsientang River in China, which reaches heights up to 8 m. Most tidal bores, however, are considerably smaller and less disruptive. Tsunamis (seismic sea waves) are unnoticed at sea because of their small amplitudes. They travel at high speeds in the range of 480 to 725 km hr^{-1} (300 to 450 mi hr^{-1}). The seismic sea waves increase in amplitude as they approach shallow water, which makes them extremely destructive to property and life should they build to 20 m or more in height as they reach shore. Overall, these various hydrospheric signals range in energy content from low to high, resulting from a spectrum of gradual to abrupt changes in the state of the ocean.

3.4 Lithospheric Signals. Mass movement in soils and seismic actions generated in the earth's crust and mantle results from the interdependence and interaction of three systems: air, water, and land.

3.4.1 Soils. Air or water acts as a medium, carrier, and sink for solid soil particles. As a dispersing medium, fluid disperses soil particles (the dispersed phase) into the atmosphere or hydrosphere to form aerosols or hydrosols, respectively.[9] The fluid acting as a carrier transports aerosols in the atmosphere through air currents and hydrosols in the hydrosphere through water currents. The sink (destination of the transport) can either be a fluid medium or a lithospheric surface (mainly terrestrial surfaces or oceanic basins). Submicron particles often remain suspended in the fluid for a long period of time, if not indefinitely. Large particles usually precipitate out of the fluid system to eventually deposit on solid surfaces.

The above process is illustrated by a number of lithospheric signaling phenomena: transport and formation of particles in sand dunes, loess deposits, and dust bowls as a result of atmospheric transport; and deposition of sediments as alluvial fans, alluvial terraces, and deltas as a result of hydrospheric transport and deposition. Much of the microscale behavior of soils in their process of formation (podzolization, gleziation, laterization, calcification, salinization, etc.) is characterized by the upward (flocculation) and downward (leaching) movement of nutrients (minerals and organic materials) and solid particles (silt, clay, and colloidal particles). The macroscale phenomena involving denudation and deposition are discussed in more detail.

(1) Dust Bowl. A dust storm consists of a column of dust extending as high as 1 km above the surface and reducing visibility to practically zero. This type of storm is caused by turbulent windflow in an arid or semi-arid region.[10] It occurs frequently in Iraq, north Africa, and northwest India. In 1902, for

---

[9] Aerosols are aggregations of minute particles of soil or other solid particles and/or liquid, suspended in the air; hydrosols are the suspended soil particles, oils, and even gaseous substances in a body of water.

[10] Sandstorms, on the other hand, are masses of coarse sand blown by strong winds to a height of no more than 15 to 30 meters and moving a distance of less than 1 kilometer from their source location.

example, dust clouds generated over Algeria were carried approximately 1700 km to the British Isles. In the United States, the dust bowl originally referred to the portion of the Great Plains where farmsteads were ruined by severe dust storms. The region has been extended in three separate periods[11] to a total of 13 states since the 1930's. It has been estimated that a quantity of about 875 metric tons of dust per cubic kilometer can be suspended in the air during one of these storms. Should there be more than one meter of drift sand and silt deposition, farmland is rendered useless.

On the other hand, loess deposition over the midwest and central states of the USA, particularly in Iowa, Illinois, and Nebraska, has been beneficial to agriculture. Loess is a porous, friable, yellow sediment of finely divided mineral fragments of silt about 4 to 62 μm in diameter. The loess deposited in Illinois, for example, is as thick as 2.54 meters; it is rich in minerals making it a high quality agricultural soil.

Measurement of the microscopic particles in the air is of importance to both agriculture and sanitary engineering. Various aerosol photometers have been designed to detect the total number of particles per unit volume for each size distribution (spectral range specification) and also to determine the total mass of particles per unit volume per unit time. Sophisticated remote-sensing techniques have been developed to measure aerosol content as a function of the amount of light scattered. (See Section 8.2.) Estimates by Peterson and Junge[12] of the natural production of particles throughout the world are cited in Table 3-9.

(2) Sediments. In the process of denudation, sediments consisting of clay, silt, sand, and even sizable rocks are continually transported by rivers and streams from the continents to the oceans. In addition, particles of continental origin, such as windblown dust, forest fire debris, and volcanic ash, are transported and deposited in the oceans (see Table 3-9), and are carried from the oceans to the continents by the wind, part of which eventually is returned to the oceans via other processes. The maximum rate of measured denudation varies between 1 m and 1.5 m per 1000 yr for high mountain regions. The actual rate would be much smaller than this maximum rate for continents as a whole. It is obvious that denudation is a rather slow process.

Judson[13] estimated the mass transport of mineral matter from the continents to the ocean basins, shown in Table 3-10. Stream transport is the largest avenue

---

[11] These were the new dust bowls of the 1950's and 1960's. Perhaps more dust bowls will occur in the future, if soil conservation practices are not reinforced.

[12] Peterson, J.T. and C.E. Junge. 1971. Man's Impact on the Climate. Mathews, W.H., W.W. Kellog, and G.D. Robonson, ed. MIT Press, Cambridge, Mass.

[13] Judson S. 1968. Erosion of the land or what's happening to our continents? American Scientist 56(4): 356-374.

Table 3-9 Estimates of Particle Production

Source	Particle Diameter > 5 microns	< 5 microns
A. Direct particle production:	Units: 10^6 metric tons yr^{-1}	
Sea salt	500	500
Windblown dust	250	250
Forest fires	30	5
Meteoric debris	10	0
Volcanoes (highly variable)	?	25
	790+	780
B. Particles formed from gases:		
Sulfates	85	335
Hydrocarbons	0	75
Nitrates	15	60
Total:	100	470

After Peterson and Junge (1971).

Table 3-10 Estimated Mass Transport of Mineral Matter
from Continents to Ocean Basins

Components of Mass Transport	Amounts
Mass derived from continents:	Units: 10^6 metric tons yr^{-1}
Stream transport	9.3
Wind transport	0.06 to 0.36
Glacier transport	0.1
Mass from extraterrestrial sources:	0.00035 to 0.14
Total entering ocean basins	9.6 approx.
Mass of sediments deposited in oceans:	
Water depth less than 3 km	5 to 10
Water depth over 3 km	1.2
Total accumulation in ocean basins	6.2 to 11.2

After Judson (1968).

of the total mineral mass entering the ocean. It must be noted that most sediments accumulated in ocean basins are found in water less than 3 km in depth, primarily close to the continental margins.

3.4.2 The Crust and Mantle. Other lithospheric signals are generated totally by forces of the interior earth: volcanic eruptions, tectonic plate movements, earthquakes, and tsunamis.

The major land and ocean basin topographies originate from volcanic activities. Such landforms as basaltic domes, basaltic cinder cones, stratovolcanoes, and explosion depressions are a result of the differential intensity of volcanic activities. To a lesser extent, volcanic ash, dust, and lapilli (cinders) produced by eruptions contribute to the landscape.

Earthquakes, on the other hand, are more likely to be of tectonic origin, probably due to the existence of faults. They are also associated with volcanic eruptions. Sudden changes of landforms related to a large magnitude quake have been reported. The birth of new islands in the ocean basins reveals the power and strength of large quakes and volcanic activity. On a global basis, the smallest detectable quake of one magnitude on the open-ended Richter Scale is a daily occurrence but is always unnoticed. Earthquakes occurring in ocean basins trigger seismic sea waves or tsunamis which are extremely dangerous to islands and coastal areas. Tsunamis very often go unobserved on the open ocean by sailors because the waves have low amplitudes on the open ocean.

Tectonic processes are extremely slow in shaping the land. They mould the earth's surface through breaking, bending, and warping of the earth's crust, and creating depressions and elevations. These processes take eons on a geological time scale to produce significant changes (exclusive of the occasional catastrophic events associated with earthquakes, etc.).

CHAPTER 4   REQUIREMENTS AND SPECIFICATIONS

This chapter covers the specifications of instruments and instrument systems
for physical environmental measurements; special emphasis is given to operational
requirements.

The designer of an instrument for a manufacturer needs a complete set of specifications describing in detail the design requirements for the potential user.
The builder, in consultation with the designer, then may modify the design to
fit production requirements. A prototype instrument is then manufactured. Performance specifications are established by calibrating and testing the prototype
through comparison with a standard instrument and by other standardized procedures. Further modifications, if required, are made and the instrument is ready
for production.

By comparing all available specifications with his own operational specifications,
the individual user searches for the most suitable instrument to meet his operational needs. The user could be the designer, but usually he is not. Therefore, the user depends upon the performance specifications supplied by the manufacturer for decision-making. Not all commercial instrument manufacturers give
reliable or correct specifications; the user often has to perform calibrations
and other tests prior to purchase. Those instruments with a certificate from
the National Physical Laboratory of Great Britain or the National Bureau of Standard in the USA are considered reliable, having been rigorously tested and compared to known standard instruments. If the user is unable to find an instrument
suitable to his needs, it is best to buy one which approximates his requirements
and modify it, with or without the assistance of the manufacturer.

4.1 Range and Limitation. The first performance specification of an instrument
to be considered is its range. The range is the difference between the two extreme measurable signal readouts as indicated by the highest and lowest scale
values of an instrument. An instrument which is capable of measuring a large
range of signals is said to have a high capacity and a small range, a low
capacity. Thus the two terms, range and capacity, have been used alternatively
and loosely in the literature. Sometimes, two instruments may have about the
same capacity but not exactly the same range. For example, if two thermometers
have a capacity of an 80C temperature differential between maximum and minimum
values, but one ranges from -80C to OC and the other between OC and +80C, then
the difference between the two can be recognized only by the range and not the
capacity. However, for some instruments, especially those which are designed
for weight or volumetric measurements, the term capacity perhaps is justifiably
employed.

The second specification for an instrument is limitation. Limitation may be
designated as the capability of an instrument to give accurate readings only
within a certain state of environmental conditions. Several examples illustrate instrument limitations. At a temperature of -5C, the wet bulb readings
from a liquid-in-glass thermometer are always too high. This is caused by
the release of latent heat of fusion by the freezing water, and a reduction of
evaporative cooling due to the presence of an ice sheet over the muslin. In
high winds exceeding 100 knots, most conventional anemometers fail to function
due either to mechanical failure or response errors. Relative humidity readings

72

over 95% are incorrectly indicated by most hair hygrometers. The response of
the hair used as the sensing element becomes grossly non-linear under these
conditions. These examples suggest that all instruments have their own environmental restrictions. Stated another way, the sensor of an instrument detects
only a certain range of its complete scale accurately. Usually, at the extreme
limits of the scale, the readings are less accurate and hence not considered to
be reliable. Therefore, the limitation of an instrument must be known to the
user.

4.2 Accuracy and Representativeness. The two terms, accuracy and representativeness, are interrelated. Without a knowledge of the accuracy of an instrument,
the time and space representativeness of an environmental variate cannot be
specified. In fact, the accuracy of an instrument has a more or less clear-cut
and generally accepted definition among scientists, but its representativeness
does not.

4.2.1 Accuracy. The accuracy of an environmental instrument is designated as
the degree to which the response of a sensor to its immediate or distant environmental signal conforms to a standard or known quantity (true value).[1] Stated
another way, during the process of calibration there is a difference in responses
between the standard and the tested instruments in a varying environment. The
difference between the two responses is a measure of the sensor accuracy. For
an immediate (or ambient) environment, an ideal sensor would react with the environmental signal in exactly the same way as the signal changes. For example,
a highly sensitive temperature sensor is always in thermal equilibrium with its
medium and changes with the changing temperature of the medium. In this case,
a close direct contact between the sensor and the medium must be established in
order to permit an accurate readout. This type of instrument is known as an
in situ or contacting instrument as opposed to a remote-sensing instrument.
For a distant environment, the sensor either has to receive energy from or
transmit energy to the environment in order to detect and synchronize with the
variation of the distant signal. A camera receives the reflected light from
an object whereas a weather radar station emits radiowaves to the cloud and
detects reflected energy. Both are examples of remote-sensing instruments.
The former, however, is a passive type and the latter, an active. In either
case, the magnitude of the energy determines the accuracy of the readout. It
is clear that the accuracy of a sensor depends upon the degree and speed of
reaction between the sensor and its environmental signal. A perfect sensing
element is theoretically impossible because the element always has a different

[1] An alternate definition of accuracy is the closeness with which an observation
of a quantity, or the mean of a series of observations, approaches the unknown
true value of the quantity. The standard of temperature, for example, on the
thermodynamic Celsius scale for the freezing point of water is 273.15K (as
recommended by the Ninth General Conference on Weight and Measures in 1948)
whereas a value of 273.16K is utilized in the United States. Although a
standard is artificially chosen, it serves as a reference point.

physical property than the medium it measures. This means that one hundred per cent accuracy cannot be obtained.

Alhtough sensor accuracy generally describes the degree of response of an instrument, instrument accuracy and system accuracy also are used to indicate the accuracy of the output of an instrument and its instrument system. Errors introduced by the mechanism of an instrument and instrument system, including electronic, optical, chemical, and mechanical devices, determine the instrument accuracy and the system accuracy, respectively.

In commercially available instruments, instrument accuracy instead of sensor accuracy usually is expressed either in terms of the absolute unit of readout (e.g., 0.1C or +0.1C) or per cent of full scale (e.g., 2% of full scale). The latter expression is not a suitable indication of instrument accuracy since it indicates the smallest scale division of the instrument that can be read.

So far, the discussion has covered the performance specifications. The operational requirements or specifications impose additional restrictions. Relative measurements, such as the vertical gradient of temperature or wind speed, require a greater accuracy (precision)[2] than absolute measurements, such as a specific value of temperature or wind speed. Subdivisions of variate ranges are another aspect to be considered in the operational requirements. For the forecasting of fog, low accuracy of humidity measurement is sufficient in the lower range between 0% and 65% relative humidity; high accuracy is needed in the upper range between 65% and 100%. For frost preventative warnings and controls, the lower end of minimum temperature measurements should be made as accurately as possible to within 0.1C.

The error, E, of an instrument, designated as the algebraic difference between the indicated value (readout), Q_i, and its real (true) value, Q_r, for the environment in situ is

$$E = Q_r - Q_i \dots \dots \dots \dots \dots \dots \dots \dots \dots \dots (1)$$

Eq. (1), the overall error of an instrument system, is closely related to resolution accuracy and representativeness. When E is divided by Q_r and multiplied by 100, the per cent error is obtained.

Resolution refers to the smallest change in a measurable property of the ambient environment that can be detected and indicated by an instrument. The meaning of the term originated in the field of optics. Resolution originally referred to the ability of an optical system to render visible separate parts of an object or to distinguish between different sources of light. Later, the usage was expanded to include electro-optical systems including radar, laser, and lidar. Now it applies to all physical and biological measurements. Strictly

[2] It should be noted that in engineering the term precision refers to the degree of agreement of repeated measurements of the same property in a controlled environment. Although the two terms are often used interchangeably, the term accuracy connotes a broader set of response conditions, including measurements in controlled and natural environments, than does the engineering use of the term precision.

speaking, the term resolution is more suitably employed for the representativeness of an instrument system while the term accuracy is used primarily for sensor accuracy.

4.2.2 Representativeness. As previously mentioned, there is no generally
accepted definition of representativeness. To most engineers, it means the
spatial and time representation of an environmental variate in terms of the output of an instrument. Often, it is referred to as system accuracy. Thus both
random errors and system errors[3] determine the representativeness through sampling
techniques. Thus, the site selection for instrument exposure, whether on the
surface or underground, is their first concern. To a statistician, it is the
manipulation of data through such techniques as filtering, smoothing, and modeling in order to obtain the best spatial and time representativeness, while disregarding any improvement of the instrument system and its sensor per se.

It is now obvious that proper application of the term representativeness is
achieved only by consideration of the interdisciplinary nature of the measurement and sampling problem. Both the agricultural engineer and the hydrologist,
for example, measure the rainfall with a gage in order to evaluate the total
fall per acre, usually expressed in units of acre-feet[4] of precipitation. The
rainfall over an acre as measured by a single gauge of 10 in. diameter is considered to be representative. However, in the United States the standard gage
density is one gage per 225 mi^2 and this, of course, is far from adequate. Many
factors with respect to representativeness must be taken into consideration when
making simple rainfall measurements. These factors include the types of precipitation, its intensity and duration, the effects of topography (including the
steepness and orientation of slopes), and the exposure of the gage (its height
and any surrounding obstructions). Some meteorological variates measured at a
station (point source) represent the state of the atmosphere at that station and
the surrounding area while others do not. Examples of representative meteorolog-

---

[3]A random error usually is caused by the sudden change of a property of the
environment being measured or an oversight by the observer. A system error is
caused by scale or zero setting errors of the instrument or errors in the instrument system besides those associated with the personal habits of the user.
System errors also can result from periodicities in weather or environmental
elements, such as a low solar angle near sunrise or sunset producing a consistent error in the measured intensity of optical or radiant energy.

[4]In the USA, the term acre-foot refers to the volume of water required to cover
one acre to a depth of one foot (equivalent to 43,560 ft^3) while ignoring dissipating factors such as runoff, percolation, and evaporation (see Figure 3-3).
The acre-foot unit is used in measuring discharge of rivers and streams, runoff
volume over land, irrigation water requirement, and reservoir capacity. Another
expression is the second-foot day representing the flow of water of 1 ft^3 s^{-1}
for a 24-hour period (or 86,400 ft^3 of water). This unit is used extensively
for runoff volume and reservoir capacity evaluations.

ical variates are atmospheric pressure and radiation, whereas non-representative ones are humidity and wind. Representativeness becomes less useful for such variates as turbulence, wind, temperature, and humidity of the upper air. These latter variates, when measured, have very complicated aspects which influence the representativeness of the measurements. The areal coverage and fluctuations associated with upper-air measurements are further complicated by the methods used to obtain them. An example of the influence of platform on the variate measured is the effect of dynamic heating on the measuring probe induced by aircraft movement.

To summarize, <u>accuracy</u> refers to how closely a sensor or instrument presents a readout which conforms to some recognized standard value. <u>Representativeness</u>, broader in meaning, includes not only the accuracy of the instrument but other features of the variate measured, instrument platform measuring and translating the input, and environment producing the conditions. These include the coupling of the instrument to the environment, smoothing performed by the sensor-transducer-recording system, spatial and temporal coverage of the sensor, and the number and distribution of the data points used to establish the central values of the quantity being measured.

4.3 <u>Sensitivity and Time Response</u>. The sensitivity of an instrument system is the change in system output to a given change of input signal. In other words, it is a measure of the accuracy of the output signal, usually expressed as the ratio of the full-scale output of the instrument to the full-scale input value. It differs from the resolution of an instrument. Thus, the output of an instrument system with high sensitivity reproduces every fluctuation, regardless of size, received as an input signal from the environment. An extremely sensitive thermometer, for example, may indicate a ±0.01C change in temperature whereas an insensitive one may only indicate changes to ± 2C or more. The term time response, on the other hand, refers to the time required for an instrument sensor to respond maximally (usually taken as a 63% response to the true value)[5] to an input signal. Stated another way, the response is the time elapsed between signal input and signal output of the sensor. An electric thermal sensor, for example, requires 1 s to register a maximum response to a temperature change whereas a mercury-in-steel thermometer requires as much as 280 s.

It is now apparent that sensitivity refers to the degree of accuracy provided by instrument output, whereas time response pertains to the speed of response of a sensor. Nevertheless, the two terms are easily confused by the beginner. For example, the smallest detectable amount of a signal refers to sensor accuracy rather than sensitivity. There are other interrelated terms such as range, limitation, accuracy, and even economy of operation in addition to sensitivity and time response. A clarification of these terms therefore is necessary.

4.3.1 <u>Sensitivity</u>. When the signal-to-noise ratio is specified, the output signal of an instrument must be large enough to be distinguished above its

---

[5] For details of response curves, see Section 4.3.2.

corresponding noise level.[6/] This means that the sensitivity of the response should be large enough so that the input signal (or infomation) passes through the system without degradation. Also, the sensitivity of response from the instrument must be considered so that it is economical for the desired application. Thus, avoidance of excessive (unwanted) sensitivity in the system is necessary. For example, if sensitivity is greater (e.g., 0.01C) than either the input signal (e.g., 0.1C) or the operational requirement (e.g., 1C), then such excessive sensitivity in an instrument system is expensive and should be weighed against the operational requirements of the system.

When a sizable instrument system involves several subsystems with varying sensitivities, the compatibility of all the subsystems must be carefully considered. The output signals from the subsystems are matched and amplified so that uniform sensitivity for the entire system is achieved. Simultaneous, compatible measurements and signal processing are possible.

Another important consideration is linearity or non-linearity of response sensitivity. Although the advantages of having linear response from an instrument sensor are acknowledged by all scientists, non-linearities in response are important for some instruments. For example, a hot wire anemometer, underheated with a constant low current, responds non-linearly to low wind speeds while becoming insensitive to high speeds. Expanding its range of sensitivity by increasing current flow to the hot wire, however, may expand the anemometer's sensitivity at higher wind speeds but destroys its response to lower speeds due to heat damage to the sensor. A dewpoint hygrometer, which responds non-linearly to humidity, is highly sensitive in the low humidity range but decreasingly so in the higher range. Its low humidity sensitivity, however, makes it very suitable for upper-air soundings of humidity. The non-linearity of response for these instruments can be used to advantage for specific applications.

4.3.2 Time Response. Among the various expressions, the term time response has a concise definition.[7/] Both laboratory and field techniques have been employed for determining the time response of an instrument sensor. The laboratory environment is easier to control than the field but the latter represents real time and place in situ. Where possible, a field experiment is better than a laboratory experiment. For the sake of convenience and control over conditions, however, the laboratory method has been widely employed. In this method, various environmental chambers provide controlled radiation, temperature, humidity, wind, precipitation, visibility, and even air composition.

---

[6] Noise is usually introduced through the transducer. To a lesser extent, it may be generated by the transmitter and other electronic components of the instrument system.

[7] Examples of the other nomenclature that have appared in the literature are response time, time constant, time transit, speed of response, and time lag coefficient. The latter is sometimes called a lag coefficient or simply time lag. Although these terms can be defined concisely and used interchangeably, their definitions differ among various specialists. In this book the term time response is adopted but other nomenclature will also be used whenever appropriate.

Most of these chamber conditions model physical properties of the atmosphere, and only occasionally conditions of the hydrosphere or pedosphere. Various sizes of atmospheric chambers are available, ranging from a small Dewar flask to the large chambers in an aerospace facility. Many of the chambers are used in special applications related to biological systems, such as the growth cabinet, phytotron, human chamber, solar house, and the biotron. The dual purposes of environmental control chambers are controlled testing of instruments and instrument systems and controlled studies of the responses of living organisms to their environment. The most difficult conditions to simulate are wind, radiation, precipitation, and cloudiness. Boundary effects on the controlled environment are of major concern, primarily in wind tunnels and cloud chambers. The proximity of boundaries in these facilities reduces the likelihood of true simulation of the natural environment. Reproducing solar radiation is hampered by divising a source whose output is equivalent to the sun's full spectrum and intensity. Likewise, the effects of natural rainfall, turbulent flow, or cloudiness are difficult to simulate. Pollutants that contribute to the pollution syndrome have not been totally duplicated by air chemists in controlled environments. Despite these limitations, valuable findings have resulted from experiments conducted in the limited facilities of control chambers.

In the procedures used to determine time response, mathematical representation is necessary. The mathematical calculations depend more upon the type of instrument and the mechanism employed than on the environmental variates per se. For example, the calculation of the time response of a psychrometer differs completely from that of a hair hygrometer. The former is based on the mechanism of evaporative cooling while the latter responds to the elongation of a human hair due to the absorption of water vapor. The physical processes and mathematic descriptions differ for the two instruments.

Two examples are given in this section to illustrate the mathematical determination of time response using laboratory techniques. They are: (1) the liquid-in-glass thermometer and (2) the cup anemometer. Only the first-order system is covered here. Second-order systems, such as damped oscillatory devices, together with sinusoidal and random forcing functions as applied to several other instruments, are discussed in the appropriate sections of the remaining chapters.

(1) <u>Liquid-in-Glass Thermometer</u>. As shown in Figure 4-1$_a$, a standard mercury thermometer graduated in 0.5F increments is placed in a Dewar flask filled with ice made from distilled water. During the experiment, distilled water is added to the flask to ensure complete contact of the water-ice mixture with the thermometer. Temperature stratification in the flask is eliminated by using a stirring device, such as a motor stirrer, to agitate the water-ice mixture. The stirrer rotates continuously while moving up and down. The stem of the thermometer is immersed as far as possible into the constant temperature container. Let T_o be the initial reading of the thermometer at time t = 0 prior to the immersion of the thermometer into the flask, and T_a be the ambient temperature of the medium (ice-water mixture). In this case the ambient temperature remains constant during the entire period of testing, so T_a = 0C. T is the instantaneous temperature reading of the thermometer at time t, and k is the constant of proportionality between the rate of change of temperature, dT/dt, and the difference in temperature between the medium, T_a, and the

thermometer, T. The time response of the thermometer is λ and its negative reciprocal, $-1/\lambda$, equals k. Thus

$$dT/dt = k(T-T_a) = -1/\lambda(T-T_a) \quad \ldots \ldots \ldots \ldots \ldots \ldots \quad (2)$$

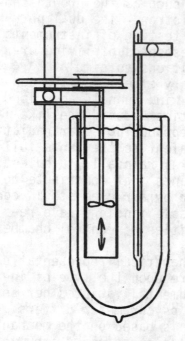

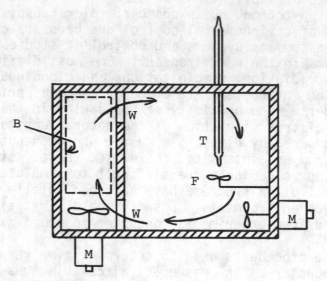

W – Window, F – Flow Meter, M – Fan Motor
T – Thermometer, B – Wire Basket for CO_2

(a) Dewar flask for measuring the time response of thermometers in liquid.

(b) CO_2 cabinet for measuring the time response of thermometers in air.

Figure 4-1 Thermometer Calibration by Dewar
Flask and CO_2 Cabinet

where t and λ are expressed in seconds. Rearranging Eq. (2) and integrating over the temperature range from T_o to T for time 0 to t seconds

$$\int_{T_o}^{T} \frac{dT}{(T-T_a)} = -\frac{1}{\lambda} \int_{0}^{t} dt,$$

or

$$\ln \frac{T - T_a}{T_o - T_a} = -\frac{t}{\lambda},$$

Therefore

$$\frac{T - T_a}{T_o - T_a} = e^{-\frac{t}{\lambda}} \quad \ldots \ldots \ldots \ldots \ldots \ldots \ldots \ldots \ldots \ldots \quad (3)$$

Illustrations of the exponential relationship can be derived from Eq. (3). For instance, when $t = \lambda$ s, Eq. (3) becomes $(T-T_a) = 1/2.718(T_o-T_a)$ which indicates that the difference in temperature between the thermometer and its surroundings $(T-T_a)$ decreases to $1/2.718$ or ~ 0.368 of its initial difference in temperature (T_o-T_a). Similarly, when $t = 0.69 \lambda$ s, then $(T-T_a) = 1/2(T_o-T_a)$; and when $t = 2.3 \lambda$ s, then $(T-T_a) = 1/10(T_o-T_a)$.

Figure 4-2 shows the exponential function by plotting the above values on a semi-logarithmic chart with t/λ as the abscissa and $\ln (T-T_a)/(T_o-T_a)$ as the ordinate.

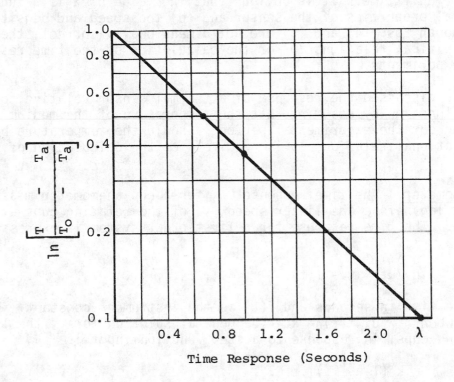

Figure 4-2   Ratio of the Instantaneous to the Initial Temperature
Difference Versus the Time Response

When the above experiment is repeated in a CO_2 cabinet (dry ice box), as shown in Figure 4-1$_b$, the time response, λ, is determined from the relationship

$$\lambda = \frac{K}{(\rho v)^n} \quad \dots\dots\dots\dots\dots\dots\dots\dots\dots \quad (4)$$

where ρ is the density and v the speed of air. K is a constant and n is a dimentionless constant[8] less than unity. For example, a mercury-in-glass thermometer with a spherical bulb 1.12 cm in diameter has a K value of 117, and n is 0.48.[9] The ventilation as measured by a flowmeter is 4.6 m s^{-1}. The λ

---

[8] The constants n and K refer to the experimental and instrumental constants, respectively.

[9] Middleton, W.E.K. and A.F. Spilhaus. 1964. Meteorological Instruments. University of Toronto Press.

value is approximately 56 s.

Factors affecting the value of λ are summarized by the relationship

$$\lambda = \frac{C}{c_p m} \quad \ldots \ldots \ldots \ldots \ldots \ldots \ldots \ldots \ldots \ldots \ldots \ldots \ldots (5)$$

where C is the thermal capacity of the sensor of a thermometer; c_p, the specific heat of air at constant pressure; and m, the mass of the medium flowing around the sensor per unit time. It is obvious that the value of λ is a function of (a) the thermal properties of the sensor and (b) the speed and density of the medium. Although Eqs. (4) and (5) are not dimensionally correct, they offer a convenient empirical relationship for the calculation of the time response in wind tunnel experiments.

So far we have discussed the response of a thermal sensor relative to the input step or step function by assuming that the temperature of the medium remains unchanged throughout the experimental period. Should the temperature be a variable, a complex functional relationship is required for the calculation of the time response.

(2) <u>Cup Anemometer</u>. The time response, λ, for a cup anemometer measures the delay time or lag before the linear speed, v, of the rotating cups attains its maximum value. The time response for a first-order system is expressed mathematically as

$$\lambda \, dv/dt + v = f(t). \quad \ldots \ldots \ldots \ldots \ldots \ldots \ldots \ldots \ldots (6)$$

where t is the time in seconds and f(t) is the instrument constant. When t < 0, f(t) = 0 and when t > 0, f(t) = k. The general solution for v, the linear speed of the rotating cups, in response to a step function input is

$$v = k \, (1 - e^{-t/\lambda}) \quad \ldots \ldots \ldots \ldots \ldots \ldots \ldots \ldots \ldots (7)$$

where λ is the time required for the rotating cups to reach the final value of (1 -1/e)v or $\sim$ 0.63v. The time response, λ, increases inversely with actual wind speed, u, assuming the frictional force of the anemometer is negligible and the second-order Reynolds number effects are ignored. The distance response, L, is defined as the length (in m or ft) of an airstream required for an anemometer to respond to 63.2% of a step change in speed or (1-1/e). It is constant for all wind speeds. Since u is the speed, then $L = u \lambda$. When wind speed changes from an equilibrium value, u, to $(u_0 + \Delta u)$, then the linear speed of the rotating cup, v, for the incremental change in wind speed, Δu, is

$$v = \Delta u (1 - e^{-t/\lambda}) = \Delta u (1 - e^{-x/L}) \quad \ldots \ldots \ldots \ldots \ldots \ldots (8)$$

where x is the distance of horizontal displacement of the wind during time, t. Figure 4-3 illustrates that the fractional change in the linear speed, v, of the cups varies as a function of both the time response and the response distance, L. Note that the <u>tangent</u> to the exponential curve at x = 0 intersects the horizontal axis (1.0 v) at x = L. At this point the linear speed of the cups has reached 0.63 of its final value. The relationship of the three variables are shown in Table 4-1.

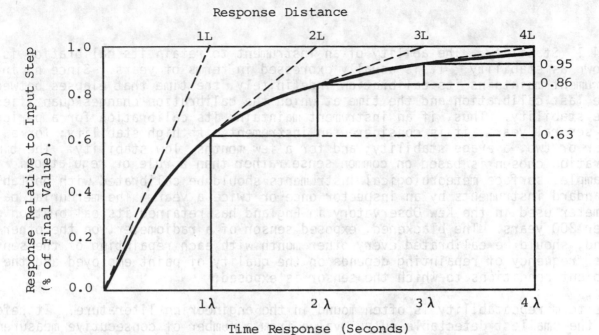

Figure 4-3 An Illustration of the Step Function of a Response Curve

The time response concept so far discussed is more applicable to a recording rather than a non-recording device. For non-recording devices, the measuring time is important. Measuring time is designated as the time required for one or more operational procedures to be executed in order to yield a discrete measurement. Since each operation depends heavily on the individual observer's skill, it becomes difficult to arrive at a consistent measuring time. Accordingly, all instruments, recording or non-recording, are conveniently classified into the fast, medium, or slow response types. At room temperature, a fast response instrument may reach 90% of the final recording value (the true ambient condition) on the order of seconds. A medium response time is on the order of minutes and a slow response time requires 30 minutes or more.

Table 4-1 First-Order System Response to a Step Input

Response relative to input step, % of the final value	63	86	95	98
Response distance, L	1L	2L	3L	4L
Time response, λ	1λ	2λ	3λ	4λ

4.4 <u>Stability, Reliability, and Ruggedness.</u> Instruments and instrument systems also are specified according to their stability, reliability, and ruggedness.

4.4.1 Stability. The ability of an instrument to retain its calibration is known as stability. It is usually expressed in terms of years. Since no instrument maintains its calibration indefinitely, the time that elapses between the last calibration and the time at which the calibration changes quantifies the stability. Thus, if an instrument maintains its calibration for a period of several years, it is considered an instrument with high stability; for a year or two, average stability; and for a few months, low stability. The time duration chosen is based on common sense rather than a rule or regulation. For example, surface meteorological instruments should be calibrated with portable standard instruments by an inspector once or twice a year. The mercurial barometer used in the Kew Observatory in England has retained its calibration for over 200 years. The blackened, exposed sensor of a radiometer, on the other hand, should be calibrated every other month with each repainting of the sensor. The frequency of repainting depends on the quality of paint employed and the ambient conditions to which the sensor is exposed.

The term repeatability is often found in the engineering literature. It refers to the smallest detectable output value from a number of consecutive measurements, each made at the same initial input value, step change, and operating conditions. It should be noted that the word repeatability implies a short term laboratory test. It differs from hysteresis, which refers to an input response, step change, and recovery measurement.

4.4.2 Reliability. This is another closely related term to stability. Reliability is designated as the probability that a system or instrument will perform satisfactorily for a period of time disregarding calibration. Satisfactory refers to freedom from one or more of such unwanted conditions as drift, friction, and hysteresis.[10] These unwanted conditions result from a number of environmental factors: shock and vibration, contamination, extreme weather conditions, and so on. Specifications, therefore, include both the operating environmental conditions and the period of satisfactory performance. A more precise definition of reliability has been given by Norton.[11]: "A measure of the probability that the instrument will continue to perform within specified limits of error for a specified length of time under a specified condition." Another related term is maintainability. It is defined as the probability that a repair can be completed within a specific time. In addition, the expected maintenance and support conditions must be specified: the skill-level of maintenance personnel, the diagnostic procedures, and required test equipment and spare parts.

4.4.3 Ruggedness. The ruggedness of an instrument is measured as the usable

---

[10] Within a specified range the hysteresis phenomenon is the maximum difference in output for any given input value. The hysteresis error is caused by the absorption of energy by a sensor; it is expressed in per cent of the full scale.

[11] Norton, H.N. 1969. Handbook of Transducers for Electronic Measuring Systems. Prentice-Hall Inc.

time of the instrument before it becomes impaired. Simply stated, it is the life-span of an instrument. Most environmental instruments built today are strong and durable so their ruggedness is generally high. Usually a quality instrument or instrument system has been subject to various controls and tests, such as vibration, shock, high temperature, and radiation tests. Therefore, it is capable of withstanding handling, packing, and transporting without undesirable effects. The terms durability and robustness are sometimes used interchangeably with ruggedness.

4.5 Portability and Simplicity. For the sake of convenience, a portable instrument that is both simple and small is highly desirable provided it meets the operational specifications of the user. Usually a portable and simple instrument is economic in terms of installation and maintenance costs. In many applications, a portable and simple instrument is required, such as those for agricultural applications, military operations, geological surveys, and many other field operations. Therefore, simplicity and portability enhance the usefulness of an instrument or instrument system.

4.6 Costs of Installation and Maintenance. A number of factors govern the cost of an instrument and its installation. Two major determining factors are the designer's requirements and the production quality. If the designer's requirements are complicated and if the demand is low, the cost of the product is extremely high. On the contrary, mass production with the choice of several alternative designs usually results in greatly reduced costs.

In the long run, the costs for purchase and installation sometimes are much less than those of maintenance. Instruments with low stability require constant calibration; those with low reliability require frequent repairing; and those with low maintainability require time for maintenance. All of these involve costs of material and labor, as well as producing lower operational efficiency. It, therefore, is necessary for the user to consider carefully the overall costs of maintenance in terms of his operational requirements prior to his selection of instrumentation.

CHAPTER 5   ENVIRONMENTAL INFORMATION SYSTEMS

Environmental information systems can have many meanings to scientists and engineers of different disciplines. In this book the term refers to instrument
systems designed to measure and deliver useful physical environmental information about the geosphere[1] to the user. In a typical instrument system of this
kind, the physical properties of the geosphere are measured and its data are
handled by means of gathering, transmitting, storing, processing, and retrieving.
Therefore, the system is considered to be an assemblage of subsystems arranged
to operate as an entire system; each subsystem is designed to work properly with
the other subsystems. The above principle is illustrated by the block diagram
of a simple electrical instrument system shown in Figure 1-2.

The natural environmental system generates signals; there is a flow of energy
and/or mass from one place to another. The energy systems have physical or
dynamical boundaries and their inputs and outputs cross these boundaries. These
systems can attain a dynamic equilibrium or a steady state. After a time interval, however, the system often undergoes a change to a different steady state
condition. The period of change between the two steady state conditions is
called the transition period. If the input of energy or mass, or both, is cut
off, the system undergoes decay. The types of energy systems which are dependent
on time variations of energy inputs include: (1) growth and decay systems, (2)
cyclic or rhythmic systems, and (3) random fluctuation systems. The simplest
form of a decay system can be illustrated by an ephmeral river system that no
longer receives water from the source region so it dries up and disappears.
This may occur by diversion, damming, evaporation, percolation, or other processes.
Cyclic fluctuations can be associated with timed water releases from a storage
dam, while random fluctuations in river flow can be caused by intense precipitation and runoff associated with a storm.

In the coordination of manmade instrument systems and natural environmental
systems, many techniques have been developed to obtain information on the physical
characteristics of the environment. Each measuring technique is associated with
an overall data acquisition system. Better known techniques include measurements
taken by atmospheric probes such as weather satellites, sounding rockets, radar,
lidar, radiotheodolites, ceilographs, radiometers, and so on. Less familiar probes
are those used to measure space electricity, magnetic storms, and atmospheric
acoustic properties. The measurement techniques, together with many others used
for surveillance of the geosphere, are described in later chapters.

This chapter discusses some fundamentals of environmental instrument systems with
special emphasis on the system components and subsystems. Environmental engineering systems, with specific applications to gathering physical environmental information, are stressed first. This is followed by a brief description of Doppler
systems, static and dynamic systems, in situ and remote-sensing systems, and
Eulerian and Langrangian systems. Additional details on these systems can be
found in the literature of the geophysical sciences.

---

[1]The discussion in this book is restricted to physical environmental systems and
physical information derived from such systems.

5.1 <u>Environmental Engineering Systems</u>. The systems described here are essentially digital and analog systems and their applications to instrument systems. The application of feedback techniques is included. A description of system reliability includes brief discussions on the failure law, reliability prediction and measurement, measurement error, and error analysis.

5.1.1 <u>System Components</u>. As mentioned in Section 1.1 the data acquisition system encompasses a sensor or transducer, amplifier, data processor and recorder, in addition to transmission and telecommunication devices. In this acquisition system the data processor receives the sensor input signals and conditions the signals to a useful form for data recording and transmission. Signal conditioning may take such forms as amplification, linearization, totalizing counts, and offsetting the zero.

There are two basic systems involving various combinations of signal conditioning, namely digital and analog. A digital system uses numerical digits expressed in a scale of notation to discretely represent all variables in a system. An analog system uses physical analogues, such as electric voltages or shaft rotations, of the numerical variables continuously occurring in a system.

(1) <u>Digital Systems</u>. As shown in Figure 5-1, the basic elements of a digital system consist of the input, processing, and output units. The input unit includes the sensor and/or transducer and the signal conditioner. The signal conditioner accepts signals from the sensor and/or transducer. As previously noted, the sensor may be a mechanical device such as an aneroid capsule or a bimetallic strip. In this case, a transducer is required to convert the mechanical energy into electrical energy (e.g., a voltage). The transducer is coupled to the signal conditioner, where the signal is amplified or otherwise modified to be acceptable to the processor. The signal, still in analog form, is digitized and modified for coupling to the output which may be paper punch tape, magnetic tape, numerical readout device, or combinations of these. The signal format often is in a form acceptable to a computer for further processing.

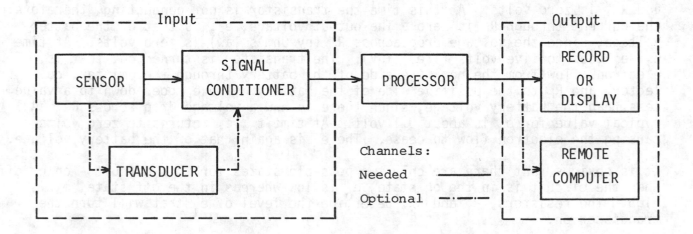

Figure 5-1 A Simple Digital Instrumentation System

At this point, it may be helpful to briefly review the operation of a transistor, since the use of both discrete transistors and integrated circuits containing large quantities of transistors has become commonplace in modern instrumentation systems.

The bipolar transistor, also known as the junction transistor, is most commonly used at this time. However, it should be mentioned that the field effect or uni-polar transistor will, in all likelihood, be making its appearance in new equip-ment as both discrete transistors and as active elements of integrated circuits. While the two types utilize different principles of operation and possess dif-ferent input and output characteristics, the end result is essentially the same. Therefore, the variations between the two types are of concern to the circuit designer rather than the instrument user.

There are some analogies between junction and field effect transistor nomenclature that should be kept in mind. These are:

Junction Transistor[2]	Field Effect Transistor
base	gate
emitter	source
collector	sink

The operation of a field effect transistor closely resembles that of a vacuum tube. The current flow from the source to sink is controlled by an internal electrostatic field developed by a voltage applied to the gate. In a vacuum tube, current flow from cathode to plate (anode) is also controlled by an electrostatic field. This field is developed by a voltage applied to the grid.

In the junction transistor current between collector and emitter is controlled by a small current flowing between the emitter and the base. As shown in Figure 5-2, the bipolar transistor has three elements: the collector, C, the base, B, and the emitter, E. The collector is connected through the load resistor, R_L, to the positive side of the battery, with the emitter connected to the negative side. At time t_1, the input emf, e_i, which is applied to the junction of B, R_1, and R_2, is zero volts. At this time the transistor is <u>not</u> conducting, therefore, the current through R_L is zero. The output voltage, e_o, is that of the battery voltage, since the voltage drop across R_L (by Ohm's law) is zero volts. At time t_2, e_i goes to five volts (i.e., +5V). The transistor is turned on, that is, electrons flow from the negative side of the battery through the emitter, col-lector, and R_L to the positive side of the battery. The e_o goes down to a value less than the battery voltage, since there is now a voltage drop across R_L. A typical value for e_o is about 0.1 volt. At time t_3, e_i returns to zero volts causing the electron flow to cease. The e_o is again that of the battery voltage.

It is obvious that there are only two possible states for this circuit - on or off. When the circuit is in the on state, e_o is low whereas in the off state, e_o is high. The resistors, R_1 and R_2, determine the level of e_i that will turn the transistor on.

---

[2] In operation the collector receives the electrons; the base controls the amount of electrons; and the emitter emits the electrons.

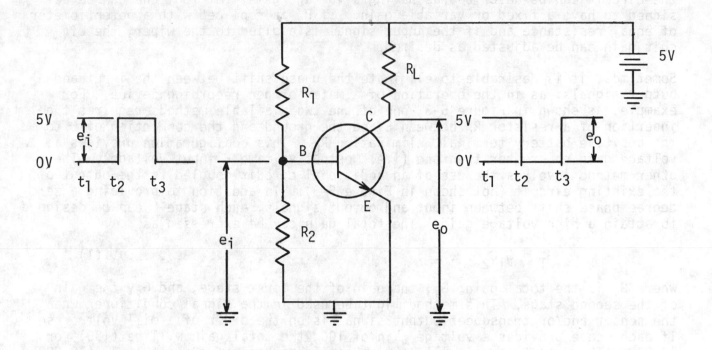

Figure 5-2 A Digital Transistor Circuit

Digital circuitry represents information in codes or digital forms by the state of groups of circuits such as described above. Present day techniques involve the use of integrated circuits whereby several hundred circuits may occupy an area about the size of a pin head. By connecting these circuits in various configurations, a variety of functions is achieved. The recent development of complex new integrated circuits have made available a variety of low cost pocket-size calculators and highly advanced desk-size computers.

The advantage of digital circuitry is that large variations in circuit parameters may occur without causing an error in operation. Some sample digital instruments, together with typical operating parameters, are shown in Table 5-1. A summary of some representative digital systems with a brief description of their functions is given in Table 5-2.

(a) Analog Systems. The basic elements of an analog system are somewhat similar to those of a digital system. For example, the sensor and/or transducer can be used with either system. The primary difference is in the operation of the processor. In the analog system the signal is not digital but is essentially a replica of the electrical output of the sensor and/or transducer.

The circuit shown in Figure 5-2 can be used by merely changing the values of R_1 and R_2 and increasing the battery voltage. Figure 5-3 illustrates an analog transistor circuit in which the input and output are clearly shown. The output signal is 180° out of phase with and about 100 times the input signal. Thus, the circuit can be described as having a voltage gain[3/] of 100, and can be designed to have a fixed or variable gain. If R_2 is replaced with a potentiometer of equal resistance and if the input signal is applied to the wiper, the circuit gain can be adjusted as desired.

Sometimes, it is desirable to eliminate the phase shift between the input and output signals, as in the operation of a meter or pen recording device. For example, as shown in Figure 5-3, one of the two available methods requires the insertion of a resistor R_3 between E and the ground and the connection of C to the positive battery terminal, eliminating R_1. This configuration provides a voltage gain no greater than one (i.e., output voltage < input voltage). The other method involves the use of an additional circuit coupled to the output of the existing circuit (not shown in Figure 5-3). In addition to providing a zero degree phase shift between input and output signals, each stage[4/] can be designed to attain a high voltage gain. The total gain can be expressed as

$$G_T = G_1 G_2 \quad \dots \dots \dots \dots \dots \dots \dots \dots \dots \quad (1)$$

where G_T is the total gain; G_1, the gain of the first stage; and G_2, the gain of the second stage. This method might be used in the signal conditioner when the sensor and/or transducer output signal is on the order of a millivolt or so. If each stage provides a voltage gain of 100, the total gain will be $(100)^2$ or 10,000. The output signal will be in the neighborhood of 10-volt level (10,000 mV). This level is sufficient to drive a pen recorder without the need of the processor.

The most important difference between the analog and digital instrumentation systems is accuracy. The digital system is capable of producing data with an accuracy several orders of magnitude greater than that of the analog system. In the digital system, accuracy is not a function of circuit characteristics but is dependent only on the number of digits used to form a data bit. Accuracy of the analog system, on the other hand, is totally dependent on its circuit characteristics which may be affected by such factors as the ambient temperature, humidity, shock, vibration, and circuit component aging. The digital system is of greater complexity than the analog, thus it is considerably more expensive.

Aside from the digital and analog processing systems, several other subsystems are required for an environmental engineering system. Of specific importance are the feedback system, signal transmission system, and remote-sensing techniques.

(1) _Feedback Systems_. Feedback techniques are frequently employed in instrumentation systems to ensure exactness of data output. These techniques involve the measurement of a set of output signals which are compared with an appropriate set of reference input signals, and generate a set of error signals. The error signals are applied to the appropriate instrument circuit in such a manner as to minimize output data errors.

(2) _Signal Transmission_. An instrumentation system can be a simple system such as a pH meter sitting on a dock, the probe in the water, and a pen recorder

---

[3] Also called amplification factor.

[4] The term stage is commonly used in place of circuit.

Table 5-1 Sample Digital Instruments

Class Name	Quantities Measured	Available Outputs	Typical Specifications
Digital multimeter analog-to-digital converter	DC volts AC volts Ohms Voltage ratio	Digital display Printer Code (The code output may be used to drive a card or tape punch or other recording device.)	Indicate 5 decimal digits Accuracy of DC voltage reading is 0.01% ± 1 digit Range = 0.0001 to 1010V with automatic ranging Input impedence = 10 megaohms
Universal counter and timer	Events per unit time or frequency Total count of input cycles Frequency ratio Period of wave Width of pulse Time between pulses	Digital display Printer Code	Accuracy of frequency and time measurement is 3 parts in 10^7 ± 1 digit
General-purpose digital computer	A program of operations to be performed is fed into the computer in code form. Capability of the computer includes addition, subtraction, multiplication, utilizing tables, and utilizing results of previous operations.	Usually punched or magnetic tape, automatic typewriter, and high speed printer	

After Susskind, C., ed. 1962. The Encylcopedia of Electronics. Reinhold Publishing Corp., New York.

attached to the meter to provide a record of variations in pH as a function of time. It can also be a subsystem of a large complex data acquisition and transmission system, as shown in Figure 5-4. Five monitors are measuring various

Table 5-2 Sample Digital Systems

Class	Function
Data reduction system	Input is data from a test such as magnetic tape recordings from a missile flight. Data are converted into digital form, then utilized directly or fed into a digital computer. The data reduction system may perform mathematical operations to derive the output data.
Tachometer system	Conditions of an engine under test are sensed by transducers and converted to meaningful numbers which may be remotely indicated and recorded.
Missile checkout system	The system operates according to a set of instructions, generally on IBM cards or magnetic tape. These instructions cause the system to apply electrical stimulus to control equipment on the missile and to measure voltages and resistances at specified points. The system records any value outside specified tolerances.
Time code generator	On demand this instrument generates a signal which is modulated to a digital code that very precisely states the time of day.

After Susskind (op cit).

environmental variates. The multiplexer or signal combiner makes it possible to transmit many signals over a single transmission line. In the mechanical commutator method of multiplexing, shown in Figure 5-5, each data source is sampled by a motor driven wiper. Each complete revolution is called a frame, and the beginning of each frame is indicated by a special pulse known as the synchronizing or sync pulse. Electronic commutators, which serve the same purpose, are available. Since they have no moving parts, they require much less maintenance, and generate much less noise than the mechanical commutator. At the user's end, a device known as a decommutator separates the signals, which can then be fed to a recorder, display unit, or computer.

Another method of signal combining is shown in Figure 5-6. In this method, the output of each monitor is fed into a sub-carrier oscillator (SCO). Each SCO generates a unique frequency, usually in the ultrasonic range (30 kHz to about 100 kHz), which is frequency modulated by the incoming signal from the monitoring instrument. The frequency modulated signals from each SCO are then fed to a summing amplifier, the output of which is fed to the transmission line. At the user's end, the incoming summed signals are applied to a device known as a discriminator. The discriminator contains tuned filters, one for each sub-carrier frequency. Each filter extracts its sub-carrier frequency from the

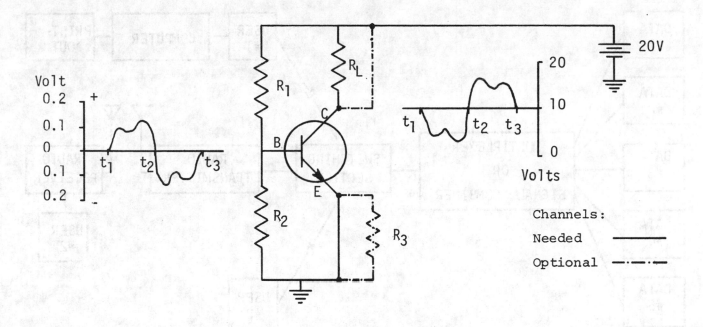

Figure 5-3 An Analog Transistor Circuit

summed signals. The sub-carrier frequencies are then demodulated, separating
the data from the sub-carrier frequencies, and the data is then fed to a recorder, display unit, or computer. As shown in Figure 5-4, this can be transmitted to the user via a transmission line or radio signal.

(3) Remote-Sensing Techniques. Several remote-sensing methods are briefly
described here. Although they are complete systems or subsystems they are primarily sensors. These include radar, infrared, and other electro-optical, and
acoustical systems.

(a) Radar. Radar (from RAdio Detection And Ranging) is a system for locating
natural and artificial objects by means of radio signals. In addition to per-
forming the functions of detection and ranging, radar can scan moving objects
and determine their number, size, and identity.

Two basic types of radar systems commonly employed are weather and tracking
radar. The former echoes weather phenomena such as clouds, precipitation, and
tornadoes, while the latter tracks the movement of airborne balloons for wind
velocity measurement.

The basic elements of a radar system include the transmitter, an antenna to
radiate energy into space, an antenna to receive energy reflected from the object in space, the receiver to process the received energy, and the output
system for display or utilization of the energy. The same antenna may be used
for transmitting and receiving by utilizing a time-sharing switch called a
duplexer. It is activated when the transmission is pulsed (instead of being
continuous). Such systems are called monostatic radar systems. A block diagram
for such a system is shown in Figure 5-7.

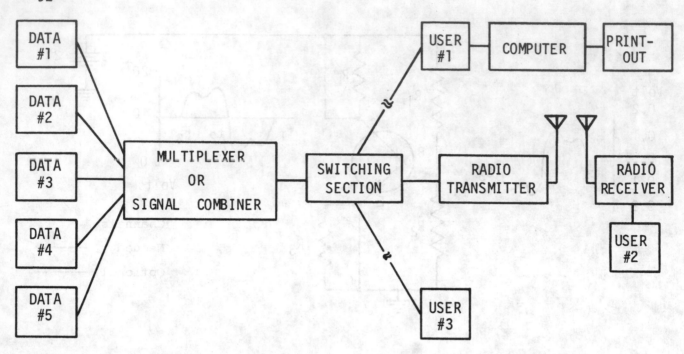

Figure 5-4 A Complex Data Acquisition and Transmission System

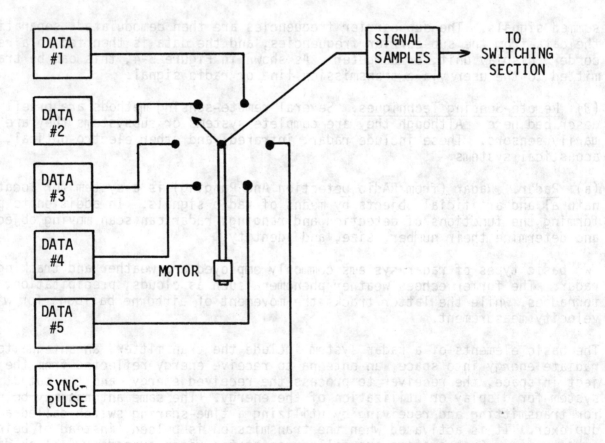

Figure 5-5 A Multiplexing Method Using a Mechanical Commutator

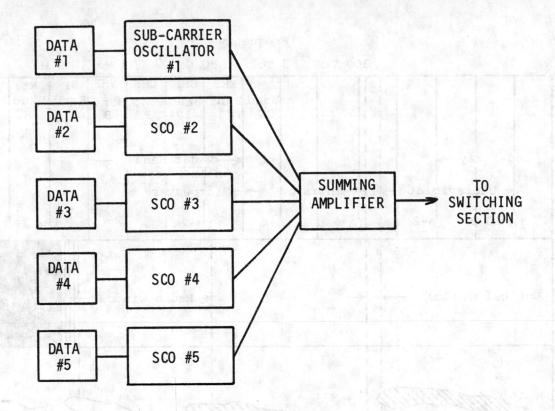

Figure 5-6   A Signal Combiner Method

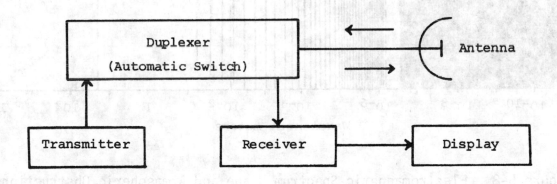

Figure 5-7   Elementary Monostatic Pulse Radar System

The transmitter sends out a series of accurately timed short pulses at radio frequencies (from 100 to several thousand megahertz). The portion of the electromagnetic spectrum used in radar systems and the atmospheric obstructions encountered are shown in Figure 5-8.

The functional parts of a typical superheterodyne radar receiver are shown in the block diagram in Figure 5-9. This receiver is used most frequently as it eliminates thermal noise from such external sources as the sky. It provides a

94

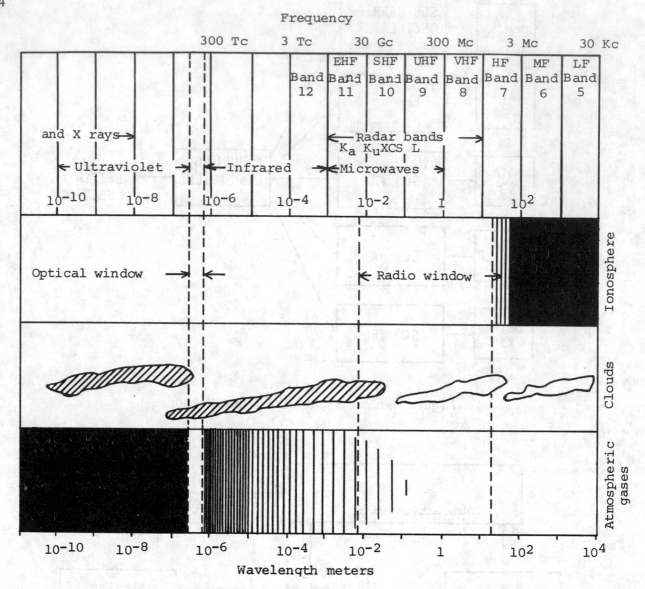

Figure 5-8   Electromagnetic Spectrum Usage and Atmospheric Obstructions

matched filter through the use of an IF amplifier[5] for optimal detection.

The output is in the form of a direct video display, or the data is processed by a computer where it can be refined and used in calculations. Multiple exposure radar photographs are used to interpret information acquired over a period of time.

---

[5] The beat frequency used in heterodyne reception results from the combination of the received radio frequency signal and a locally generated signal. The intermediate frequency (IF) signal is usually the difference between the above signals, and is employed to avoid the direct amplification of radio frequency signals.

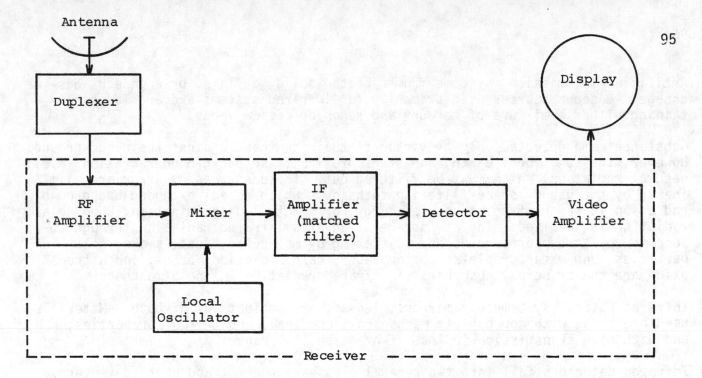

Figure 5-9 Typical Radar Receiving System

(b) _Infrared_. The basic elements of an infrared (IR) system include the energy source, the radiation background against which the source must be detected, the atmosphere or environment which affects the infrared signal, the optical system that collects the signal, the detector, and the electronic system for signal processing and display. A generalized infrared remote-sensing system is shown in Figure 5-10.

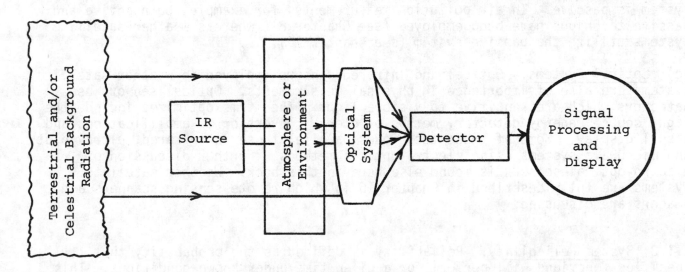

Figure 5-10 Generalized Infrared System

The radiation from solids and liquids usually contains a broad continuous band of wavelengths, while gases emit and absorb radiation at discrete frequencies or bands.

Artificial IR emission may come from a heated solid, a flame, or a gaseous discharge. Gaseous sources used primarily for infrared systems are arc lamps containing either a mixture of mercury and xenon or cesium vapor.

Substances are detected only by virtue of their contrast against their background. However, the sun, moon, earth, and stars provide false background radiation (i.e., natural radiation) which must be filtered out. These sources of error are eliminated by the use of space-filtering techniques which reject extended background radiation in favor of point source energy, or spectral discrimination, when solar radiation is involved. The atmosphere attenuates infrared radiation through absorption by atmospheric gases and scattering by molecules, dust, smoke, salt particles, and water droplets. Water vapor, carbon dioxide, ozone, and nitrous oxide are the principal absorbers of infrared radiation in the atmosphere.

Infrared optical systems use mirrors, lenses, or combinations of both. Materials used are those with appropriate mechanical, chemical, and thermal properties, and with high transmission in the infrared spectral range.

Infrared detectors fall into two general classes - thermal and photo detectors. Thermal detectors have an active element of low thermal capacity coated with a black absorbing material. The thermistor and thermocouple are examples of thermal detectors. Photodetectors depend on individual photons activating electron flow within a sensitive film or crystal. The photon either ejects the electron (photo-emission) or raises it to an excited state, as in photoconductive or photovoltaic detectors. These electronic systems include detectors, amplifiers, and readout devices similar to those described in previous applications.

The type of systems described are considered to be active systems. When natural radiation is utilized as the energy source for environmental measurements, the system is passive. In air pollution measurements, for example, both active and passive technique have been employed (see Chapter 8), whereas weather satellite systems utilize the passive system (see Chapter 16).

(c) Optical Systems. Optical and infrared sensors employed in weather satellite systems are also of importance in physical measurements. Optical sensors use detectors which are sensitive to visible light. The items optimized include the light source, energy detector, method of energy conversion or amplification, and visual display. Some of these techniques parallel those in infrared, ultraviolet, and microwave systems using electro-optical methods. Further discussion of electro-optical sensors is found elsewhere in this book. Weather satellite systems are fully described in Chapter 16 in which remote-sensing scanners and imagers are discussed.

5.1.2 System Reliability. Reliability is defined as the probability that specified functions will perform for a given time under known conditions. This definition requires the following specifications: the required time or life, the environmental conditions, and the operational requirements. From these specifications, a quantitative value is determined for reliability and used for prediction purposes.

As shown in Figure 5-11, the first stage is the early life of the equipment where

short-lived and marginal components fail and are removed by debugging processes.[6]
The second stage is a period of constant failure rate over the main operating life
of the equipment when failures occur randomly and infrequently. The curve in the
figure is broken at the central portion to indicate that this stage lasts much
longer than either of the other stages. In the third stage, failures begin to increase because of wear after an extended life. Since the failure rate in the
middle portion of the curve is fairly constant, the problem of reliability design,
prediction and measurement is simplified. By using statistical theory, the basic
reliability equation or failure law is obtained

$$R = e^{-\lambda t} = e^{-t/m} \quad \ldots \ldots \ldots \ldots \ldots \ldots \ldots \ldots \ldots \quad (2)$$

where R is reliability expressed as a probability of successful operation for
time, t; λ, failure time; m, inverse of the failure rate or Mean Time Between
Failures (MTBF); and e, base of the natural logarithm. Note that this law applies
to the second stage only.

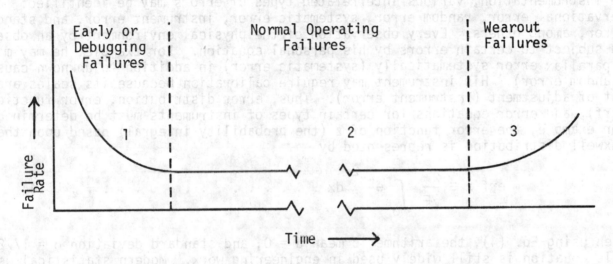

Figure 5-11 Typical Failure Behavior Curve

(1) Reliability Prediction. Since a sample of equipment consists of many components, each with its own failure rate, the overall failure rate is simply the sum
of the individual failure rates. Therefore, the equipment designer must know the
history and/or testing of the failure rate of the individual components. If the
total system then does not meet the requirements for an overall MTBF, more reliable components must be used or the stresses on the components reduced. For
a high reliability system, a parallel redundancy is required so that if a component fails, an alternate equivalent path is provided for continued successful
operation.

(2) Measurement. Most time-based reliability test data are mathematically based
on the constant failure rate shown in Stage 2 of Figure 5-11. If one has a large
number of samples, and these are tested for a long period of time, the MTBF of
the equipment is determined by

$$MTBF = \frac{\text{Total operating time of all samples}}{\text{Total number of failures}} \quad \ldots \ldots \ldots \quad (3)$$

---

[6]Debugging involves detecting and removing defects or errors from a system.

As this cannot be done in actual practice, the determination is made on a limited number of samples.

(3) <u>Measurement Errors</u>. In any measurement, errors are attributed to known and unknown causes. The errors from known causes can arise from test conditions (environment), measurement equipment errors (calibration against standard equipment), mistakes, systematic errors (test procedure, equipment or measured object), and visual errors (human errors, fatigue, etc.). Errors due to unknown causes are random in nature and are related to transient changes in environmental conditions or in the observer.

(4) <u>Error Analysis</u>. In any experiment, one can reduce the errors due to known causes by taking the proper precautions. For the determination of the probable error from unknown causes, however, it is necessary to apply the law of probability after taking a number of measurements.

In instrumentation, various interrelated types of errors may be identified: observational error, random error, systematic error, instrument error, and standard error, among others. Every observation of the physical environment by an observer is subject to certain errors by his personal equation. For example, he may make a parallax error systematically (systematic error) in addition to unknown causes (random error). His instrument may require calibration because its scales are out of adjustment (instrument error). Thus, error distribution, error function (erf), and error equations for certain types of instruments must be determined. For example, the error function of z (the probability integral) based upon the Maxwell distribution is represented by

$$\text{erf } z \equiv \frac{2}{\sqrt{\pi}} \int_0^z e^{-x^2} \, dx \quad \ldots \ldots \ldots \ldots \ldots \ldots \ldots \quad (4)$$

When using Eq. (4), the arithmetic mean $\mu = 0$, and standard deviation $\sigma = 1/\sqrt{2}$. This equation is still widely used in engineering work. Modern statistical usage favors the unit normal variate, u, in which $\mu = 0$ and $\sigma = 1$. The relationship between the probability integral (or erf z) and the distribution function F(u) of the unit normal variate, u, is specified as

$$u \text{ positive: } F(u) = \frac{1}{2} + \frac{1}{2} \text{ erf } (u/\sqrt{2}) \quad \ldots \ldots \ldots \ldots \quad (5)$$

$$u \text{ negative: } F(u) = \frac{1}{2} + \frac{1}{2} \text{ erf } (-u/\sqrt{2}) \quad \ldots \ldots \ldots \ldots \quad (6)$$

Error equations for instruments are discussed below in association with each instrument system.

5.2 <u>Doppler Systems</u>. The Doppler effect is the change in the apparent frequency of a wave train produced either by the motion of the source toward or away from an observer, or by motion of the observer toward or away from the source, or both. The Doppler effect is of great importance in the cases of light, sound, and radio waves. For sound waves, the observed frequency, f_0, in hertz is given by

$$f_o = \frac{v + w - v_o}{v + w - v_s} f_s \dots \dots \dots \dots \dots \dots \dots \quad (7)$$

where v is velocity of sound in the medium; v_o, velocity of the observer; v_s, velocity of the source; w, velocity of the wind in the direction of the sound propagation; and f_s, frequency of the source.

For optical waves

$$f_o = f_s \sqrt{\frac{c + v_r}{c - v_r}} \dots \dots \dots \dots \dots \dots \quad (8)$$

where v_r is velocity of the source relative to the observer; and c, the speed of light.

In applications to radar, the shift in frequency, f_d, is proportional to the relative radial velocity, v, and the original frequency, f_o, such that

$$f_d = \frac{2v}{c} f_o \dots \dots \dots \dots \dots \dots \dots \dots \dots \quad (9)$$

where both f_d and v are a function of the difference between the transmitted and received signals.

5.3 <u>Static and Dynamic Systems</u>. Recognizing that the air, water, and land are continuously in motion, two basic concepts have been developed in the science of physics: <u>statics</u> and <u>dynamics</u>. Several coordinate systems have been established as frames of reference for the motion of mass and energy of the geosphere. This section elucidates the various applications of the two concepts to the real time and space of geospheric instrumentation. This two concept application deals with the moving system and, consequently, modifies the conventional concept of statics to an appreciable extent.[7] The term static system is here designated as the system of steady motion (i.e., steady state) or the system of equilibrium.

5.3.1 <u>The Systems of Statics</u>. This system deals with conditions under which the continuous flows of mass and energy remain unchanged when acted upon by various forces. Examples are the rotation of the Earth (the geosphere), the motion of plate tectonics (the lithosphere), the permanent oceanic currents (the hydrosphere), and the planetary circulation (the atmosphere). In general, the mean speed and direction of these motions remain constant in time and space. A bal-

---

[7] According to the law of conservation of momentum (the First Law of Motion as formulated by Newton in 1687), if a body is <u>not</u> acted upon by an external force, its momentum remains constant. This means that a <u>rigid</u> body remains either in constant motion or at rest. But such an ideal rigid body is non-existent. Furthermore, several external forces may act upon a rigid body which still conserves its momentum because these forces may balance one another. With respect to spatial and temporal variations, a dynamic system may refer to a varying fluid flow (e.g., turbulence flow), whereas a static system, an orderly flow (e.g., laminar flow).

ance of forces is established in this system so that an orderly and steady flow is observed. In fact, the entire geosphere is governed by a system of statics. This makes it possible for geophysicists to recognize the many orderly flow patterns of moving air, water, and land.

The application criteria of the static system are classified into the following three categories: (1) those which are restricted to the threshold value of one or more variables, (2) those which are restricted to time factors, and (3) those which are restricted to the dimensions of the geosphere. Once the criteria are known, the design and performance of observation and measurement readily become possible. It is, therefore, important to understand the above criteria. Brief illustrations of these criteria are in order.

(1) <u>Those Which Are Restricted to the Threshold Values</u>. With a wind speed of 15 knots or more, standing waves are generally stationary; they are commonly observed in association with airflows over and/or behind a mountain barrier (mountain waves), behind the mountains and at a distance between 4 to 40 km (lee waves). The inertial flows, a balance of Coriolis and centrifugal forces with a measurable amount of pressure-gradient force and frictional force, are found both over the ocean, and in the atmosphere. This flow occurs only in low latitudes.

(2) <u>Those Which Are Restricted to Time Factors</u>. The daily phenomena which are commonly observed are the land-sea breeze, the lake breeze, valley winds, and upslope or downslope (drainage air) flows. Common seasonal phenomena are the monsoons and the diurnal waves of atmospheric pressure, temperature, and radiation at a given season. The flow of the solar radiant flux during a cloudless day over a given season of the year, for example, has its own flow pattern. All of these may also be called the rhythmic systems.

(3) <u>Those Which Are Restricted to the Dimensions of the Geosphere</u>. In this case, specifications of the time and spatial dimensions are necessary. When a general circulation is recognized as having regular and continuous flow patterns (such as the prevailing westerlies in the middle latitude and the easterlies of the tropics), the global and annual dimensions are taken into consideration. In view of the various sizes of turbulent flows within the general circulation, the latter cannot be considered as a systematic and orderly flow but rather a random motion. Another typical example of this nature is the oceanic tide. Examples of microscale phenomenon are water circulation of a pond over the year in a cold region and the formation of saline soil in an irrigated land over a century or more.

The above three categories involve one or more of the following systems: the growth, decay, cyclic and rhythmic systems, as mentioned earlier. They definitely do not belong to the random fluctuation system.

5.3.2 <u>The System of Dynamics</u>. This system deals with conditions under which the flows of mass and energy are continuously and constantly in alteration. It is a typical random fluctuation system. All of the hazardous phenomena are created randomly by the dynamic system. Examples of some of these phenomena are: earthquakes and volcanic eruptions (geologic), dust bowls and erosion (pedologic), floods and tsunamis (hydrologic), and blizzards, tornadoes, and hurricanes (atmospheric). Although the dynamic system of the geosphere may or may not be governed by these random phenomena, they have appreciable influence on the biosphere and sociosphere in addition to the geosphere itself.

Consequently, there have been intensive studies on the development of instruments and instrumentation to measure each of these phenomena. Each phenomenon or event has its own system including such processes as its formation, movement, and dissipation. For instance, the initiation of a tropical cyclone, its track or path, and its decay are processes of nearly all severe storms. Most of the geologic hazardous events, however, are more or less localized. Earth tremors as a result of earthquakes may continue for some time and cover a huge geographical area, but they usually do not have destructive power over areas the size of a continent. At any rate, the prediction and control of all of these hazardous events are the ultimate goals of geophysicists. In the study of instrumentation, an understanding of the dynamic system of each of these events is of paramount importance. The remaining chapters of this book will contain discussions of the measurements of the various physical properties of these phenomena.

5.4 _In Situ and Remote-Sensing Instrument Systems_. In an _in situ_ or contacting instrument, the sensor must be in close contact and equilibrium with the medium from which it obtains the input signal. Examples of contacting instruments are a mercury thermometer for measuring soil temperature, a hair hygrometer for measuring relative humidity of the air, an aerovane for measuring wind speed and direction, and a secchi disk for estimating the transparency of water. In a remote-sensing or non-contacting instrument, the sensor obtains its input signal from a distance. Examples of remote-sensing instruments are a human eye for visibility observations, a camera for taking cloud pictures, radar for detecting the location and size of thunderstorms, a sonar detector on board ship for determining the depth of ocean basins, and an IRIS (_InfraRed Interferometer Spectrometer_) for measuring the vertical temperature profile of the earth's atmosphere from a weather satellite. More instruments, both prototype and commercial, are now available for non-contacting types of instrumentation.

Different systems of measurement should be assigned to each type due to the differences in specifications of the above two types of instruments in spatial and time coverage, representativeness, time response, sensitivity, and accuracy. Of course, the use of platforms and the different environmental parameters measured also affect the choice of a system. By and large, a ground mobile unit system is favored for use with _in situ_ instruments so that a better area coverage will be attained. For aircraft measurements, remote-sensing instruments are more suitable. A radar system, for example, is usually installed on a ground platform, but because of its light weight, side-looking radar[8] can be installed on aircraft.

5.5 _Eulerian and Lagrangian Systems_. When a sensor of an instrument is placed at a fixed point in space, it senses the time rate of the local change, $\partial/\partial t$, of the signal, Q. Thus, the measurement is expressed as $\partial Q/\partial t$. For example, a series of temperature records, T, registered by a thermograph at a local weather

---

[8] Side-looking airborne radars (SLAR) emit pulses sequentially along the line of flight. Return signals from a target are stored along with phase information and then the returns are combined inphase. Film is moved across the image line, formed as a recombined beam on a CRT, producing a photographic image. (Wolff and Mercanti, 1974)

station in time, t, can be expressed as $\partial T/\partial t$. This frame of measurement is called the Eulerian Reference System or Eulerian coordinates. In fact, in the past most observations and measurements in the geosphere were obtained from fixed points in space. This is particularly true in meteorological and hydrological surveillances. It is usually, though by no means always, more convenient to use Eulerian systems. However, Eulerian systems are unsuccessful in the study of diffusing particles or turbulence, particularly in fluid mechanics. An alternative system, the Lagrangian Reference System, is required.

When an airborne or waterborne sensor of an instrument is moving with the air or water current, it senses the physical property, Q, of the individual fluid parcel all the time, D/Dt. The individual change is expressed mathematically as DQ/Dt. This frame of measurement is called the Lagrangian Reference System or Lagrangian Coordinates. Mathematically, the relationships between the Lagrangian and the Eulerian changes are expressed by

$$\frac{DQ}{Dt} = \frac{\partial Q}{\partial t} + u\,\frac{\partial Q}{\partial x} + v\,\frac{\partial Q}{\partial y} + w\,\frac{\partial Q}{\partial z} \cdot \ldots \ldots \ldots \ldots \ldots \quad (10)$$

where u, v, w are the velocity components of the fluid passing through a fixed point. Equation (10) can be simplified by using the vector operator, or

$$\frac{DQ}{Dt} = \frac{\partial Q}{\partial T} + V \cdot \nabla Q \ldots \ldots \ldots \ldots \ldots \ldots \ldots \ldots \quad (11)$$

In practice, the most successful experiments using Lagrangian frames of reference are the applications of tracer techniques and constant-level balloon techniques. Various tracers, natural and artificial, have been used in the atmosphere and over the ocean. Tracers are relatively stable substances such as dyes, fluorescent particles, stack plumes, radioactive fallout, volcanic dust, clouds, and pollen. By assuming these tracers are weightless in comparison with their fluid medium, the velocity of the tracer represents that of its medium. The position of the tracer at any instant in time is then detected by such remote-sensing instruments as radar, laser, lidar, and even vidicon cameras from satellites. The tracer technique is discussed later in Sections 6.4.1 and 7.2.6. In the constant-level balloon technique, the position of balloons at any time can be tracked by the detector of a space vehicle. The balloon carries miniature meteorological instruments to measure pressure, temperature, and humidity as it moves with a certain air parcel. Constant-level balloon techniques are further elaborated in Sections 7.1 and 16.4.2.

In summary, the Eulerian approach has been employed in all measurements of the geosphere, whereas the Lagrangian has been applied only to the atmosphere and hydrosphere. Most of its applications have been in the determination of the flow pattern of fluids. There are fewer studies of the physical properties of individual fluid parcels. The limitations of balloon techniques and aircraft samplers in measuring these properties are discussed in detail in Chapter 7.

PART B

ATMOSPHERIC INSTRUMENTATION

At present, instruments for routine observations and special measurements of atmospheric variables are in the general areas of:

1. Surface meteorological instruments which customarily are installed in weather shelters at 1.5 m above the ground and on masts, towers, and buildings at various heights above, on, or under the ground surface.

2. Upper-air instruments installed on the earth's surface, on airborne balloons, and in aircraft, covering a vertical range of about 35 km.

3. Research or special meteorological instruments for mesoscale meteorological measurements[1] installed on a variety of platforms and covering a vertical range of about 0.5 to 2 km.

4. Upper atmospheric instruments installed on such spaceborne platforms as rockets, satellites, high altitude balloons, and U-2 aircraft, in addition to ground-based platforms. Both operational and research studies of these instruments, however, are discussed in Part E.

The first three areas of measurements constitute the next three chapters of Part B.

Chapter 6 describes the principles, mechanisms, and operation of instruments for surface air measurements. The meteorological variables measured are the electromagnetic field, thermofield, pressure field, wind field, and hydrofield. Both conventional and research instruments are described. Because large numbers of instruments are now available, illustrations are given only for a few major types. However, other important instruments are listed and their references cited.

Chapter 7 covers upper-air instruments. Since most upper-air soundings are performed by a variety of instrument systems and platforms, discussions follow the techniques of measurement rather than meteorological variables. Thus, the order

---

[1] Special meteorological instruments pertain to those instruments which are not used by national weather services, but are employed by such organizations as agricultural, health, military, and other special research institutes.

of presentation is balloon techniques, ground platform techniques, and aircraft techniques, and their system reliabilities. Space vehicle techniques involving satellites and rockets are given in Part E. Because of the great advancement in upper-air instrumentation, there are too many instruments and instrument systems to be adequately described in this chapter. Therefore, only a few major, commonly used instruments are described in detail. As in the previous chapter, references are cited for other important instruments.

Chapter 8 summarizes the research and special mesoscale measurements of such atmospheric variables as aerosols and particulates, atmospheric ions, radioactive fallout, gaseous chemicals, airborne organisms, and noise. Most of these measurements are conducted at ground-level, but some are monitored by airborne balloons or aircraft. While some of the above measurements are performed by the Environmental Protection Agency (EPA), the Nuclear Regulatory Agency (NRA), and a number of other federal, state, regional and local governmental agencies in the USA, others are conducted by research institutions.

Part B emphasizes the physical and operating characteristics of selected instruments, comparison of techniques in using these instruments, and alternative instrumentation methods available.

CHAPTER 6 SURFACE AIR MEASUREMENTS

For our purposes, surface meteorological instruments may be divided into two categories: Category A, which consists of the conventional instruments employed by national weather services in the USA and abroad; and Category B, which includes special prototype instruments developed and used in research. Table 6-1 classifies these instruments.

Table 6-1 Classification of Surface Instruments

Collective Nomenclature	Meteorological Variables Measured	Representation in Category A	B
Radiometry	Solar radiation, infrared radiation, net radiation, sometimes ultraviolet radiation, and other spectral values.	no	yes
Photometry	Sunshine,* starshine, visual range,* and sometimes transparency of the atmosphere.	*	sometimes
Thermometry	Air temperature* and vertical temperature gradients.	*	yes
Barometry	Atmospheric pressure	yes	yes
Anemometry	Wind velocity* (direction and speed), sometimes three dimensional wind components as well as turbulence and diffusion.	*	yes
Atmometry	Evaporation,* transpiration, and evapotranspiration.	*	yes
Hygrometry	Atmospheric humidity.	yes	yes
Nephometry	Clouds (types, amount, height, etc.).	yes	yes
Hyetology	Precipitation* (rainfall, snowfall, hailstone, etc.), and condensation (dewfall, fog drips, etc.).	*	yes

*The asterisks indicate the only meteorological variables measured within each type of observation (collective nomenclature) by Category A. (e.g., In thermometry, Category A instruments measure air temperature but not vertical temperature gradients.)

6.1 <u>Radiometry and Photometry</u>. While radiometry is the science of measurement of radiant energy in general, photometry is concerned with the luminance or brightness which exists only in the visible portion of the electromagnetic spectrum. An instrument designed to measure this visible radiation is called a photometer, while one that measures radiant energy in general is known as a radiometer.

In photometry, visual observation must be weighted to take into account the variable response of the human eye as a function of the wavelength of light. The photo-unit of luminous intensity is the candle which is designated as one-sixtieth of the luminous intensity of one square centimeter of a blackbody radiator at the temperature of 2,046K.[2]   One candle per square centimeter (cdl cm^{-2}) is the brightness or luminance equivalent to 3.142 lamberts, another expression of brightness. Several derived units of brightness are given in Appendix 1. In radiometry, the thermal energy emitted by the sun and the earth is measured. Over the surface of the earth, the spectral range of radiation covers the near ultraviolet, the visible, and the infrared radiation, or 0.3 to 4.0 µm. Below the near ultraviolet the radiant energy does not reach the earth's surface; beyond the near infrared the radiant energy is too weak to be measured by conventional radiometers. The common units for radiometric measurements are either calories per square centimeter per minute (cal cm^{-2} min^{-1}) or milliwatts per square centimeter (mW cm^{-2}).[3]   One cal cm^{-2} is referred to as a langley (ly). Hereafter, langley instead of calories per square centimeter is employed.

6.1.1 <u>Radiometry</u>. As shown in Table 3-5, there are six major types of radiometers. In this section, however, only the following three types of radiometers are discussed:

> (1) the pyrheliometer, as represented by the Ångström compensation pyrheliometer;

> (2) the pyranometer, as represented by the Eppley pyranometer; and

> (3) the net pyrradiometer, as represented by the Gier and Dunkle net exchange radiometer.

The last two of the above instruments can be used to make albedo measurements. Some simple, inexpensive but useful radiometers are also briefly discussed. The spectral range for the above instruments is about 0.3 to 4.0 m. Radiant energy rather than duration or wavelength of radiation is measured. Specifications of various radiation devices are summarized in Table 6-2 below.

(1) <u>Ångström Compensation Pyrheliometer</u>. About 1890 K. Ångström of Sweden invented the well-known Ångström compensation pyrheliometer. It has since been

---

[2]This is the solidification temperature of platinum.

[3]One 15C g-cal cm^{-2} min^{-1} is equal to 69.75 mW cm^{-2}. Sometimes mW cm^{-2} and W m^{-2} are employed.

Table 6-2 Specifications of Radiation Instruments

A. Pyrheliometer (direct radiation 0.30 to 4.0 μm)

1. Angstrom Pyrheliometer (see text)

2. Abbot Silver Disk Pyrheliometer (sensor: blackened silver disk; readout: mercury thermometer scale unit, 0.5C; accuracy: 0.5%; mechanism: Q_n = C d $\Delta T/\Delta t$, where C is the thermal capacity of the disk and d is its thickness, $\Delta T/\Delta t$ is the rate of temperature change between the alternately shaded and unshaded disk as it receives solar radiation at regular 2 min. time intervals.) Used principally in the USA as standard instrument.

3. Michelson Actinometer (sensor: blackened bimetallic strip; readout: microscopic readings of the deflecting angle; time response: 20 to 30 s; accuracy: 1%; mechanism: 3C of strip temperature equivalent to 1 ly min^{-1}.) A portable instrument used as a secondary standard.

4. Eppley Normal Incidence Pyrheliometer (sensor: thermopile; readout: potentiometer readings; sensitivity: within 1% for a temperature range of -30 to +50C; time response: 20 s for 98% of signal.) Follows the same principle as the silver disk pyrheliometer.

B. Pyranometer (global radiation 0.30 to 5.0 μm)

1. Eppley Pyranometer and Black and White Pyranometer (see text)

2. Moll-Gorczynski Solarimeter (sensor: 14 alternate manganin and constantan strips forming 14 thermojunctions; readout: potentiometer readings; sensitivity: 8 mv per ly min^{-1} for 10 ohms resistance; time response: about 10 s. Uses black and white surfaces like the Eppley pyranometer.) Commonly used in Europe.

3. Robitzsch Bimetallic Actinograph (sensor: blackened bimetallic strips; readout: paper strip chart on a drum; mechanism: lever principle; accuracy: 5 to 10%; time response: 10 to 15 min. for 98% response). Recommended for field use without power supplies.

4. Others: Among several others, the Kipp and Zonen Solarimeter and Linke and Feussner Actinometer deserve special attention. Both of these are built and manufactured by P.J. Kipp and Zonen, Deft, Holland. The latter is the actinometer at its best. It measures daylight, sunlight, and nocturnal radiation. Both instruments employ a manganin-constantan thermopile similar to the Moll-Gorczynski solarimeter. The sensitivity of the Kipp and Zonen solarimeter is about 7.9 mv per ly min^{-1}.

C. Pyrradiometer and Net Radiometer (global and terrestrial radiation at various wavelengths)

1. <u>Ventilated</u> <u>Exposed</u> <u>Surface</u> (all are similar in mechanism; see Gier and Dunkle instrument in the text).

   a. Franssila-Suomi Net Radiometer (prototype instrument).
   b. Gier and Dunkle Net Exchange Radiometer (manufactured by Teledyne Geotect).
   c. Kew Observatory Net Radiometer (prototype instrument).
   d. Courvoisier Balance Meter (manufactured by the Physikalisch-Meteorologischen Observatorium, Davos, Switzerland).

   Among the above four, the Courvoisier Balance Meter is the best.

2. <u>Non-Ventilated</u> <u>Exposed</u> <u>Surface</u> (see references 2 and 4)

   a. Hofman Heat-Compensated Net Radiometer (prototype).
   b. Monteith Net Radiometer (prototype).

3. <u>Non-Ventilated</u> <u>Enclosed</u> <u>Surface</u> (see references 2 and 7)

   a. Suomi Economic Net Radiometer (prototype).
   b. Funk Net Radiometer (manufactured by C.S.I.R.O., Division of Meteorological Physics, Melbourne, Australia).
   c. Fritschen and van Wijk Net Radiometer (prototype).

References:

1. Air Ministry Meteorological Office. 1956. Measurement of the duration of sunshine and starshine and of the intensity of solar radiation, In: Handbook of Meteorological Instruments, Part I: Instruments for Surface Observations. HMSO, London. 458 pp.

2. Gates, D.M. 1962. Radiation instruments, In: Energy Exchange in the Atmosphere. Harper and Row, New York. 151 pp.

3. Marchgraber, R.M. 1970. The development of standard instruments for radiation measurements, In: Meteorological Observations and Instrumentation. Meteorological Monographs 11(33). 455 pp.

4. Monteith, J.L. 1972. Survey of Instruments for Micrometeorology. Blackwell Scientific Publications, Oxford. 263 pp.

5. Robinson, N. 1966. Solar Radiation. Elsevier Publishing Co., New York. 347 pp.

6. Suomi, V.E. 1962. Instrumentation, In: Bibliography of Agricultural Meteorology. Wang, J.Y. and G.B. Barges, ed. The University of Wisconsin Press, Madison, Wisconsin. 673 pp.

7. Wang, J.Y. 1972. Instrumentation and observation, In: Agricultural Meteorology. Milieu Information Service Inc., San Jose, California. 537 pp.

8. Coulson, K. 1975. Solar and Terrestrial Radiation. Academic Press, New York. 332 pp.

improved such that the instrument has an accuracy on the order of ± 0.5 per cent. This instrument now is used as the primary standard pyrheliometer for calibrating secondary pyrheliometers and pyranometers, particularly in European countries.

As shown in Figure 6-1$_a$, the pyrheliometer is first leveled by screws G, H, and I. Its tube, blackened inside, is oriented toward the direction of the sun by siting through diopters C and D, mounted on top of the tube. The tube is fine adjusted by knobs A and B so that the sun's beam will fall at an incidence angle normal to the surface of the manganin strips[4] L and M, shown in Figure 6-1$_b$. The two manganin strips, which are mounted in the rear of the tube and tightened by screw K of the strip-holder, are identical, thin, blackened sensing elements (see Figure 6-1$_c$). While one of the strips is exposed to the sun, the other is both shaded by the radiation shield T and heated by an electric current. The current is regulated by thermocouples Q and R (see Figure 6-1$_d$) so that the heating of the two strips is identical. This can be done by the null balance method, using a sensitive galvanometer.[5] Both the heating current and thermocouple circuits are given in Figure 6-2. By placing shield T in front of the tube, either one of the two strips can be shaded from the sun. At the same time, the heating circuit is coupled only to the shaded strip by means of either platinum contact N or P.

The theory of operation of the pyrheliometer is as follows. Let Q_n be the radiant energy (ly s^{-1}); b, the width of the strips (cm); and α, the absorptivity.[6] Then one unit length of the strip receives $\alpha b\, Q_n$ ly s^{-1}. If Ω is the electric resistance in ohms per unit length of manganin strip, and i is the electric current in amperes, then the radiant energy absorbed by the strip per unit length (equal to $\alpha b\, Q_n$) is $\Omega i^2/4.187$ (in which 1 cal s^{-1} equals 4.187 watts). Thus

$$Q_n = \frac{\Omega i^2}{4.187\ \alpha b}\ (\text{ly s}^{-1})\ \dots\dots\dots\dots\dots\dots\dots\ (1)$$

or

$$Q_n = \frac{60\ \Omega i^2}{4.187\ \alpha b}\ (\text{ly min}^{-1})$$

When watts per square centimeter are employed

$$Q_n = \frac{\Omega i^2}{\alpha b}\ (\text{W cm}^{-2})$$

---

[4] Normally, the strip has dimensions of 20 x 2 x 0.01 mm. Manganin is an alloy consisting of copper, manganese, and nickel, with excellent electrical conductivity.

[5] At the beginning of the operation, the shield is turned to 90° so that both strips are exposed to the sun; there is no heating current applied to either one of the strips. The reading of the galvanometer connected with the thermocouple circuit in series is taken as the zero position. One strip is exposed to the sun by turning the shield 90° to the left or right. The other strip is connected to the heating current circuit (i.e., the compensating current circuit) and heated to the same temperature of its counterpart (the latter produced by solar heating). The total time required for switching the circuit from one strip to another is a minute or two.

[6] Absorptivity, α, is the ratio of radiant energy absorbed to the total amount incident upon the strip. The magnitude of absorption is, therefore, a function of the color and size and, in turn, the thermal capacity of the manganin strips.

The values of α, b, and Ω are determined in the laboratory.

In practice, the following equation is used

$$Q_n = k\, i^2 \quad\ldots\ldots\ldots\ldots\ldots\ldots\ldots\ldots\ldots\ldots\ldots\ldots\ldots (2)$$

where $k = \Omega/\alpha b$. The value of the pyrheliometric constant, k, is obtained by comparison with a standard pyrheliometer.

The great advantage of the Ångström pyrheliometer is that it uses the principle of thermal compensation, making it unnecessary to measure the temperatures of individual strips. This increases the accuracy and eliminates errors in making individual measurements. Other advantages are its high temporal stability, fast response, high sensitivity, and ease of operation. Most of the drawbacks of this pyrheliometer are common to all radiometers used in absolute measurements: (a) the incomplete elimination of traces of sky light entering through the window; (b) the imperfect simulation of a blackbody with respect to absorption and uniformity over the entire sensing surface; and (c) the difficulty in constructing two strips that are identical in shape and thermal capacity. The last drawback is a typical problem of dual measurement (i.e., making a comparison of two simultaneous measurements). This problem can be solved by making alternate exposures of each strip. The mean value of the electric current from several successive measurements is employed as the output signal of the pyrheliometer. In summary, the Ångström pyrheliometer has been used for over eight decades and its performance is well documented. It is believed that, in the future, this pyrheliometer will serve as a standard radiometer.

Aside from the Angstrom pyrheliometer, the Smithsonian silverdisk pyrheliometer (i.e., the modern version of Abbott Pyrheliometer of 1902) is another long standing standard instrument. Several other pyrheliometers, prototypic or commercial, used as standard or secondary instruments, are now available. Their specifications are summarized in Table 6-2.

(2) Eppley Pyranometer. For measuring the global radiation on a horizontal plane near the earth's surface, the Eppley pyranometer has been extensively employed as a secondary instrument in the United States and some countries abroad. An improved version, the Eppley black and white pyranometer, has replaced the described device below. However, the older version still is used widely; hence, a description of it is appropriate.

The Eppley pyranometer consists of an outer whitened ring coated by magnesium oxide (the cold junction) and an inner blackened ring coated by British Parson's optical black lacquer (the hot junction)[7] as shown in Figure 6-3. While the black coating has uniform absorptivity for the measurable spectral range, the reflectivity of the white coating is constant up to 2 μm. The undersides of the two rings are attached to a pair of thermopiles which are made of 60% gold/40% palladium and 90%

---

[7]This lacquer absorbs the wavelength between 1 and 15 μm with a mean absorptivity of 0.98. A new product known as NEXTEL Brand velvet coating optical black paint, manufactured by 3M Company, has a mean absorptivity of 0.98. Nextel, however, is more durable than Parsons' paint, lasting about 3 years per coating.

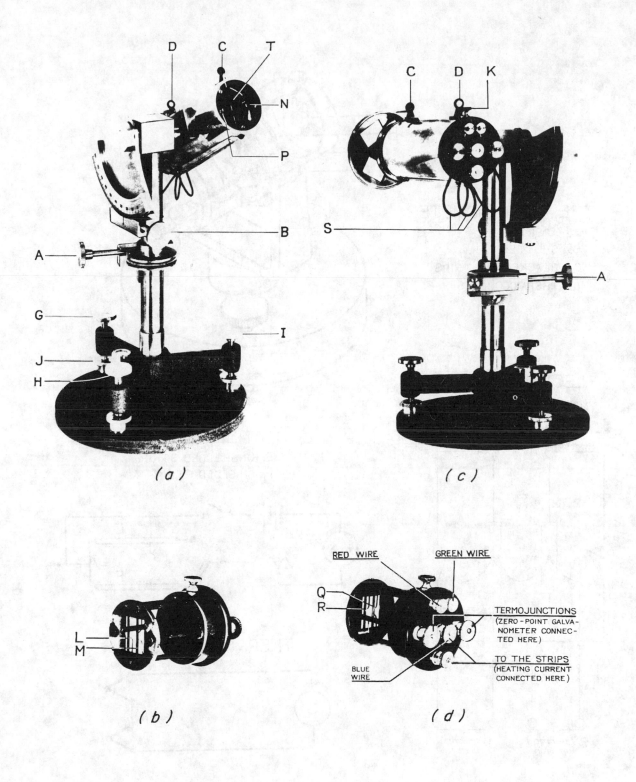

Figure 6-1 Ångström Compensation Pyrheliometer

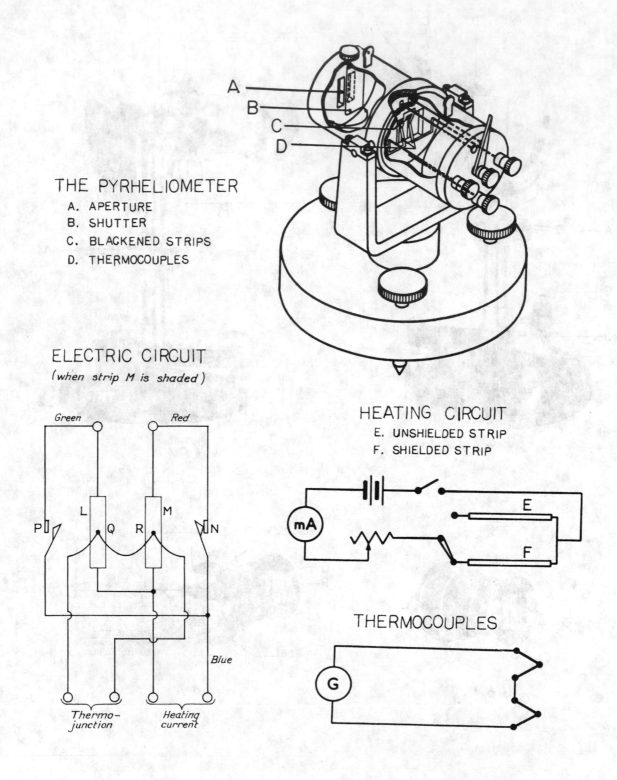

112

THE PYRHELIOMETER
  A. APERTURE
  B. SHUTTER
  C. BLACKENED STRIPS
  D. THERMOCOUPLES

ELECTRIC CIRCUIT
(when strip M is shaded)

Green   Red

P   L   Q   R   M   N

Blue

Thermo-junction   Heating current

HEATING CIRCUIT
  E. UNSHIELDED STRIP
  F. SHIELDED STRIP

mA   E   F

THERMOCOUPLES

G

Figure 6-2 Ångström Compensation Pyrheliometer

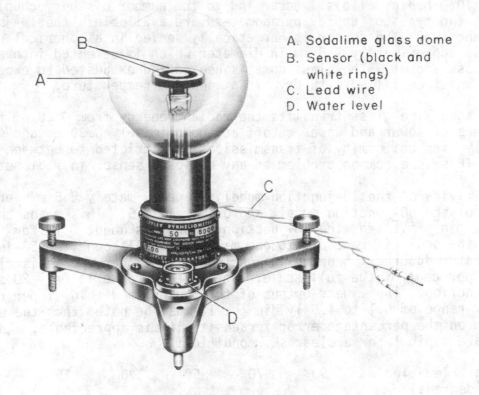

A. Sodalime glass dome
B. Sensor (black and
   white rings)
C. Lead wire
D. Water level

A. Shadow band
B. Pyrheliometer

Figure 6-3 Eppley Pyranometer

platinum/10% rhodium alloys. According to the number of thermocouple junctions employed, two types of Eppley pyranometers are available: the 10-junction and the 50-junction. The sensor is hermetically sealed in a spherical dome of soda lime glass approximately 7.5 cm in diameter which is cemented into a chromium-plated brass socket. The glass dome is heated and exhausted to expel atmospheric moisture and then filled with dry air at ambient temperature.

Although soda lime glass transmits the radiant energy from 0.28 μm to 5 μm with the centers of lower and upper cutoff at approximately 0.30 μm and 4.5 μm respectively, its uniformity of transmission is restricted to between 0.35 μm and 2.5 μm. This is a common problem of any enclosed sensor in radiometry.

The sensitivity of the 10-junction model is approximately 2.5 mV per ly min^{-1}, and that of the 50-junction model, approximately 7.5. This means that the output is 0.4 ly min^{-1} mV^{-1} for the 10-junction, and it is about 0.13 for the 50-junction, provided the scale is readable to one millivolt. Although the 50-junction Eppley has a greater accuracy than the 10-junction, its time response is also high, 30 s for a 98 per cent of the full scale. The time response is only 20 s for the 10-junction. The linear output of the Eppley is within ±1 per cent, over the intensity range of 0.1 to 1.5 ly min^{-1}. It must be noted that the effect of solar elevation on the percentage error in sensitivity is appreciable. The following figures are typical for a clear sky condition:

Solar Elevation (degrees)	90	70	50	30	15
Percentage Error (%)	0	+1	+2	+3	+5

In practice some accessories must be provided for the improvement of the performance of the Eppley pyranometer. They are a platform for exposure, a temperature compensation device, and a built-in integrator. These are described below.

(a) <u>The Platform</u>. For a good exposure the instrument platform should be high enough to be free from any viewing obstructions. Surrounding objects, if any, must not exceed an elevation angle of 5° above the plane of the sensor.

(b) <u>The Temperature Compensation Device</u>. Over the temperature range of -50 to +40C, the instrument sensitivity is not constant and lies between -0.1 and -0.15 per cent per 1C in ambient temperature change. The instrument sensitivity can be improved considerably by inserting a bead type thermistor-resistor inside the socket of the pyranometer. For the standard range of temperature (-20 to +40C), the sensitivity is within ±1 per cent. For a larger range of ambient temperature of -70C to +50C, it is about ±2 per cent.

(c) <u>Built-in Integrators</u>. When the total radiation intensity for an interval of time (hourly, daily, or weekly) is of interest, a specially built integrator is available. Thus, the integrated readings are printed out on a paper strip chart or stored in any recording device.

(3) <u>Eppley Black and White Pyranometer</u>. The thermopile and optical glass dome of the Eppley pyranometer have been improved significantly in the version that

replaces the older device described above. The thermopile utilizes a wire-wound, copper-on-constantan receiving surface, with the junctions located radially. The surface is divided into three blackened hot junctions and three whitened cold junctions alternating around the circular thermopile. Parsons' optical black paint is used on the black segments and barium sulfate on the white segments. Built-in temperature compensation of the 48 plated junctions provides overall ±1.5% compensation from -20C to +40C. The cosine response of the instrument is within ±2% for zenith angles between 0° and 80°. The glass dome protecting the sensor surface is a precision-ground Schott WG7 optical glass which is transparent between wavelengths of 0.28 μm and 2.8 μm; a quart dome may be substituted to increase the glass transmittance to wavelengths between 0.2 μm and 4.5 μm. Typical response features of the 48 junction type are: approximately 7.5 mV/cal cm^{-2}min^{-1}, impedance of 300 ohms, linearity of ±1% from 0 to 2 cal cm^{-2}min^{-1} (0 to 136.0 mW cm^{-2}), 3-4 s response time to 66% change in intensity. (It should be noted that the use of pie or star-shaped black and white sensing surfaces for pyranometers is not new. Dirmhirn and Sauberer used such a pattern on their star pyranometer as reported in 1958.)

The same basic design of domed glass covers (half hemispheres) is used for the Eppley precision spectral pyranometer. The circular wire-wound thermopile, however, is totally coated with Parsons' optical black paint. Two glass domes provide for selectivity in wavebands sensed by the instrument. The inner dome is clear WG7 Schott glass, as used in the black and white pyranometer. The outer dome is selected for particular wavebands: WG7 between 0.285 and 2.8 μm, yellow GG14 with lower cutoff of 0.5 μm, orange OG1 at 0.53 μm, red RG2 at 0.63 μm, and dark red at 0.7 μm. Instrument sensitivity is 5 mV/cal cm^{-2}min^{-1}; impedance, 300 ohms, temperature compensated to ±1% between -20C and +40C; linear response up to 4 cal cm^{-2}min^{-1} (272 mW cm^{-2}); 1 s response time to 66% change of intensity.

(4) Gier and Dunkle Net Exchange Radiometer. As shown in Figure 6-4, the Gier and Dunkle net exchange radiometer[8] consists of sensor A, a small air blower (enclosed in box C), and built in bubble level B. The sensor is a 480-junction silver-constantan[9] thermopile which is wound on a 115 mm^2 bakelite plate E. The surface of sensor D is an aluminum sheet F whose outer portion is blackened by Parsons' optical black lacquer and whose inner portion is polished. The AC blower provides a constant 20 mph air stream distributed equally across the top and bottom of the sensor plate. When the natural wind speed equals or exceeds 20 mph from any direction, the indicated radiation decreases steadily with the increase of wind speed. The exception is the case of a wind directly opposing the ventilated air stream in the range of 25 to 35 mph.[10] Aside from the minimization of wind effects, artificial ventilation is necessary for the reduction of noise signals from evaporative cooling of moisture deposits (e.g., dew, fog, rime, and to a lesser extent moisture from sorption processes), as well

---

[8] This radiometer was originally designed by Gier and Dunkle at the University of California at Davis. See Gier, J.T. and R.V. Dunkle. 1951. Total hemispheric radiometers. Trans. Amer. Inst. Elect. Engrs. 70:339.

[9] An alloy containing 60% copper and 40% nickel used for electrical resistance heating and thermocouples.

[10] See Portman, D.J. and D. Fleming. 1959. Influence of wind and angle of incident radiation on the performance of Beckman and Whitley total hemisperical radiometer. The Univ. of Michigan Res. Inst., Ann Arbor, Michigan. 388 pp.

116

as dust accumulation on the surface of the sensor. The bubble-type level is installed so that the observer can verify that the sensor remains in a horizontal position. When it rains, the radiometer fails to function and the sensor should be covered for protection.

The theory of operation of the radiometer is based upon the output of a differential thermopile measuring the vertical temperature gradient across a horizontal bakelite sheet. The recorder (any type of potentiometer) gives direct millivolt readings which can be converted into ly min^{-1} as follows. Let R_i be the total

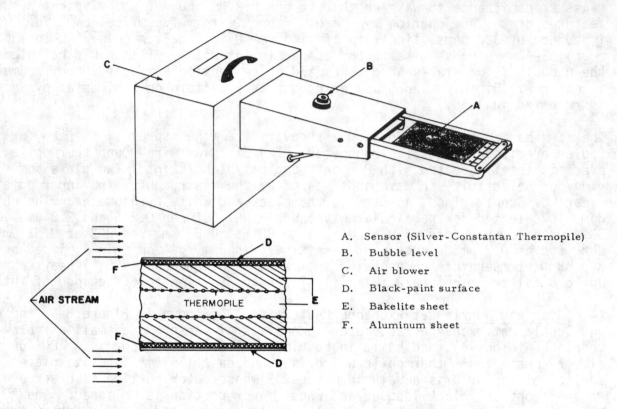

A.   Sensor (Silver-Constantan Thermopile)
B.   Bubble level
C.   Air blower
D.   Black-paint surface
E.   Bakelite sheet
F.   Aluminum sheet

Figure 6-4 Gier and Dunkle Net Exchange Radiometer

global radiation on a horizontal surface from above; R_o, the total terrestrial radiation from below; and R_n, the net radiation. Thus

$$R_n = R_i - R_o = Kv \quad\dots\dots\dots\dots\dots\dots\dots\dots\dots\quad (3)$$

where K is the instrument constant with an order of magnitude of about 0.11 ly mV^{-1} and v is in millivolts. The physical specifications of this radiometer are (a) range: -0.56 to 1.68 ly min^{-1}; (b) accuracy: ±2%; and (c) time response: 12 s to reach a 98% steady state condition. The general accuracy is about 0.05 mV min^{-1}.

The total hemispheric radiometer (Teledyne Geotech) measures direct solar and sky radiation. The structure is identical to that of the net exchange radiometer (see Figure 6-4), except that it has a thermal radiation shield at the bottom of the thermopile transducer. The equation for the total hemispherical radiation on a horizontal surface is

$$R_t = Kv + \sigma T^4 \quad \ldots \ldots \ldots \ldots \ldots \ldots \ldots \ldots \quad (4)$$

where R_t is the total hemispheric radiation on a horizontal surface in ly min^{-1}; K is the instrument constant as in Eq. (3) above; v is in millivolts; σ is 0.813 x 10^{-10} min^{-1}K^{-4} and T is 491.69 + C (Rankine temperature scale). A temperature-sensing device should be provided to compute the last term of Eq. (4).

(5) Albedo Measurements. The albedo of shortwave radiation can be obtained by alternately exposing the sensor of the total hemispheric radiometer just described to incoming and outgoing radiation. This is done by taking a reading with the black surface up first, and then by taking another reading with the black surface down. The radiation in both cases is computed from Eq. (4). Obviously, the albedo is obtained from dividing the outgoing value by the incoming value. The albedo so measured is the hemispherical albedo, not the beam albedo. The total hemispheric radiometer cannot be used for obtaining beam albedo or the albedo in the infrared or ultraviolet range.

The beam albedo is measured by inverting the Eppley pyranometer as shown in Figure 6-5$_a$. However, a better method is to place the Eppley pyranometer upright in the prime focus of a parabolic mirror. This permits a parallel beam to reflect the visible and infrared rays from the surface of bare soil, crown of vegetation, or water surface onto the sensing surface of the Eppley without allowing any diffuse light to enter from surrounding objects (see Figure 6-5$_b$). Thus, the measurement of albedo at low solar altitudes becomes possible. It is necessary to minimize the effect of shadows cast by the instrument by raising the pyranometer to an appreciable height.

When measurement of the infrared or ultraviolet albedo is desired, an infrared light or ultraviolet light sensor should be placed in the prime focus of the parabolic mirror. Several infrared and ultraviolet radiation instruments are commercially available.

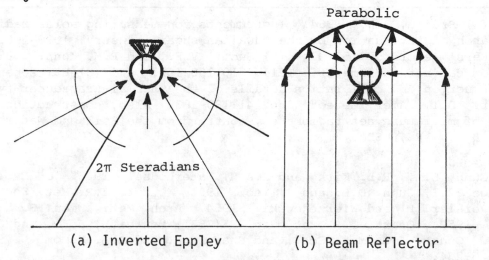

(a) Inverted Eppley   (b) Beam Reflector

Figure 6-5　Albedo Measurements with Eppley Pyranometer

(6) <u>Simple Radiometers</u>. Various simple and inexpensive but useful radiometers have been designed[11] and some of them have sufficiently high accuracy to warrant practicability. Among them, the flat or the spherical Bellani cup radiometer is widely used. The former is generally known as the Bellani atmometer, and the latter is called the Gunn-Bellani radiometer. The Bellani atmometer, which will be elaborated in Section 6.5.1 and Figure 6-54$_b$, consists of a blackened, porous, flat porcelain cup which is connected by glass tubing to a graduated cylinder. After the cup and glass tubing are filled with distilled water in a pneumatic trough, the other end of the glass tubing is submerged in the graduated cylinder which has been pre-filled with distilled water. As only the top plate of the porcelain cup is porous, it serves as the evaporative surface or sensor. By a siphon mechanism, the water in the cylinder is drawn to the plate for evaporation. The entire instrument is airtight, with the exception of the porous top plate. The amount of water consumed in a given time is measured by the graduated cylinder and indicates the amount of evaporation from the porous cup. The rate of evaporation is a function of the wind speed, humidity gradient, incident radiation, and the temperature of the evaporating surface. Under stagnant weather conditions with low humidity, solar radiant energy alone (which determines the temperature of the evaporating surface) becomes the dominant factor. In many regions a highly significant correlation has been established between the amount of water evaporated from the Bellani sensor and the total incident radiation within a given time interval. The coefficient of correlation, r, runs as high as $r = 0.98$. Under these circumstances the Bellani atmometer may serve as an integrator of solar radiation. In regions where dew deposition is high a mercury-wool valve, which acts as a check valve to retard the flow of water from the Bellani plate back into the tube, is inserted in the upper section of the glass tubing.

Another simple device, which has extensive applications in agriculture, hydrology, and meteorology, is the century-old Gunn-Bellani radiometer. Its thermal efficiency has been improved from 4% to over 75%, first by Gunn, Kirk, and Waterhouse, and later by Pereira.[12]

As shown in Figure 6-6, the Gunn-Bellani radiometer consists of two spherical concentric bulbs attached to a 40-ml graduated burette. The space between the two bulbs is evacuated to retard convective heat exchanges between the inner

---

[11] See (1) Brodie, H.W. 1964. Instruments for measuring solar radiation: research and evaluation by the Hawaiian sugar industry, 1928-1962. Hawaiian Planters' Record 57:159-97. (2) Suomi, V.E., and R.M. Kuhn. 1958. An economical net radiometer. Tellus 10:160-163. And (3) Fritschen, L.J. 1960. Construction and calibration details of the thermal-transducer-type net radiometer. Bull. Amer. Meteor. Soc. 41:180-183. Also, construction and evaluation of miniature net radiometer, contr. from the Soil and Water Conservation Res. Div., ARS, USDA.

[12] See Gunn, D.L., R.L. Kirk, and J.A.H. Waterhouse. 1945. J. Exp. Biol. Cambridge 22:1; Gunn, D.L. and D. Yeo. 1951. Quart. J.R. Met. Soc. 77:293; Courvoisier, P. and Wierzejewski. 1954. Arch. Met., Wien, B. 5:413; Pereira, H.C. 1959. Quart. J.R. Met. Soc. 85(253); Monteith, J.L., and G. Szeicz. 1960. Quart. J.R. Met. Soc.; and Shaw, R.H. and A.L. McComb. 1959. Forest Sci. 5:234.

bulb and the ambient air. The inner bulb of copper, blackened with Parsons' black paint, forms part of the alcohol reservoir. As the copper bulb is heated by the sun, the alcohol in the reservoir evaporates. Some of the alcohol vapor condenses and enters the burette through the capillary tube. Thus, the recording of the burette is a measure of the integrated solar radiation within a given time interval. The radiometer can be placed in a sunken container, with its sensor just above the ground level, so as to avoid heating from the terrestrial surface. Two types of Gunn-Bellani radiometers are manufactured by Baird and Tatlock Ltd. (London), with and without a condenser. The latter, although less expensive, is less accurate than the condenser version and is recommended only for replicated field experiments. Both alcohol and water have been tested as the fluid in the radiometer, with water giving poorer results than alcohol. The overall accuracy of the alcohol condenser type radiometer is between ±2 and ±3 per cent.

6.1.2 <u>Photometry</u>. As mentioned earlier, photometry is the science of visible light measurement. The major application of photometry in physical environmental instrumentation is in the measurement of visibility. Another area of application is measurement of the response of the human eye, green plants, and animals to visible radiation. One of the great achievements is the measurement of photosynthetically active radiation (PAR). A portable instrument that performs this function is commercially available from the Lambda Instruments Co., Inc. A brief description of their Quantum/Radiometer/Photometer is in order.

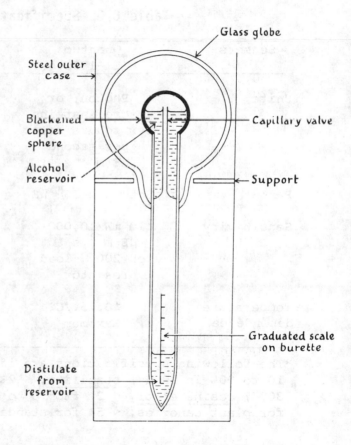

Figure 6-6 Gunn-Bellani Radiometer

(1) <u>Quantum/Radiometer/Photometer</u>. By utilizing three separate sensors, the Quantum/Radiometer/Photometer is capable of measuring PAR for plant response studies, global radiation for environmental studies, and illumination for industrial lighting, outdoor illumination, and biological applications. These three sensors are the quantum, the pyranometric, and the photometric sensors, respectively. They are illustrated in Figure 6-7. The performance specifications of

the three sensors are listed in Table 6-3.

In order to have a better understanding of the specifications, it is necessary to clarify the mechanism in terms of the silicon solar cell in general, the filter system, and the circuitry of the Lambda meter.

(a) <u>Silicon Solar Cell</u>. The silicon solar cell generates its own electrical current from the impact of solar radiation on the photoreceptor, unlike pyranometric sensors which have a thermal receiving and responding surface, a photoelectric cell, or a photochemical reagent tube.[13] In fact, the current of the silicon cell is directly proportional to the intensity of radiation and is almost independent of temperature within the surface meteorological range.

Table 6-3 Specifications of the Lambda Meter*

Sensors	Quantum (L1-190S)	Pyranometer (L1-200S)	Photometer (L1-210)
Units	Photon, or microeinsteins per square meter per second ($\mu E\ m^{-2}\ s^{-1}$)	Watts per square meter ($W\ m^{-2}$)	Lux
Range	3 to 3 x 10^4 μE	0.3 to 3000 $W\ m^{-2}$	30 to 3 x 10^5 Lux
Sensitivity	10 mW/10,000 $\mu E\ m^{-2}\ s^{-1}$ for 200 Ω load resistor	5 mV/1000 $W\ m^{-2}$ for 100 Ω load resistor	5 mV/100,000 Lux for 100 Ω load resistor
Temperature dependence	±0.15%/C Maximum	±0.1%/C Maximum	±0.15%/C Maximum

*The following specifications are identical for the three sensors: <u>time response</u>, 10 to 90% in 10 µs; <u>stability</u>, < 2% change in 1 year; <u>azimuth error</u>, < 1% over 360°; <u>cosine error</u>, < 2% from 0 to 82° angle of incidence; <u>relative error</u>, < 1% for plant canopies, < 5% for standard growth chamber lighting.

Whillier has described the use of the conventional silicon solar cells as radiation detectors, and Collins has improved the integrator for recording global radiation

---

[13] A photochemical reagent is a chemical which is sensitive to radiant flux density; the photocell is a device which converts radiant energy into electrical energy. Examples of the former are anthracene in benzene, and uranyl oxalate in sulphuric acid; that of the latter is a selenium cell.

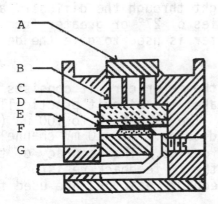

(a) Quantum/Radiometer/Photometer

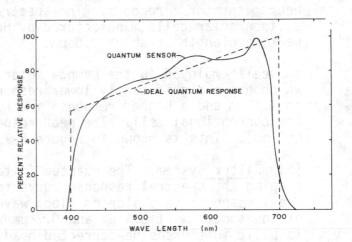

(b) Spectral response of the Quantum sensor and that of an ideal quantum sensor

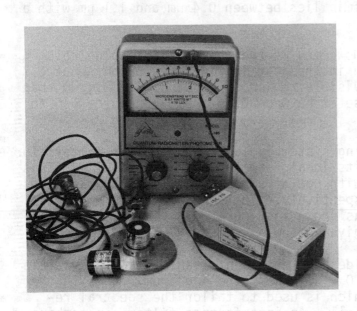

A. Diffuser, B. 3/32 holes,
C. Interference filter and 1.5
mm of Chance-Pilkington HA3 glass,
D. Kodak Wratten CC10M gelatin
filter, E. Black acetal delrin
case, F. Sharp SBC-255 silicon
blue cell, G. Bakelite disc.

(c) Cosine-corrected photosynthesis
response sensor

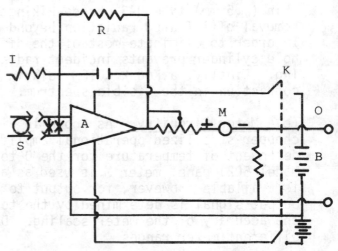

A. Chopper stabilized operation amplifier
B. Battery (14 volts); I. Millivolt imput
K. Key; M. Panel meter (0 - 100 microamp.)
O. Alternative output (0 - 10 mv)
R. A series of different feedback resistors
S. Photosynthesis light sensor

(d) Meter Circuit

Figure 6-7 Quantum/Radiometer/Photometer

over a period of one day.[14] Twenty-two cells were connected in parallel by Collins. Using low resistance leads, he wired them directly to the DC ampere hour meter for a readout. The spectral response of the standard commercial grade silicon solar cells manufactured by Hoffman lies between 0.4 μm and 1.1 μm with a peak wavelength at about 0.8 μm.

The cell employed in the Lambda meter, however, is the Sharp silicon blue cell which has a significantly lower response in the near infrared region of about 0.7 to 1.1 μm and a higher response in the visible region of about 0.4 to 0.7 μm than the conventional cell. The peak response is between 0.55 and 0.65 μm (or 550 to 650 nm). This is shown in Figure 6-7$_b$.

(b) Filter System. The quantum filtering system includes a number of filters for shaping the spectral response curve to its ideal form shown in Figure 6-7$_b$. This ideal response is a sloping block wave with cut-offs for the lower and upper ends of the spectrum at 0.4 μm and 0.7 μm respectively. The quantum filtering system is built in the cosine-corrected head, as shown in Figure 6-7$_c$. A brief explanation of the function of each filter is given below.

A diffuser (the diffusing plastic) provides for excellent transmission of visible radiation and a suitable cut-off at the 0.4 μm end of the spectrum. This further reinforces the 0.4 to 0.7 μm bandpass which is used to tailor the spectral response curve of the Sharp silicon blue cell. An interference filter, in combination with heat absorbing glass, makes a clear cut-off at the 0.7 μm end. This thin (1.5 mm) type HA3 Chance-Pikington heat absorbing glass provides for the removal of infrared radiation beyond 1 μm transmitted by the interference filter. In order to eliminate most of the diffuse sky light through the diffuser, a 7-hole cylinder prevents incident radiation at angles of 27° or greater from entering. Finally, a Kodak CC10M Gelatin Wratten filter is used to give the desired correction in the visible spectrum.

(c) Meter Circuitry. As shown in Figure 6-7$_d$, the meter circuit consists of a chopper-stabilized operational amplifier which makes the circuit practically independent of temperature for the 0 to 70C range. The 5-inch, 0 to 100 μa (API Model 502) panel meter M is used as a readout indicator. A 0-10 mV channel is available, however, for output to a recording device. The accuracy of the output signal is determined by the tolerance of the 1% feedback resistors and the accuracy of the meter scaling. Different feedback resistors are used for different meter ranges.

The principal application of photometry in physical environmental instrumentation, as mentioned earlier, is the measurement of visibility. Among the many types of visibility instruments,[15] the transmissometer is the most widely used. It is

---

[14] Whillier, A. 1964. A simple accurate cheap integrating instrument for measuring solar radiation. Solar Energy, 8(4): 134-136. Collins, B.G. 1966. Silicon solar cell radiation integrator, In: Conference on Instrumentation for Plant Environment Measurement. The Soc. of Instr. Techn., Australia.

[15] Examples of visibility instruments are several visibility meters including fog or haze detectors. For details, see Middleton, W.E.K. 1958. Vision Through the Atmosphere. The University of Toronto Press, Toronto, Canada; and Löhle, F. 1941. Sichteobacktungen vom Meteorologischen Stunpkt. Julius Springer, Berlin.

now installed in nearly every major airport in the world. Essentially, the instrument measures the space rate of diminution of light (attenuation) from a projected light source at a given distance (i.e., the preset baseline). It directly indicates the RVR (the runway visual range) as required for aviation synoptic reports for clearance in aircraft landings and take-offs. The transmissometer, however, has been employed occasionally for measuring long range visibility outside airports, and for detecting the presence of dust storms. Three sets of transmissometers were installed outside the Copenhagen Airport to supply information on visibility in the 10 km to 50 km range.

Although all transmissometers consist of a projector with a light source and a receiver with a photo-sensitive element, each varies greatly in the optical systems employed. A brief review of a commonly used optical system in a modern transmissometer illustrates one such system.

(1) Transmissometer. As shown in Figure 6-8, the beam of the xenon spark lamp L is located at the focal point of and reflected by a parabolic mirror M. The parallel beam so produced is focused through a double convex lens 1 at a photo-emissive tube P, which is the receiving unit connected to the amplifier A. Since complete stabilization of light sources is difficult to arrange, a dual receiver system is often employed. In the dual system a second receiving unit, consisting of a photoemissive tube, a lens, and a plane reflecting mirror m, acts as the comparison receiver located near the projector. This comparison unit is built into the amplifier console. If the projector light source provides a stabilized output, the dual system is not needed. The output signal of the transmissometer is displayed by both the indicator I and the recorder R. Other features of the transmissometer, as shown in Figure 6-8, are explained below.

(a) Xenon Spark Lamp. The xenon spark lamp in the projector unit provides luminous output that is independent of changes in ambient temperature and voltage of the power supply. It also has a greater longevity than an incandescent filament lamp. (The luminous output of the latter increases about 1% with increase of 0.3% in voltage or a rise of 6 to 7C in ambient temperature.) Mercury or sodium lamps, on the other hand, generate unwanted stray light and also are subject to slow and continuous change in output similar to the filament lamp. The xenon spark lamp produces high intensity pulses of about 1 μs duration at intervals of one second. The modulated light so produced by the spark lamp thus eliminates the daylight effect on the photoemissive tube. This couples with the specially built amplifier which is receptive only to the pulsation of the preset frequency produced by the spark lamp. Some transmissometers have a mechanical chopper to pulse the light at a fixed frequency. The chopper requires a lens system which reduces the intensity of light source output. In addition to all the above favorable aspects, a spark lamp emits at maximum luminous efficiency corresponding to the human eye (between 0.38 μm and 0.76 μm).

(b) Photosensitive Element. The photoemissive tube is the best photosensitive element. However, there is no ideal element for the detector. The selenium barrier-layer cell, for example, is stable but has a very high electrical capacity (about 1 microfarad per square inch). Thus, this cell is unsuitable for use in the AC circuit required to produce a modulated light source. Also, it is sensitive to variations in temperature and humidity. At temperatures over 50C,

Projector

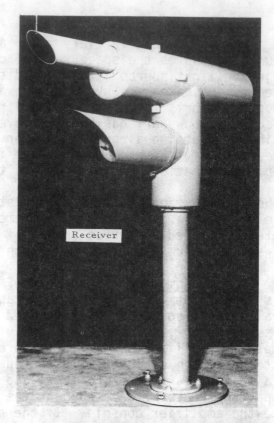

Receiver

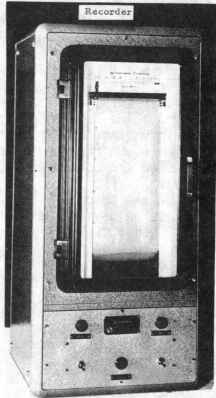

Recorder

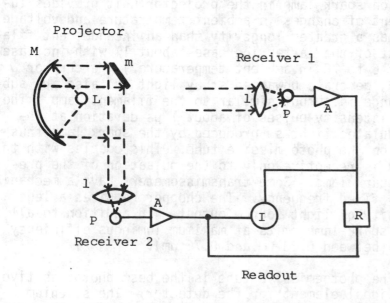

Figure 6-8 Transmissometer

a selenium cell is permanently damaged. The cadmium sulfide cell is suitable for use with a modulated light source but its sensitivity varies with the ambient temperature and illuminance. In all respects, the photoemissive tube is a better detector for the transmissometer. Its characteristics include high sensitivity, stability, linearity, and suitability in the spectral range sampled.

(c) <u>Noise Avoidance</u>. Avoidance of various types of noise is necessary. Stray light coming from the sky behind the projector unit is one source of noise. Another is direct daylight entering the receiver. All of these can be eliminated by designing the amplifier to respond only to a modulated frequency. In addition, a small aperture restricted to less than 10 milliradians for the receiver helps to reduce unwanted stray light. In this connection, a perfect alignment of the entire optical system is required.

(d) <u>Visibility Measurements</u>. A long baseline is needed for visibility measurements outside an airport. The baseline of a transmissometer can be extended from the runway length (usually about 100 meters) to 50 km or more provided the light source is sufficiently intense, the detector is highly sensitive, and the signal-to-noise ratio is high. Here again, the xenon spark lamp serves the purpose. It emits a 300 million candle power impulse light beam. This permits operation in exceptionally low visibility and at remarkably long baselines. Low visibility detection is required for visibility forecasting rather than traffic safety. An automatic warning system has been installed in some transmissometers to indicate when visibility drops below a preset value.

6.1.3 <u>Duration of Sunshine</u>. A sunshine recorder measures the daily duration of direct sunshine in the daytime whereas the starshine recorder registers duration of Pole starlight at night. The latter is designed specifically to measure the cloudiness of the night sky and, therefore, is discussed in Section 6.7.1. Duration of moonshine is not customarily measured.

All sunshine records utilize the heating power of the sun's radiation. However, the intensity and quality of radiation usually are not expressly measured. Because of the great variation in atmospheric turbidity and in optical thickness of clouds, the measured duration of sunshine is determined mainly by the type of instrument chosen. Unfortunately, the distinction between sunshine and lack of sunshine is arbitrary.

Sunshine recorders are classified into two groups according to the recorder time scale, the chronograph and the sundial. The former uses a clock-driven device for recording the time of occurrence of sunshine as represented by the Marvin sunshine recorder and Maurer sunshine chronograph. In the latter class of instruments, the time scale is obtained from the motion of the sun in the manner of a sundial. Examples are the Campbell-Stokes recorder, Jordan sunshine recorder and pers sunshine recorder. In this section, however, only the Marvin and Jordan recorders are discussed.

(1) <u>Marvin Sunshine Recorder</u>. As shown in Figure 6-9$_a$, the Marvin sunshine recorder utilizes a chronometric type of recording. It consists of two bulbs; black bulb A absorbs more radiation energy than clear bulb B. The narrow glass tube C joining the two bulbs extends nearly to the bottom of bulb A and is partially filled with mercury and a small amount of methyl alcohol. Two platinum

wires are sealed through tube C at D, with a small gap in between. The whole instrument is enclosed in a vacuum tube E.

When the temperature in bulb A is higher than bulb B, the air in bulb A expands,
and the vapor pressure of alcohol increases more in bulb A than in bulb B. When
the mercury is forced up tube C high enough to reach the contact D, the electric
circuit is closed and a record is made on a chronograph by a magnetic pen. The
instrument is adjusted so that the light inclination of the tube in the plane
of the meridian permits the mercury column to just close the electric circuit
during times when the disc of the sun can be seen faintly through the clouds.
Vacuum tube E is used to prevent dynamic and/or evaporative cooling as a result
of wind and/or moisture deposits, and to protect the black bulb from decoloring.
Alcohol is used to provide uniform pressure on the mercury. Although the Marvin
recorder gives an easily deciphered record at a distance, it exaggerates the
duration of sunshine because of its great sensitivity to diffuse skylight. For
example, there may be more energy coming from an overcast sky at noon than from
the unobscured sky in the early morning. Furthermore, the instrument is fragile.
Its mounting and standardization are subjective.

(2) Jordan Sunshine Recorder. As shown in Figure 6-9$_b$, the Jordan recorder consists of two semi-cylinders each having a short, narrow slit in its flat side.
A piece of sensitized blueprint paper lines the inner curved side of each semi-
cylinder. As sunlight passes through one of the slits, it makes a trace on the
blueprint. One slit covers the morning hours, the other the afternoon hours. At
the end of the day, the blueprint is removed and washed briefly in water to render
the record permanent. Some Jordan recorders are designed with only one cylinder
but with two slits on opposite sides; a curved metal piece is built between the
two on the top to keep rainwater from entering the slits. In this case, a single
blueprint paper suffices.

Advantages of the Jordan recorder are its ruggedness and ease of operation. Its
disadvantages are the varying sensitivity of the blueprints with time, and the
shrinkage or expansion of the blueprints with the presence of moisture. When a
comparison of the records of the Jordan with those of the Campbell-Stokes sunshine
recorder (see Section 6.7.1) is made, there is inconsistency in the daily values
but not for the average value obtained from a long period record.

6.2 Thermometry. The science of temperature measurement is called thermometry.
In the study of the thermofield of a region, we are concerned with both latent
and sensible heat. However, temperature rather than heat is measured by a thermal
sensor.

Three major problems associated with thermal sensors and their measurements must
be solved. These three problems are (a) response time, (b) accuracy, (especially
radiation error), and (c) sensitivity. These problems have already been discussed
in sections 4.2.1, 4.3.1, and 4.3.2. One point that should be stressed here, however, is the operational specification of requirements. If the user prefers a
slow rather than fast response thermal device (e.g., a liquid-in-steel thermometer
vs. a thermistor or thermocouple sensor), he should choose the former. The time
response of the thermistor or thermocouple is on the order of 1 or 2 s versus
280 to 300 s for the mercury-in-steel thermometer. The slow response sensor may

act as an integrator in a Eulerian system. However, for turbulence studies, particularly on fast moving platforms, a highly sensitive and quickly responsive sensor is required. It must be noted that an additional error, known as dynamic heating, occurs in a fast-moving aircraft. This, however, is discussed in Section 7.3.2.

Radiation error is the most serious problem to be solved for all thermal sensors. To avoid radiation error, the weather shelter commonly has been employed the world over and is still in use. On a sunny afternoon, however, the shelter temperature may be 2 to 3F too high; conversely, on a clear, calm night it is as much as 1F too low. As has long been recognized, the weather shelter is not an ideal shield for thermometer exposure. Although many types of weather shelters have been designed, there has been no substantial reduction of radiation error. Consequently, various kinds of radiation shields have been used when taking temperature measurements. An early type of radiation shield is still used in the Assmann psychrometer. A pair of mercury-in-glass thermometers are suspended in a chromium-plated, highly polished frame which serves as the radiation shield (see Figure 6-10). Air is drawn through the two coaxial metal tubes by means of a clock-driven fan which is located in the hollow top of the psychrometer. Many psychrometers have been constructed using similar principles but with electrically-driven ventilation fans.

A more sophisticated radiation shield is the motor aspirated temperature shield (Climet Model 016-1). It is designed to house both air temperature and dewpoint sensors. It limits the radiation error to 0.2F or less, and has an operating temperature range between -60F and +150F. The aspirator motor requires a power source of 12 watts at 115 volts and 60 Hz. Shown in Figure 6-11$_a$ are the fan assembly A, the primary shields B, the secondary shields C, and the front shield D. The sample air inlet E, the cooling air inlet F, the spacer G, the sensor H, the drainage hole J, and the air outlet K, are illustrated in Figure 6-11$_b$.

(a) Marvin Sunshine Recorder   (b) Jordan Sunshine Recorder

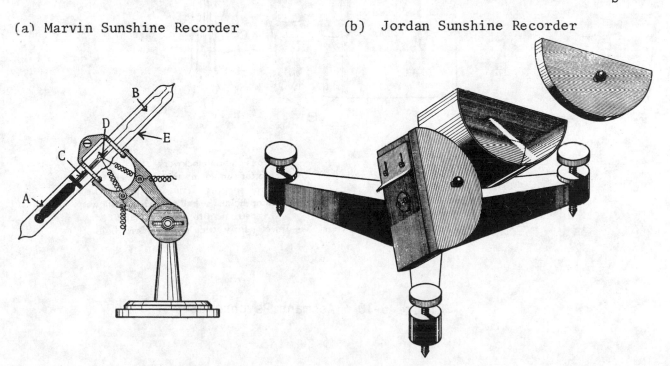

Figure 6-9 Sunshine Recorders

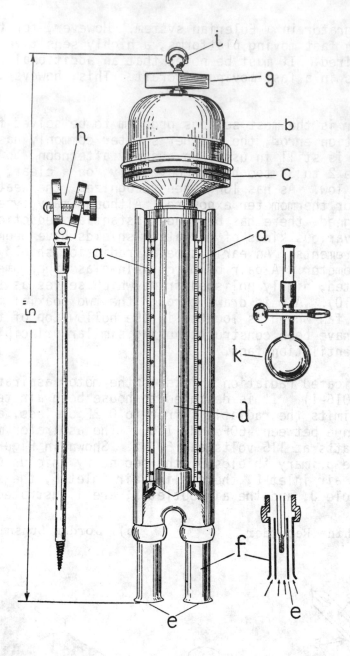

a—Thermometers
b—Dome containing clockwork
c—Fan and air outlets
d—Main air duct
e—Air inlets
f—Polished tubes protecting thermometers

g—Key for winding clockwork
h—Clamp for supporting the instrument
i—Point of support of the instrument; the clamp holds the ball securely but allows the instrument to hang vertically
k—Injector for wetting muslin of wet bulb

Figure 6-10   Assmann Psychrometer

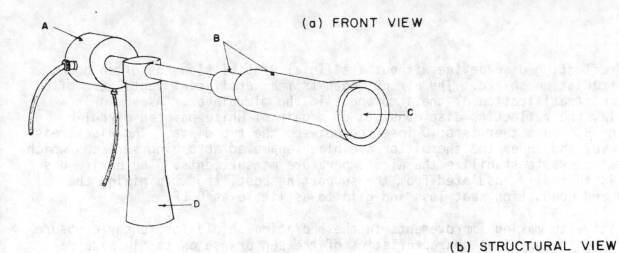

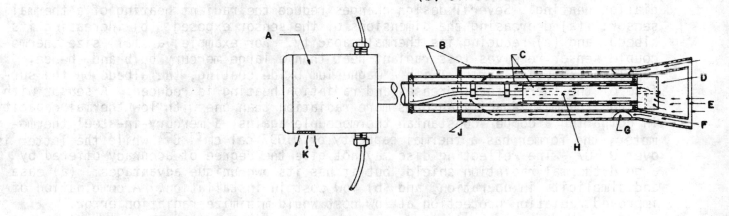

Figure 6-11   Motor-aspirated Temperature Shield

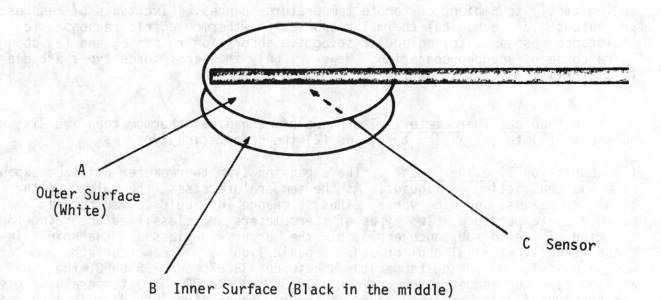

A
Outer Surface
(White)

C Sensor

B   Inner Surface (Black in the middle)

Figure 6-12   Reflecting Disc Shield

A simple reflecting disc device without artificial ventialation has often been used as a radiation shield. The air currents from an artificial ventilator often disturb the stratification of the atmosphere in the microlayer. As shown in Figure 6-12, the reflecting disc consists of a pair of white painted circular discs A and B, and a thermistor C inserted between the two discs. The disc areas directly over and under the thermistor are black enameled absorbing surfaces which act as heat sinks to stabilize the air temperature measurements. The entire disc assembly is thermally insulated from the supporting post, thus minimizing the radiation and conduction heat loss and gain to as little as 0.5F.

An alternative to making improvements in the radiation shield for sensor exposure is to improve the physical characteristics of the sensors so as to minimize radiation heating. Several design changes reduce the radiant heating of a thermal sensor: (a) decreasing the dimensions of the sensor exposed, (b) increasing its albedo, and (c) reducing its thermal capacity. For example, a small size thermocouple sensor receives less radiant heat than a large mercury bulb and, hence, has less radiation error. With a magnesium oxide coating, the albedo on the surface of the thermistor increases and radiation heating is reduced. A sensor with a large thermal capacity absorbs more radiation than one with low thermal capacity. In comparing a copper-constantan thermocouple against a mercury-in-steel thermometer, the former has a thermal capacity of 0.097 cal cm^{-3} C^{-1} while the latter is over 0.107. The reflecting disc may not give the degree of accuracy offered by a good thermal radiation shield, but it has its own unique advantages: (a) ease and simplicity in operation, and (b) low cost in installation. A combination of improved radiation protection at low cost would minimize radiation error.

In view of the three major problems in temperature measurements discussed so far, the choice of thermometer type is essential. Two main types of thermometers are recognized: the mechanical and the electrical (or electronic). The latter is far superior to the former with respect to all performance specifications. Classification of a thermal instrument is based upon the response of its sensor directly or indirectly to ambient or remote temperature changes. Five types of responses are noted: (a) mechanical thermal response, (b) thermoelectric response, (c) resistance response, (d) molecular selective absorption response, and (e) changes in response of sound propagation. However, only the first three types are elaborated here.

6.2.1 Mechanical Thermometer. Two kinds of mechanical thermometers are distinguished: (1) the expansion type, and (2) the deformation type.

(1) Expansion Type Thermometer. The expansion type thermometer utilizes expansion and contraction of liquids. As the temperature rises, the volume of the liquid increases, and vice versa. Thus, a change in liquid volume becomes a measure of temperature. Two types of thermometers are classified according to the kind of liquid substance employed: the mercury-in-glass and the spirit-in-glass (usually ethyl alcohol or other organic liquid). The mercury thermometer may be used for measuring temperatures between -38.87C and +356.58C, the ethyl alcohol type for temperatures between -117.3C and +78.5C. The temperatures given are the freezing and boiling points of the two respective liquids. In other words, where temperatures are expected to go below the freezing point of mercury, the spirit-in-glass thermometer is employed. For this reason, a spirit thermometer is

normally used as a minimum thermometer and mercury as a maximum.[16]

All liquid-in-glass thermometers consist of a glass tube (the stem) with an extremely small but uniform capillary bore, a mercury or spirit reservoir (the bulb) at the bottom, and a small vacuum space at the sealed top end. The extra space is needed for expansion of the liquid in case of overheating. The bulb and part of the tube are filled with fluids such as mercury, spirit, or organic liquids. The glass bulb is made as thin as is consistent with strength to facilitate heat conduction to and from the bulb; the tube wall is made as thick as possible for added strength and still permitting visibility of the liquid column. The thick tube wall also acts as a magnifier with a contrasting background usually furnished by a white enamel strip. Without radiational heating the liquid-in-glass thermometer has minimal error. Various instrument errors are listed in Table 6-4.

It must be noted that most of the errors listed in Table 6-4 may be eliminated by regular calibration and observer's attention to proper usage. The degree of accuracy depends upon operational requirements, given in Table 6-5. As air temperature fluctuates considerably with time and space (within minutes, or within 100 m), the accuracy can vary as much as ±2F. In general, accuracy of more than one tenth of a degree is meaningless. Usually, the spirit thermometer is much less accurate than the mercury thermometer. The time response of spirit bulk is larger than that of mercury. In addition, the difference in the coefficient of expansion of spirit with respect to glass is very much larger than that of mercury to glass.

(2) Deformation Type Thermometer. The deformation type thermometer utilizes the principles of thermal elongation and contraction of metals in temperature measurement. There are three common types: the bimetallic thermometer, the Bourdon tube thermometer, and the mercury-in-steel thermometer.

(a) Bimetallic Thermometer. The sensor of a bimetallic thermometer consists of two dissimilar metal strips, such as brass and iron, which are firmly fastened together. The linear expansion of brass is 19.9×10^{-6} C^{-1} cm^{-1} and that of iron is 11.7×10^{-6} C^{-1} cm^{-1}. This means that an increase of temperature causes brass to expand almost twice as much as iron. This bimetallic strip is bent in a circular form with the brass on the outside, and the whole unit is insulated to keep thermal conduction to a minimum. If one end of the strip is firmly fastened, the

---

[16] The maximum thermometer is similar to an ordinary mercury thermometer except that it has a constricted bore below the lowest graduation. This constricted bore permits the mercury to rise only. In order to reset the mercury column, it is necessary to swing the thermometer steadily back and forth. A minimum thermometer is similar to a spirit thermometer except that a glass index is inserted inside the bore below the liquid surface. Because of surface tension of the liquid, the index is carried downward with the liquid surface as the temperature decreases. Therefore, the top of the index always remains at the lowest temperature reading, thus recording the minimum temperature. To reset, the observer turns the thermometer upside down. Both maximum and minimum thermometers are mounted about 2 degrees from a horizontal position with their bulbs downward.

132

Table 6-4 Errors of the Liquid-in-Glass Thermometer

Sources of Error	Order of Magnitude
A. Common errors of mercury and spirit thermometers:	
1. Reversible error due to a sudden change in temperature*	For ordinary glass: ±2F (max.) For best glass: 0.05F
2. Irreversible error due to secular exposure of the thermometer to light	For the first year: 0.02F for all thermometers.
3. Emergent stem error due to the difference in temperatures between the upper and lower portions of the stem*	For all thermometers: 0.2 to 0.3F approx.
4. Parallax error due to the incorrect line of sight of the observer (increases with an increase of the angle formed by the actual line of sight and the correct line of sight, and with an increase of the stem thickness*)	For all thermometers: 0.5F to a few degrees
5. Incorrect divisions due to manufacturing defects*	For usual calibration: ±0.1F
B. Additional errors of spirit thermometers only:	
1. Secular change due to polymerization of organic compounds	It can be as much as 1F in the first year
2. Adherence of spirit to the glass due to rapid temperature change	As much as ±0.2F.
3. Breaking of the liquid column*	5 to 10F

*
Asterisks identify those items that can be corrected.

Table 6-5 Summary of Operational Requirements for
Air Temperature Measurements

Types of Operation	Requirements (sources)
Climatology	
land surface	1F (b); ±0.1C (a)
up to 200 mb	±0.5C (a)
above 200 mb	±1.0C (a)
Aeronautics	
land surface	±1.0C (a)
Synoptic	
land or sea surface	±0.1C (a)
upper air	±0.5C (a)
Marine	
sea or land surface	±0.1C (a)
Hydrology	
for psychrometric evaluation	0.1F (b)
dew point temperature	
(climatology)	±0.5C (a)
(aeronautics	±1.0C (a)
(marine & agriculture)	±0.1C (a)
Agriculture	
land surface	±0.1C (a); 0.1F (b)
4 ft. below	0.02F (b)
Automated station	
land and sea surface	±1C (a)

Sources: (a) Mani, A. 1970. Meteorol. Monogr. 11 (23).
         (b) Stringer, E.T. 1972. Techniques of Climatology. W.H. Freeman
              and Company, San Francisco. 244 pp.

free motion of the other end is an indication of temperature change. As shown in Figure 6-13, this motion is then magnified by a system of levers which in turn moves the recording pen over a rotating drum. The temperature sensing and recording device combination is called a thermograph. However, some of these are built for measuring more than one meteorological variable,[17] such as the hygrothermograph of Figure 6-13, which is used for measuring humidities and temperatures.

[17] An instrument which gives an automatic record of two or more of the meteorological elements is called a meteorograph. This term refers to a now-obsolete instrument attached to a kite or balloon to measure the pressure, temperature, and humidity of the upper atmosphere. A similar instrument used in aircraft is called an aerograph.

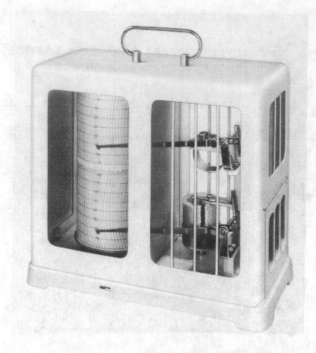

Figure 6-13 Hygrothermograph

(b) <u>Bourdon Tube Thermometer</u>. The sensor for the Bourdon tube thermometer is a hollow flexible metal tube which contains such organic fluids as glycerin, xylene, creosote, or alcohol. Instead of being helical or circular in shape as the bimetallic strip, this sensor is in the form of a flattened curved tube. As the temperature rises, the organic fluid expands and the tube tends to straighten out producing a movement. The recording system of the Bourdon tube thermometer is the same as the bimetallic thermometer.

In his book <u>Flüssigkeitsthermometrie</u>, Grundman has made an interesting comparison of performance error remedies[18] for the bimetallic and Bourdon tube thermometers and some points are detailed in Table 6-6. In view of the comparisons in Table 6-6, bimetallic thermometers are preferred over Bourdon tube thermometers.[19]

(c) <u>Mercury-in-Steel Thermometer</u>. A mercury-in-steel thermometer generally is not classified as a deformation type; however, its sensor operates under the same principle as the Bourdon tube. A 13-inch long hollow steel rod with a mercury-filled capillary acts as the sensor. When the expansion and contraction of mercury take place inside the capillary tubing, the increase (or decrease) of fluid pressure is transmitted to a Bourdon tube through long, extended, flexible capillary tubing. The Bourdon tube is usually built into the recording or the indicating device. The extended capillary tubing may be as long as 150 feet. Beyond 150 feet

---

[18] See Grundmann, W. 1938. Flüssigkeitsthermometrie. Weimar, R. Wagner Sohn. for a detailed study. Also see Middleton, W.E.K. and A.F. Spilhaus. 1953. Meteorological Instruments. The University of Toronto Press. 80 pp.

[19] The Bourdon tube discussed here is different from the Bourdon tube used as a barometric sensing element. For details see the discussion of pressure measuring instruments.

Table 6-6 Comparison of the Bimetal and Bourdon Tube
Thermometers

Bimetal	Bourdon Tube
Secular changes take place but can be corrected by thermal massage through cyclic warming and cooling.	Secular changes take place on both the metal and the fluid. They can be reduced only through very great care in manufacturing.
Time response of bimetal is controlled to an appreciable extent by proper fabrication.	Time response is greater than bimetal for the same stiffness.
Upon abrupt cooling to very low temperatures, the zero setting is made only once. There is no effect at high temperatures.	Easily damaged by moderately high temperatures because of excess internal pressure; may not respond properly to changes at low temperatures.
Both steel and brass are subject to corrosion.	Non-ferrous and flexible metals are generally employed and have a high resistance to corrosion.
Bimetallic sensors are generally stiffer than Bourdon tubes. They are very insensitive to rough handling.	Moderate stiffness; may be easily damaged by slight bending.
Cost of installation and maintenance is low.	Costs are high.

the fluid pressure is too weak to be recorded accurately.

The advantages of the mercury-in-steel thermometer are: (a) it serves as a telemetering thermometer; (b) it has wide application to measurments in air, water, and soil, in addition to maximum and minimum temperatures; and (c) the ruggedness of its sensor structure is convenient for field usage. The disadvantages are: (a) it has a slow response time; (b) it has a large exposure surface which leads to a large radiation error when unshielded and used for air temperature measurements; (c) when a long extended capillary tubing is employed, the temperature may vary along its length, causing errors in recording; and (d) this thermometer is not suitable for measuring temperatures below freezing.

136

6.2.2 <u>Electrical Thermometers</u>. Based upon the principle of thermoelectricity, three common types of electrical thermometers are recognized (see Table 6-7). Positive and negative thermometers measure changes in electrical resistance as a function of <u>absolute</u> temperature. Thermocouple thermometers measure changes in electrical <u>current</u> or voltage as a function of <u>differential</u> temperature. While the latter provides an <u>indirect</u> indication of differential temperature between two junctions of dissimiliar metals, the former gives the <u>direct</u> measurement of ambient temperature.

Table 6-7 Electrical Thermometers

Types of Thermometers	Examples
Positive Resistance* (i.e., resistance thermometers)	Copper, platinum, iron, and nickel.
Negative Resistance* (i.e., thermistors)	Type A thermistor and Type B thermistor.
Thermocouples*	Copper-constantan and Iron-constantan

*
 As will be explained shortly in the text

In this section, discussions on each of the three types of electrical thermometers are followed by a comparison of their performance characteristics.

(1) <u>Resistance Thermometers</u>. The electrical resistance of most pure metals and alloys increases with ambient temperature (positive resistance), although a few demonstrate a decrease (negative resistance). Metals commonly employed in temperature measurements are silver, copper, platinum, nickel, and iron. Examples of common alloys include: constantan (60% copper and 40% nickel), manganin (a mixture of copper, manganese, and nickel), and nichrome (a nickel-based alloy containing chromium and iron).

Among various metals and alloys, platinum wire has been found to be the most accurate sensor for a resistance thermometer. In addition, platinum has high sensitivity, linearity, stability, and durability. It resists contamination and corrosion, and thus does not drift with age and use. A strain-free wire of 99.99% pure platinum is commercially available. However, due to the high cost of platinum, other metals and alloys are still used in resistance thermometers.

Platinum wire, 0.05 mm or more in diameter, is coiled around an insulating core and protected by a weatherproof sheath. A small electric current is passed through the wire and its resistance, R_T, is measured by a potentiometer at the ambient temperature, T, of measurement. It should be noted here that a small current flow is necessary to reduce wind effects on the sensor. Otherwise the sensor acts as a hot wire anemometer. A predetermined reference resistance, R_o, is referred to as temperature T_o. For example at the reference temperature $T_o = 18C$, the specific resistance is $R_o = 11 \times 10^6$ ohm cm^{-1}. Then the ambient temperature, T, can be determined by the functional relation as

$$R_T = R_o (1 + \alpha T + \beta T^2) \ldots \ldots \ldots \ldots \ldots \ldots \ldots (5)$$

where α and β are determined for one of the fixed points OC, +100C, or -78.51C through laboratory calibration.[20]

Applying Eq. (5) to the surface meteorological range of temperatures, it is re-written and simplified as

$$R_T = R_o [1 + A(T - T_o)] \ldots \ldots \ldots \ldots \ldots \ldots (6)$$

where A is the calibration constant. This establishes the linear relation between resistance and temperature.

As shown in Figure 6-14, the Wheatstone bridge has three fixed resistors R_1, R_2, and R_3. R_T is the resistance thermometer and R_v, a variable resistor that is the rheostat. A battery is connected to the two opposite terminals of the bridge and the galvanometer G to the other two. The positive electrode is in series with the rheostat. The rheostat is used to balance the circuit in the bridge and thus ensure zero current flow through the galvanometer (i.e., null-balance). Instead of manually adjusting the null-balance, a two-directional motor is employed to position the slider on the rheostat to balance the circuit. A chopper converts the DC voltage of the Wheatstone bridge into AC voltage, which is amplified to drive the two-directional motor.

Both copper and platinum wires are the most suitable thermal sensors for resistance thermometers as they show excellent linearity between temperature and resistance. Heat treatment is necessary to relieve all strains caused by winding and to ensure the zero reading of the thermometer. For a platinum sensor, the wire is electrically heated to approximately 1000C for about 30 minutes before the winding, and to 700C several times after. The standard temperature range of the platinum resistance thermometer, for example is -270C to +660C. Within a 500C range, platinum wire has a drift of about ±0.05C, and an accuracy of about ±0.25C at OC.

---

[20] The resistance thermometer is calibrated by determining its resistance at the fixed points of OC (the melting point of ice), 100C (the boiling point of water) and -78.51C (the sublimation point of carbon dioxide).

For interpolations at temperatures below OC, Eq. (5) becomes

$$R_T = R_o [1 + \alpha T + \beta T^2 + \gamma (T - 100) T^2]$$

where γ is determined by calibration at the temperature of equilibrium between liquid oxygen and its vapor at atmospheric pressure (i.e., the oxygen point).

138

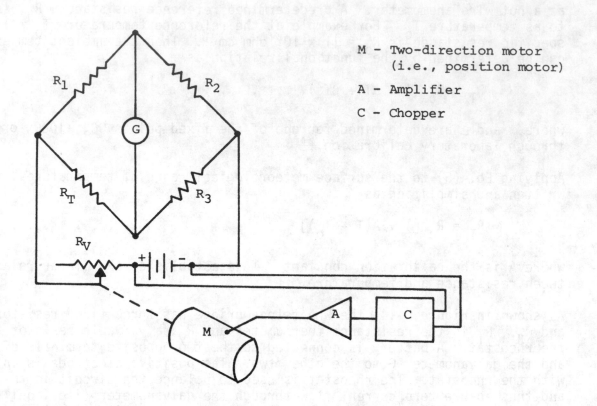

M - Two-direction motor
(i.e., position motor)

A - Amplifier

C - Chopper

Figure 6-14 Wheatstone Bridge

(2) _Thermocouples_. Because of their fairly high accuracy, sensitivity, stability, and durability, thermocouples are unique in their contribution to temperature measurement and control. They operate over a useful range of temperatures
(∿90 to 3000K), depending upon their types and sizes. In this respect, they
offer a greater variety of choices than any other type of thermometer. Other
criteria for selection are based on the desired limits of error in temperatures,
time response, and voltages to be measured. A discussion of these criteria is
given shortly.

The time response of thermocouples often is measured in seconds. Aside from temperature measurement, other applications of thermocouples are for radiation and
humidity measurement, and refrigerating and heating control.

The most common thermocouple wires employed in surface air temperature measurements are combinations of copper-constantan, iron-constantan, and manganin-
constantan,[21] although other combinations may be used depending upon the operational
specification. The relationship between EMF and temperature is non-linear and is

---

[21] It is a general convention that the first metal mentioned is the positive wire
and the second is the negative. For example, in copper-constantan thermocouples,
copper is positive and constantan is negative.

closely approximated by

$$E = \alpha T + \beta T^2 \quad . . . . \quad (7)$$

where the coefficients α and β are constant for a given metal with reference to platinum; T is the temperature of the measuring junction in C; and the temperature of the reference junction is 0C. Such coefficients for a few selected metals are listed in Table 6-8. These coefficients are used for determining the resultant EMF from Eq. (7) when the temperature of the measuring junction is specified. In the temperature ranges of concern in meteorology, the departure from linearity as given in Eq. (7) is small enough to be negligible.

Table 6-8 Coefficients* for Thermocouple EMF

Metal	α (μv C^{-1})	β (μv C^{-2})
Iron (Fe)	+ 16.7	− 0.0148
Copper (Cu)	+ 2.7	+ 0.0040
Constantan (CU + Ni)	− 34.6	− 0.0279

*With reference to platinum.

A description of an elementary thermocouple circuit follows. An iron-constantan thermocouple forms an electrical circuit with junctions that have temperatures T_1 (the measuring junction) and T_2 (the reference junction), and in which iron is the positive metal and constantan, the negative. As shown in Figure 6-15, the electric current produced tends to flow from the positive metal, through the hot junction, to the negative metal. The magnitude of the EMF will depend on the metals used, the temperature difference (T_1-T_2) and the specific values of the actual temperatures $(T_1$ and $T_2)$.

When two metals other than platinum are used, the coefficients α and β (see Table 6-8, Eq. (7)), for each individual metal with respect to platinum, are algebraically combined. This is illustrated by the following example. Let the iron-copper thermocouples have temperatures of 0C and 200C at the reference junction and the measuring junction, respectively. If their combined coefficients are α_{Fe-Cu} and β_{Fe-Cu}, then

$$\alpha_{Fe-Cu} = \alpha_{Fe} - \alpha_{Cu} = 16.7 - 2.7 = 14 \ \mu v \ C^{-1}$$

$$\beta_{Fe-Cu} = \beta_{Fe} - \beta_{Cu} = -0.0148 - 0.0040 = -0.0188 \ \mu v \ C^{-1}$$

From Eq. (7),

$$E = 14(200) + (-0.0188)(200)^2$$

$$E = 2048 \ \mu v$$

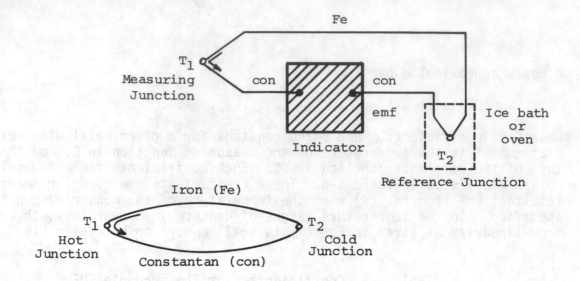

Figure 6-15 Elementary Thermoelectric Circuits

In actual engineering practice, the EMF values are obtained from the National Bureau of Standards (NBS) Thermocouple Reference Tables in which the EMF is listed with respect to individual thermocouples rather than individual metals, as shown in Table 6-8. The reference junction temperature is OC or 32F. Letter symbols are used in the Reference Tables to identify commonly used thermocouples. The standard and special limits of error and the temperature range for each type are listed in Table 6-9.

Table 6-9 Limits of Error of Thermocouples

Thermocouple Type	Temperature Range, F	Limits of Error	
		Standard	Special
J (Iron vs Constantan)	32 to 530	±4F	±2F
	530 to 1400	±3/4%	±3/8%
K (Chromel vs Alumel)	32 to 530	+4F	±2F
	530 to 2300	±3/4%	±3/8%
	-300 to -75	-	±1%
	-150 to -75	±2%	±1%
T (Copper vs Constantan)	-75 to 200	±1-1/2F	±3/4F
	200 to 700	±3/4F	±3/8%
E (Chromel vs Constantan)	32 to 600	±3F	-
	600 to 1600	±1/2%	-
S(Pt - 10% Rh vs Pt)*	32 to 1000	±5F	±2.5F
	1000 to 2700	±1/2%	±1/4%

*
 The values of Type S also apply to Type R (Pt - 13% Rh vs Pt).

As shown in Figure 6-16, the electrical bridge facilitates automatic EMF measurement and direct temperature reading. In this system, the reference temperature is controlled by a cold junction compensator[22] at a temperature $T_2 = 0C$. This compensator incorporates a temperature-sensitive resistance element, R_S, which is located in one arm of the bridge network and is thermally integrated with the reference junction. The bridge is powered by a mercury battery B. The output voltage is proportional to the imbalance created between the pre-set equivalent reference temperature at 0C and the temperature of the measuring junction, T_1. Integrating copper leads with the cold junction isolates the thermocouple material from the input terminal of the measurement device, eliminating secondary errors.

A combination of several thermocouples of the same metals connected in series is called a thermopile. The output of the thermopile is equal to the sum of the outputs of each individual thermocouple. Thermopiles have been used as sensors and/or components in radiometers, dewpoint hygrometers, psychrometers, and thermometers, in addition to the heating and cooling controls of instrument systems.

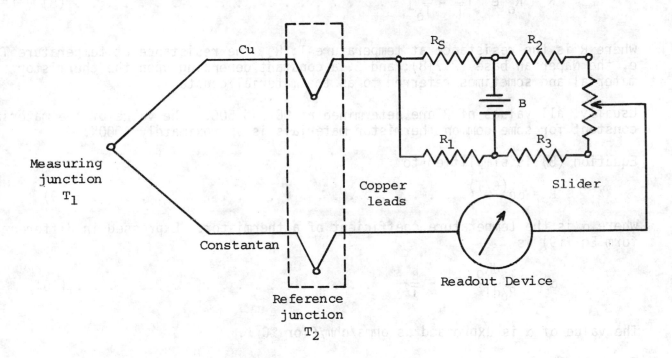

Figure 6-16 A Self-compensating Electrical Bridge

---

[22]This compensator, which is a self-compensating electrical bridge shown in Figure 6-16, introduces an equal and opposite voltage if a thermally generated voltage created by the change of the ambient temperature arises in the reference junction. Thus a variation of ambient temperature, which would cause an error in the output, is automatically cancelled by this bridge device.

(3) Thermistors. The name thermistor comes from the term thermally sensitive resistor. Thermistors are hard, ceramic-like, electronic semiconductors, usually made from a mixture of metallic oxides. Great varieties of structures and mountings are available. Simple types are beads, discs, wafers, rods, or washers with leads attached. Complex types include beads or rods sealed in glass probes, beads in glass bulbs, and glass-coated beads. Three basic electrical characteristics of thermistors are noted: resistance-temperature, current-time, and voltage-current.

(a) Resistance-Temperature Characteristic. The resistance of a thermistor is solely a function of its absolute temperature. Since electrical power dissipated in a thermistor might heat it above its ambient temperature and therefore reduce its resistance it is necessary to use a very small amount of power when measuring its resistance.

The equation which relates the electrical resistance, R (Ω), and the temperature, T (K), is given as

$$R = R_0 \, e^{\beta} \left[\frac{1}{T} - \frac{1}{T_0}\right] \quad \dots \dots \dots \dots \dots \dots \quad (8)$$

where R is the resistance at temperature T; R_0, the resistance at temperature T_0; e, the Naperian base (2.718); and β, a constant depending upon the thermistor material and sometimes referred to as the material constant.

Usually, all values of β are determined at 0C and 50C. The value of the material constant for some common thermistor materials is approximately 4000K.

Equation (8) is simplified to

$$R = \alpha e^{(\beta/T)} \dots \dots \dots \dots \dots \dots \dots \dots \dots \dots \quad (9)$$

where α is the temperature coefficient of a thermistor. Expressed in differential form Eq. (9) is

$$\alpha = \frac{1}{R_0 dT} \frac{dR_0}{} \simeq -\frac{\beta}{T^2} \dots \dots \dots \dots \dots \dots \quad (10)$$

The value of α is expressed as ohms/ohm/C or $\%C^{-1}$.

Figure 6-17 shows the resistance variation with respect to temperature for a typical thermistor material compared to platinum. Between the temperatures of -100C and 400C, there is a change of ten million to one in the resistance of a thermistor, whereas platinum changes its resistance by only ten to one.

(b) Voltage-Current Characteristic. If the voltage is increased gradually, the current increases and the heat generated in the thermistor eventually raises its temperature above that of its surroundings. The resistance then is lowered and current flow is greater than if the resistance had remained constant.

(c) Current-Time Characteristic. If a voltage is applied to a thermistor and resistor in series, current flow is determined by the voltage and the total circuit resistance. As more and more heat is generated in the thermistor, the process of lowering the resistance will continue until the thermistor reaches the <u>maximum</u> temperature possible for the amount of power available in the circuit. At this time a steady state exists. A thermistor with a certain mass takes time to heat to its maximum value. The response time is a function of the mass of the thermistor, the resistance of the circuit, and the amount of voltage applied. By a suitable choice of thermistor and its associated circuitry, it is possible to produce time delays from 0.001s to several hours.

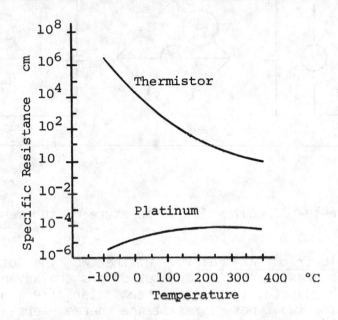

Figure 6-17 Resistance-Temperature Response of Typical
Thermistor Material Compared with Platinum

For temperature measurements three common circuit designs are available, depending upon the objectives of the user. As shown in Figure 6-18$_a$, a simple circuit provides for telemetering temperature readings at a great distance. In this circuitry, the thermistor T is connected in series with a battery B and a micro-ammeter M. Because of the high resistance of the thermistor (in the order of 100,000 Ω or more), any change in the resistance of the transmission line (e.g., copper wire) due to ambient temperature changes is negligible.

More sensitive temperature measurements require the Wheatstone bridge design shown in Figure 6-18$_b$. In this circuit, the thermistor occupies one arm of the bridge. A sensitive galvanometer G is connected to the two opposite terminals of the bridge, and a battery B to the other two terminals. The meter is a center galvanometer. As the meter sensitivity increases, its full-scale temperature indication decreases. Thus, it is possible to achieve a full-scale reading of 2F with this type of circuit.

For measuring horizontal or vertical temperature gradients, two thermistors (T$_1$ and T$_2$) are connected to the two arms of the bridge, as shown in Figure 6-18$_c$. When the two thermistors are placed in different horizontal or vertical locations, the balance of the bridge is dependent upon the difference in temperature of the two thermistors. When this circuit is used with alternating current with a high

144

gain amplifier on the output of the bridge, temperature differentials of 0.001F can be readily measured.

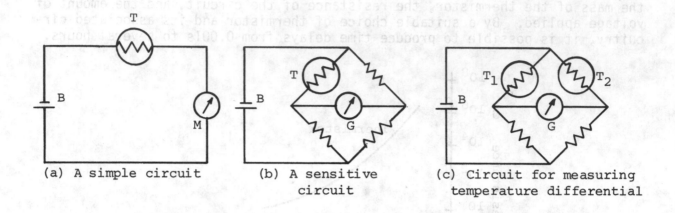

(a) A simple circuit   (b) A sensitive   (c) Circuit for measuring
                            circuit   temperature differential

Figure 6-18 Circuitry for Temperature Measurements

(4) Comparison of Electric Thermometers. Of the two types of resistance thermometers, the platinum resistance thermometer offers the advantage of a large temperature range, whereas the thermistor thermometer offers high precision. The performances of the two types of resistance thermometers are compared in Table 6-10.

When resistance thermometers and thermocouples are compared with respect to the types of measurements, temperature range, accuracy, sensitivity, and application, the following differences are noted.

(a) Types of measurement. Both platinum wires and thermistors are used as thermal sensors for direct measurements, whereas the thermocouple requires a reference junction. Thus, a more complicated circuit is needed.

(b) Temperature range. Temperature spans are as follows: thermocouples have the largest; platinum resistance thermometers, the next; and thermistors, the least. The normal range of the three are specified below.

		Lowest		Highest
Positive resistance thermometers	Platinum	-297F	to	+2000F
	Irridium	-297F	to	+4000F
Negative resistance thermometers	Thermistor	-475F	to	+ 600F
Thermocouples		-475F	to	+6000F

Table 6-10 Performance Specifications of Platinum and Nickel
Resistance Thermometers Versus Thermistors*

Specifications	Thermistors		Platinum	Nickel
	400 probe	700 probe	RTD-800	RTD-810
Temperature Range	-35 to 140C in 4 steps	0 to 100C	-350 to 1350F	-100 to 600F
Resolution	0.1C/0.01C or 0.001C	0.1C	0.1F	0.1F
Accuracy	±1% full	0.2C full	±2F	±5F
Sensor Size	5/32" dia. 4 1/2" l.	3/32" dia. 6 1/2" l.	1/4" dia. 11 1/2" l.	1/4" dia. 4" l.
Recorder Output	10 mv full scale	10 mv full scale	1 V full scale	1 V full scale
Time Response	3.4s	3.6s	5.0s	---

*
Source: Science Essentials Company, A Division of Beckman Instruments, Inc.,
Anaheim, California.

It is clear that positive resistance thermometers have a range about two to four-fold larger than the thermistors, and thermocouples are almost six and one-half times larger in range.

(c) Accuracy. Of all the electric thermometers the thermistor has the greatest accuracy (up to as high as 0.01C), and hence high resolution; platinum ranks next (about ±0.1C); and the thermocouple is the least accurate (about ±1C for ordinary commercial-grade thermocouples). At extremely high and low temperatures, all thermometers are usually limited in their accuracy.

(d) Sensitivity. The sensitivity of a resistance thermometer is designated either by the per cent change in resistance, or the output resistance per degree F or C, over the selected temperature span. Thus, its sensitivity may be expressed either by $\% \ C^{-1}$ or $\Omega \ C^{-1}$, respectively. A typical platinum resistance bulb will exhibit a change of less than 0.1 ohm F^{-1} at room temperature. Thermistors at the same temperature demonstrate a 10 to 10^5 ohms F^{-1} change. The relative sensitivities of the sensors of electrical thermometers are given in Table 6-11.

(e) Applications. Thermocouples have greater usage than thermistors or platinum resistance wires in environmental instrumentation. Aside from use as sensors of thermometers, both thermistors and thermocouples can be employed as components of radiometers, dewpoint hygrometers, and psychrometers. In addition, thermistors

Table 6-11 The Relative Sensitivities of Electrical Thermometers*

Sensor	Microvolts per Degree Celsius (μv C^{-1})	
	at 25C	at 300C
Thermistor in low temperature bridge circuit providing 0.01C readout accuracy	10^4 to 10^6	2×10
Platinum resistance bulb 30 Ω readout accuracy 0.01C	3×10^2	2×10^2
Copper vs. constantan T.C.**	4×10	6×10
Iron vs. constantan T.C.	5×10	5.5×10
Chromel vs. alumel T.C.	4×10	4×10
Platinum vs. platinum-10% rhodium T.C.	6×10^0	9×10^0

*
 Source: The 1973 Temperature Measurement Handbook. Omega Engineering, Inc., Stamford, Conn.

**
 T.C. refers to Thermocouple.

have been used as components of anemometers and bolometers. The potential uses of thermistors are expanding.

(5) Integrated Circuit Temperature Transducers. Aside from the three common thermometers, a new more sophisticated IC temperature transducer is now available. Rapid evolution of integrated circuitry (IC) has produced new generations of electronic devices that are applicable to meteorological measurements. The integrated circuit temperature transducer is representative of monolithic integrated circuitry that measures temperature. One device representative of this type of IC packaging is the two terminal IC temperature transducer, AD590, manufacturered by Analog Devices (Norwood, Massachusetts). The device in chip form provides temperature measurement, temperature compensation for other devices, or biasing proportional to absolute temperature. The physical dimensions of the entire package, shown in Figure 6-19, make it suitable for remote or instantaneous measurements where package size is not critical. The temperature transducer operates over a temperature range between -55C and +150C, using DC voltages between +4V and +30V with a typical power requirement of 1.5 mW (5V, +25C).

The temperature transducer uses two silicon transistors operated at a constant ratio of collector current densities. The difference in their base-emitter voltages is $(kT/q)(\ln r)$; where k is Boltzmann's constant; T is temperature (Kelvin); q is the charge of an electron; and r is current density. The resulting voltage of this tandem operation is directly Proportional To Absolute Temperature (abbreviated PTAT). The integrated circuit takes PTAT voltage and converts it to a PTAT current by low temperature coefficient, thin film resistors. The total current of the device is calibrated at +25C. The PTAT current regulator has an output current equal to a scale factor times the temperature of the sensor in degrees Kelvin. The scale factor is trimmed to 1 μA/K so the output current agrees with the actual temperature (25C or 298.2K at +5V DC). (See Figure 6-20 for effects before and after trim on accuracy.)

Non-linearity (Figure 6-21) over the -55C to +150C range is rated as superior to all conventional electrical temperature sensors such as thermocouples, RTD's, and thermistors. Circuit trimming of two resistors (illustrated in Figure 6-22a) at two reference temperatures (such as OC and 100C) further reduces error due to non-linearity (Figure 6-22$_b$). A summation of the transducer's specifications is given in Table 6-12.

The IC temperature transducer is ideally suited to remote sensing over hundreds of feet from the receiving circuitry. Its calibration of 298.2 μA output at 298.2K makes it well-suited for direct readout of temperature.

6.3 Barometry. The science of pressure measurement, barometry, is one of the most important and early-established areas in atmospheric instrumentation. The instrument used in barometry is generally known as the barometer, which means pressure-measure. Almost all the principles and mechanisms of barometric designs developed by the old-world scientists[23] are still in use today. The barometer remains one of the most accurate among all meteorological instruments. Unlike those involved in designing instruments for temperature measurement, the inventors of the barometer knew what is being measured and what errors are involved.

There are two basic approaches to barometric measurement: (1) measuring the pressure (force per unit area) exerted by the weight of the atmosphere, and (2) measuring the boiling point of a liquid at its corresponding atmospheric pressure. The first is measured by such barometer types as the mercurial and the aneroid (i.e., without fluid), and the second is measured by the hypsometer (GK hypsos = height). While barometers are direct-measurement instruments, the hypsometer applies the principle of indirect measurement.

6.3.1 Direct Method. Two major mechanisms are recognized in the direct measurement of atmospheric pressure: (1) balancing the weight of the atmosphere against a column of mercury, as represented by the mercurial type barometer; and (2) balancing the weight of the atmosphere against a known spring force, as represented by the elastic type barometer.

(1) Mercurial Barometer. The use of mercury for the measurement of atmospheric pressure was introduced over 300 years ago, and is extensively practiced today in barometric instrumentation. The main advantages of mercury as the barometric fluid are:

---

[23] The Italian scientists Gasparo Berti (?-1642) and Evangelista Torricelli (1608-47), the British Robert Boyle (1627-91), Nicholas Fortin (1750-1831), and John Dalton (1766-1844), the French Blaise Pascal (1623-62), and others offered valuable contributions to barometry. Among them Pascal invented the aneroid barometer. Fortin made the first Fortin Barometer, and Dalton, the Kew-pattern.

1. Its high specific gravity (13.5939 g cm^{-3} at OC) which enables the instrument to be of a manageable length;

2. Its low vapor pressure (0.00021 mb at OC and 0.00343 mb at 30C) which makes possible a nearly perfect vacuum in the space at the top of the barometric column;

3. Its small linear coefficient of expansion (0.000182 cm C^{-1}) and low specific heat (0.0335 cal g^{-1} C^{-1} at OC) which minimize thermal effects on the pressure readings; and

4. Its fairly low freezing point (-38.87C) and high boiling point (356.58C) which permit the measurement of atmospheric pressure at a fairly low or a high temperature, respectively.

Table 6-12 Performance Specifications of a PTAT Temperature Transduces, AD590L

OUTPUT	
Nominal current output (at 25C, +5V DC)	298.2 µA
Nominal temperature coefficient	1 µA/deg C
Calibration error (at 25C, +5V DC)	±1.0C max.
Absolute error (-55C to +150C)	
Without external calibration	±2.4C max.
With +25C calibration error set to zero	±1.0C max.
Non-linearity (-55C to +150C)	±0.5C max.
Electrical turn-on time	20 µs
RESPONSE TIME	
Moving air (9 ft s^{-1}) with heat sink	5 s

Specifications from Two Terminal IC Temperature Transducer AD590, 1978: Analog Devices, Route 1, Industrial Park, P. O. Box 280, Norwood, Massachusetts.

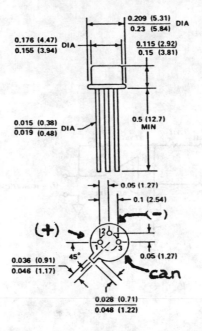

BOTTOM VIEW

Dimensions shown in inches and (mm).

Figure 6-19 Dimensions of AD590

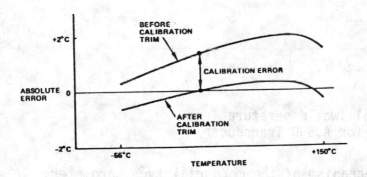

Figure 6-20 Effect of Scale Factor Trim on AD590 Accuracy

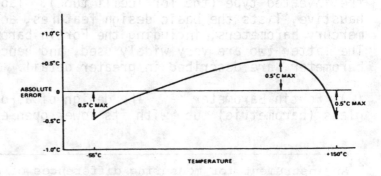

Figure 6-21 Nonlinearity of AD590

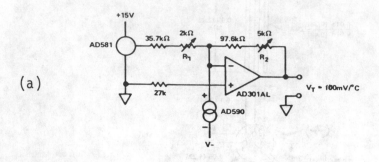

(a)

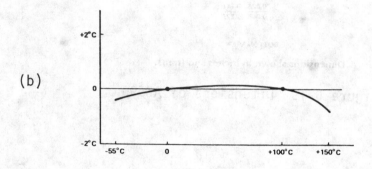

(b)

Figure 6-22 Typical Two Temperature Trim
Specifications for AD590 Transducer

Figure 6-23 illustrates the basic mechanism of the mercurial type barometer, i.e., the weight of a column of mercury is balanced against the atmospheric pressure. Of the many kinds of mercury barometers available, the least complicated device is known as the manometer.[24] The simplest form of manometer is the inverted type (the Torricelli tube). Table 6-13, which is by no means exhaustive, lists the basic design features, accuracy, and sensitivity of several mercury barometers, including the Fortin barometer and the Kew-pattern barometer. The latter two are very widely used, and hence these, in addition to the servo-barometer, are described in greater detail.

(a) Fortin Barometer.[25] The design of a Fortin barometer includes a closed glass (barometric) tube with its lower open end immersed in a cistern of mercury.

---

[24]An instrument for measuring differences of pressure.

[25]Named after the inventor, a French physicist, Nicolas Fortin (1750-1831).

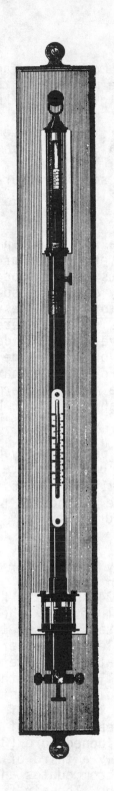

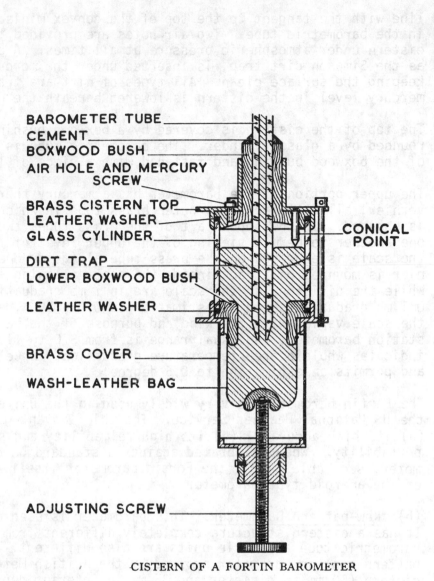

BAROMETER TUBE
CEMENT
BOXWOOD BUSH
AIR HOLE AND MERCURY
SCREW

BRASS CISTERN TOP
LEATHER WASHER
GLASS CYLINDER

CONICAL
POINT

DIRT TRAP
LOWER BOXWOOD BUSH

LEATHER WASHER

BRASS COVER

WASH-LEATHER BAG

ADJUSTING SCREW

CISTERN OF A FORTIN BAROMETER

Figure 6-23 Fortin Barometer

The mercury level in the cistern can be changed by an adjusting screw located at
the bottom of the leather bag that contains mercury (see Figure 6-23). The Fortin
barometer operates on the principle that when the level of the mercury in the
cistern is adjusted to the zero point (i.e., when the tip of the conical pointer
meets the mercury surface in the cistern), the upper mercury surface in the barometric tube indicates the atmospheric pressure reading. The scale reading is in

line with the tangent to the top of the convex miniscus of the mercury surface in the barometric tube. Two air holes are provided to keep the mercury in the cistern under atmospheric pressure at all times. A perforated glass disc, known as the Simpson dirt trap, is inserted under the mercury surface and assists in keeping the surface clean. All types of dirt are trapped under the disc when the mercury level in the cistern is lowered beneath the disc and raised up again.

The top of the cistern is covered by a boxwood bushing and the sides are surrounded by a glass cylinder. The barometric tube is cemented firmly at the top of the boxwood bushing and fastened with a piece of leather.

The upper portion of the barometer has a mercury thermometer, a scale, and a vernier, all attached to an outer brass tube in which the barometric tube proper is encased. In the upper part of the brass tube, two long slots are cut opposite one another to permit viewing of the upper level of the mercury column. Usually the scale is engraved on the brass tube alongside one of the slots, and the vernier is mounted on a rack inside the slot, allowing it to be raised or lowered. While the divisions of the scale are in 1 mb graduations, the vernier provides 0.1 mb graduations and an estimation to 0.05 mb. The millibar or inch range of the scale varies with the kind and purpose of the barometer. For the ordinary station barometer, the usual range is from 870 to 1100 mb. The mercury thermometer indicates whole degree temperature graduations in Celsius or Fahrenheit, or both, and permits an estimation to 0.5 degree.

The Fortin barometer is very widely used in the United States, particularly by the US National Weather Service. The chief advantages of this barometer are: (a) its high accuracy, (b) its high reliability and sensitivity, and (c) its portability. When calibrated against a standard barometer (i.e., a normal barometer, see Table 6-13), the Fortin barometer itself may be used for the calibration of the aneroid type barometer.

(b) Kew-pattern Barometer. This barometer is used extensively in Great Britain. It has a cistern structure completely different from the Fortin barometer. Its barometric tube and scale units are also different. Figure 6-24 shows the Kew-pattern station barometer adopted by the British Meteorological Office. The cistern (50 mm in diameter and 35 mm in interior depth) is a stainless steel vessel whose vented top is covered by a boxwood bushing and a leather washer to prevent dust and other contaminants from entering. Some Kew-pattern barometers have a flange construction in the cistern to damp out the oscillations of the mercury when being transported. Others do not have the vent holes, but instead utilize the pores of the boxwood to let air pass through. The lower end of the barometric tube is firmly cemented into the boxwood, with a greater part of its length having a constricted bore (1.6 mm in diameter), and a widened bore (8 mm in diameter) in the visible portion near the top. Approximately a third of the way up the tube is another widened portion of the bore which accommodates an air trap to prevent air from entering the vacuum chamber at the top. The air trap, as seen in Figure 6-24, is a simple inverted-pipette arrangement.

The scale of the Kew barometer, as adopted by the regular British meteorological stations, is graduated in whole millibars with a short-range scale of 870 to 1100 mb. For mountain stations, a long-range scale of 700 to 1100 mb is used. Similar to the Fortin barometer, the Kew barometer is equipped with a vernier to facilitate reading to 0.1 mb and estimation to 0.05 mb.

Table 6-13 Specifications of Some Mercury Barometers*

Instrument	Advantages	Design Features and Specifications
Fortin Mercurial Barometer	1. Fairly portable. 2. Permits inspection of the free mercury surfaces whose difference in level is to be measured.	Adjustable cistern with high accuracy (0.1 mb) and sensitivity (estimated to 0.05 mb); uses a laborious procedure for making an observation. Recommended for use at land stations.
Kew-pattern Barometer	1. Portable and rugged; ease in making observations. 2. Reduction of the pumping of mercury when used on board ship.	Somewhat similar to the Fortin barometer in design, except it has a 1.6 mm constriction in the tube which damps out oscillation caused by oceanic waves. It has a large response time (5 min. or more) and a probable error of 0.15 mb. Recommended for use on board ship.
Normal Barometer	1. Extremely high accuracy. 2. Can be used to calibrate other barometers.	Similar to Fortin Barometer, with an adjustable cistern and a very high accuracy (0.02 mb); it is used as the standard barometer in England.**
Non-adjustable	1. Very slight error due to capillary action. 2. Capable of providing a high accuracy. 3. Adaptable to a recording device.	Areas on the upper and lower mercury surface of the manometer tube are made equal. Engraved scales are provided on both ends of the tube.
Adjustable Siphon	1. Generally has the same advantages as the non-adjustable type. 2. Initial zero point setting renders measurements independent of volume of mercury within instrument.	Similar in construction to the non-adjustable siphon type, with an additional zero setting device like the Fortin Barometer. A greater accuracy than the non-adjustable type is achieved.
Weighing Mercury Barometer	1. High sensitivity and accuracy. 2. Built-in recording device.	Either the mercury tube or the cistern is weighed by a sensitive balance. With the aid of levers, recordings of less than 1 mb are achieved.

*For details, the reader may refer to the following literature:
(1) Middleton, W.E.K. and A.F. Spilhaus. 1953. The measurement of atmospheric pressure, In: Meteorological Instruments. The University of Toronto Press. 286 pp.
(2) British Meteorological Office. 1956. Measurement of atmospheric pressure, In: Handbook of Meteorological Instruments, Part I, Instruments for Surface Observations. HMSO, London. 458 pp.
(3) U.S. Government Printing Office. 1964. Manual of Barometry (WBAN). Supt. of Documents, U.S. Govmt. Printing Off., Washington, D.C.
**At the National Physical Lab., Teddington, England (see Sears, J.E. and J.S. Clark. 1933. Proc. Roy. Soc., London, A 139: 130-146.)

Unlike the Fortin barometer, the Kew barometer has a fixed cistern (i.e., it does not require the zero-point setting). It operates on the principle that the loss of the mercury volume in the cistern is always equal to the gain of mercury volume in the tube, and vice versa. Accordingly, the change of volume in both the tube and the cistern may be expressed as $Ah = aH$, or

$$h = Ha/A \qquad \ldots \ldots \ldots \ldots \ldots \ldots \ldots \ldots \ldots \quad (11)$$

where h is the fall of mercury level in the cistern; H, the rise of mercury column in the tube; A, the area of the cistern minus the tail area of the tube; and a, the area of the tube (i.e., the mercury column).

When atmospheric pressure increases, the corresponding rise of the mercury column (in mm) will be

$$H + Ha/A = H(A+a)/A. \qquad \ldots \ldots \ldots \ldots \ldots \ldots \ldots \quad (12)$$

In other words, a rise of 1 mm Hg in atmospheric pressure will produce a rise of $A/(A+a)$ mm in the scale. Thus, the scale of the Kew-pattern barometer is graduated in $A/(A+a)$ standard millimeters. The same principle is also applied to the milli-bar or inch scales. The main advantages of the Kew-pattern barometer are its ease of operation and its ruggedness. However, it is comparatively less accurate and perhaps less sensitive than the Fortin barometer.

(c) Servo Barometers. To eliminate the need for temperature and gravity corrections and to permit continuous recording of the readings, several types of so-called servo barometers are now available. Two of these independently developed instruments are the Model 1100 Servo Barometer of Henry J. Green Instruments, Inc., and the 500 Series Model of Exactel Instrument Co., Division of Statham Instruments, Inc.

As shown in Figure 6-25, the 500 Series Model Servo Barometer consists of primary and secondary differential transformers, an analog temperature compensator, an index sensor, and a barometer tube in which a float and an armature containing magnetic material are inserted. The barometer tube has a very large internal diameter (19.05 mm) with a thick wall (3.18 mm) and is made from either stainless steel or pyrex glass. The large diameter tends to reduce the capillary error to a minimum and the thick wall minimizes the ambient temperature effect on barometric readings.

Somewhat similar to the Dines floating barograph[26] which uses two pulleys to mechanically activate the recording pens, the servo barometer's index sensor, which is outside the barometer tube, is continuously moved to a position corresponding to the location of the armature inside the tube. This is done electrically by means of the differential transformer. The precision of determining the change of barometric pressure is as accurate as 0.002 mb. With the aid of a servo-amplifier (either vacuum tube or transistor type) and a two-way motor, the output signals are transmitted to a digital encoder. They may be modified

_____

[26] Dines, L.H.G. 1929. Q.J. Roy. Meteorol. Soc. 55: 37-53. (See the lower right hand photograph of Figure 6-25.)

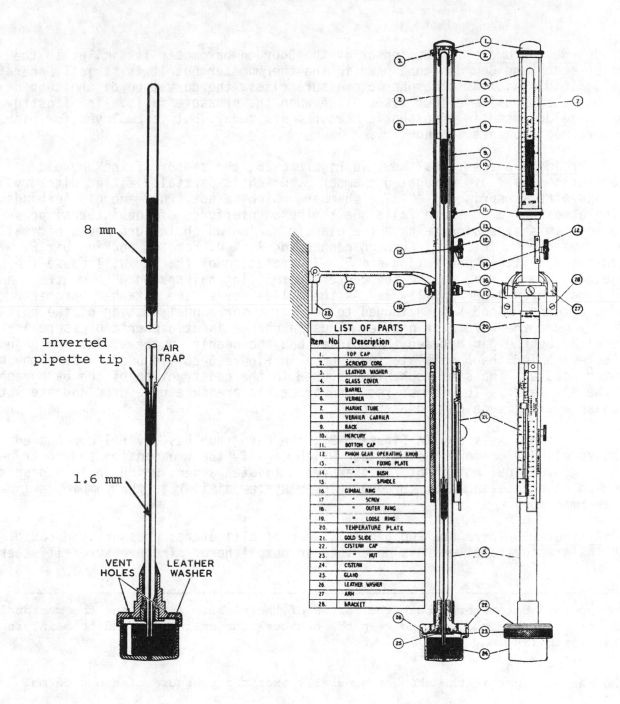

8 mm

Inverted
pipette tip

AIR
TRAP

1.6 mm

VENT
HOLES

LEATHER
WASHER

LIST OF PARTS	
Item No.	Description
1.	TOP CAP
2.	SCREWED CORE
3.	LEATHER WASHER
4.	GLASS COVER
5.	BARREL
6.	VERNIER
7.	MARINE TUBE
8.	VERNIER CARRIER
9.	RACK
10.	MERCURY
11.	BOTTOM CAP
12.	PINION GEAR OPERATING KNOB
13.	" " " FIXING PLATE
14.	" " " BUSH
15.	" " " SPINDLE
16.	GIMBAL RING
17.	" SCREW
18.	" OUTER RING
19.	" LOOSE RING
20.	TEMPERATURE PLATE
21.	GOLD SLIDE
22.	CISTERN CAP
23.	" NUT
24.	CISTERN
25.	GLAND
26.	LEATHER WASHER
27.	ARM
28.	BRACKET

Figure 6-24   Kew Barometer

by an analog converter and used for either teletype transmission or paper chart
registration. The temperature compensator, its thermal sensor coordinated with
the differential transformer, makes temperature corrections continuously.

(2) Elastic Barometer. Similar to the mercury barometer, the elastic barometer
is a direct-measuring instrument. In this type, the sensor is a highly flexible
tube or bellows made from metals or alloys such as stainless steel and phosphor
bronze. There are several kinds of elastic barometers including the Bourdon and
the aneroid types.

156

(a) Bourdon Barometer. The sensor of the Bourdon barometer is similar to the earlier described Bourdon tube used in the thermometer but it is strictly aneroid (i.e., fluidless). As atmospheric pressure rises, the curvature of the tube decreases; the opposite effect takes place when the pressure falls. This instrument is seldom used in atmospheric measurements today, but is employed for high pressure measurements in industry.

(b) Aneroid Barometer. As shown in Figure 6-26, the sensor of the aneroid barometer consists of corrugated chamber A, which is partially filled with dry air, and elastic spring B.[27] The chamber, which is not truly aneroid, expands as the atmospheric pressure falls due to the counterforce of the internal pressure.[28] This expansion relaxes the elastic spring which in turn moves bimetallic arm C upward.[29] This arm, through connecting link D, rotates rocking bar E, which has a built-in projecting arm F. The rotation of the bar will cause the projecting arm to move toward the center, permitting hairspring G to contract and wind arbor chain H upon pulley I. As the pulley moves, the attached barometer needle J turns. Knob K is attached to pointer L for manual setting of the pointer directly over the barometer needle. Thus, a change in atmospheric pressure is determined by the gap between the pointer and the needle on the scale. An adjusting screw behind the base plate (not shown in Figure 6-26) is used to zero the barometer needle. The same mechanism is used in the construction of the barograph and the altimeter. The former records barometric pressure on a drum and the latter displays altitude on a dial face.

(c) Barograph. As shown in Figure 6-27, the barograph has several corrugated chambers aligned on top of each other which magnify their expansion and contraction. The vertical motion of the chambers activates a pen, which in turn graphs a record on a rotating drum through a lever system similar to that employed in the thermograph.

(d) Altimeter. There are two general types of altimeters, pressure and radio.[30] Only the pressure altimeter is described in detail here. The pressure altimeter

---

[27] For aneroid barometers, both the forces of the elastic spring and the residual gas pressure are utilized to keep the chamber from collapsing and to maintain its flexibility.

[28] The residual gas in the chamber generally exerts a pressure of about 66 mb.

[29] The bimetallic arm acts as a component of the lever system and at the same time compensates for temperature effects on the chamber and the spring.

[30] The radio altimeter is an electronic device used in determining the height of an aircraft above the earth's surface. One device employed is a pulse radar which measures the range (height) of the transit time of the radar pulse. The other is a continuous wave radar which measures the height in terms of the phase differences between the transmitted and the received signals. The third type measures the variations in electrical capacities between the aircraft and the ground.

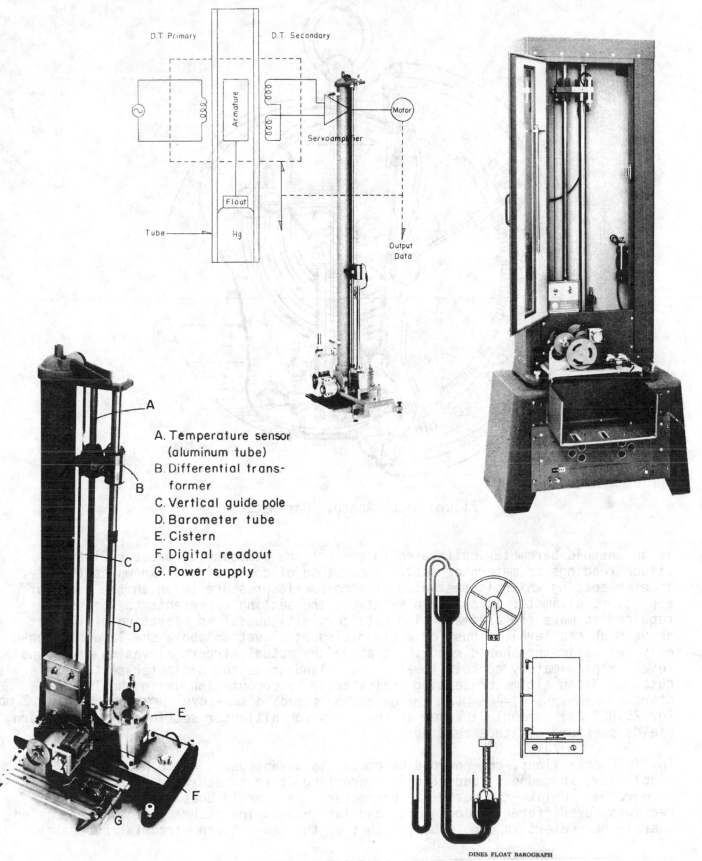

D.T. Primary   D.T. Secondary

Armature

Servoamplifier

Motor

Float

Tube

Hg

Output
Data

A. Temperature sensor
(aluminum tube)
B. Differential trans-
former
C. Vertical guide pole
D. Barometer tube
E. Cistern
F. Digital readout
G. Power supply

DINES FLOAT BAROGRAPH

Figure 6-25   Servo Barometer

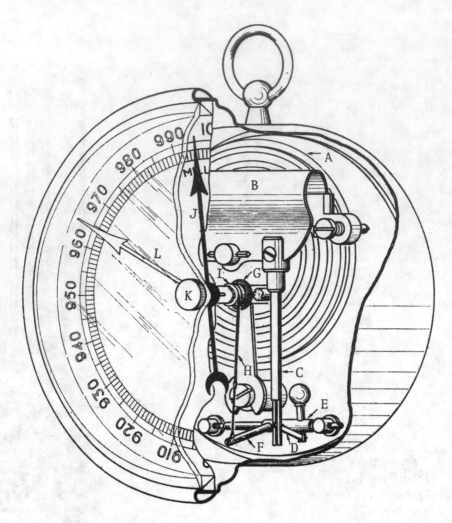

Figure 6-26 Aneroid Barometer

is an aneroid barometer calibrated to convert the atmospheric pressure into altitude readings in meters or feet. The method of conversion is known as the altimeter setting which is the value of atmospheric pressure to which the scale of
a pressure altimeter is set. In aviation, the setting represents the pressure
required to make the altimeter indicate zero altitude at an elevation of 10 feet
above mean sea level. Thus for a given airport elevation above sea level a properly set altimeter should read 10 ft above the actual airport elevation at ground
level (approximately cockpit level). Over land areas the altimeter setting is
obtained by an altimeter setting indicator or by computation using the ICAO
standard atmosphere.[31] Over the ocean the standard sea-level pressure of 1013.2 mb
(or 29.92" Hg) commonly is used as the basis for altimeter settings which, in turn,
yields pressure altitude readings.

In field operations, the aneroid barometer is advantageous as to portability,
simplicity, ruggedness, and ease in converting it to a recording device (the
barograph). Unlike the mercurial barometer, the aneroid does not require corrections for different temperatures and latitudes. The accuracy of some precision
aneroid barometers is comparable to that of the Kew-pattern mercurial barometer.

---

[31] In 1952, the International Civil Aeronautical Organization (ICAO) adopted and
modified the U.S. Standard Atmosphere of 1925. Since then, several Standard
Atmospheres, up to an altitude of 200 km, have been established. For details,
see ESSA, NASA, and USAF. 1966. U.S. Standard Atmosphere Supplement. U.S.
Government Printing Office. 289 pp.

Figure 6-27 Barograph

The main disadvantage of an aneroid barometer is the necessity of frequent calibration against a standard barometer. An ordinary aneroid barometer with a
higher sensitivity may not have greater accuracy than a mercurial barometer. The
aneroid is subject to secular changes due to sudden or gradual change in pressure.
This is known as hysteresis. Also, a rise in air temperature may cause an increase in the residual gas pressure as well as the expansion of the chamber itself despite the temperature compensating device in the bimetallic arm.

6.3.2 <u>Indirect Method</u>. The indirect measurement of atmospheric pressure utilizes
the principle expounded in the Clausius-Clapeyron equation

$$\frac{d \ln P}{dT} = \frac{L}{R} \frac{1}{T^2} \quad\ldots\ldots\ldots\ldots\ldots\ldots\ldots\ldots\ldots \quad (13)$$

where L is the latent heat of vaporization; R, the universal gas constant; and T,
the temperature of the liquid in question. At a given temperature, T, the corresponding pressure is determined while L and R are held constant. The relationship between temperature and pressure is interpreted as follows:

a. The vapor pressure of a liquid in a boiling state is the same
    as the pressure on the surface of that liquid.

b. The vapor pressure has a one to one relationship with the vapor
    temperature.

c. Therefore, if the vapor temperature can be measured, it should
    give a good indication of the atmospheric pressure.

The instrument which measures the vapor temperature of a liquid (i.e., measures
atmospheric pressure indirectly) is known as a hypsometer. For surface pressure

160

measurement, distilled water is used as the liquid; and for upper air, either carbon tetrachloride or freon are employed.

As shown in Figure 6-28, the hypsometer consists of heater H, double tube T, sensitive mercury thermometer t, and manometer M. The double tube, through which water vapor escapes, is used to minimize radiational heating from the tube wall to the thermometer. The attached mercury manometer is used to register any excess of pressure over than of true atmospheric pressure within the apparatus. This excess in pressure within the apparatus could be caused by resistance offered by the tube walls to the passage of the vapor. Taking the temperature reading and using Eq. (13), the atmospheric pressure is computed.

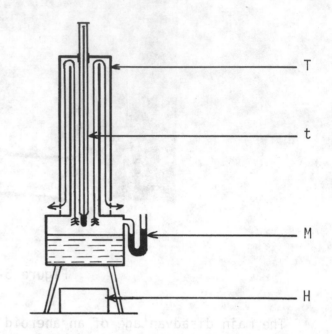

Figure 6-28 Simple Hypsometer

The principle advantages inherent in the hypsometer are compactness, light weight, simplicity, low maintenance, and low cost. Its main disadvantage is the time consumed in making observations. However, the automatic recording hypsometers employed in radiosondes at pressures of 300 mb and lower are extremely useful. A thermistor is used as the temperature sensing element in the hypsometric fluid.

6.4 Anemometry. The science of measuring and recording the direction and speed of wind is called anemometry. Instruments for surface wind speed measurements are known as anemometers; those which make a record are anemographs. Devices which measure and record wind direction alone are known as wind vanes. Some types of anemometers, however, include a vane while others do not. Also some surface wind measurements are made without a vane or anemometer such as a sonic anemometer. As all of these pertain to anemometry, they are properly and collectively called wind devices. In addition to monitoring the wind speed and direction, the device should respond to fluctuations of wind which are characterized by gusts and lulls. Both the speed and direction aspects of fluctuating winds are specified by the gustiness factor.[32] For a better understanding of anemometry, we begin with a discussion on

---

[32] The gustiness factor is calculated as the percentage ratio of the difference between the maximum and minimum horizontal wind speeds to the mean wind speed recorded in a given period. Gust shapes generally recognized are flat-topped, sinusoidal, sharp-edged, and triangular. Additional specifications are: (a) angular width of gusts, expressed in radians; (b) gust length, the distance occupied by the ascending part of the gust profile expressed in meters; and (c) gust intensity or gustiness (dimensionless).

systems for wind measurement.

6.4.1 _Systems for Wind Measurement_. As mentioned in Section 5.5, both the Eulerian and Lagrangian systems have been applied extensively to instrumentation for both upper-air and surface wind measurements. In the Eulerian system, velocity and gusts are measured by a wind device based on a fixed platform with respect to the earth's surface. In the Lagrangian system, the changing positions of a moving air parcel are continuously monitored by a wind device which follows the moving air current. These two systems are used as criteria for the classification of wind measuring instruments.

(1) _Eulerian Wind System_. In the Eulerian wind system, _fixed_ coordinates are chosen to specify wind speed and direction. Usually, orthogonal Cartesian coordinates are employed, but polar or cylindrical coordinates (curvilinear coordinates) may also be used. In the Cartesian coordinate system, a wind device is considered to be located at the center of the system at all times. Thus, two-dimensional horizontal and one-dimensional vertical winds are identified as the wind passes over the device.[33] When many wind devices are installed, vertically or horizontally, at different locations over a large geographical region on the earth's surface, each device has its own orientation for coordinates. This is due to the random motions of eddies under canopies, for instance, and differences in curvature of the earth's surface.

A rational classification of Eulerian type anemometers facilitates a better understanding of their operational principles and mechanisms. The following classification[34] is based upon the utilization of wind energy by anemometers.

    A. Aerodynamic forces utilizing the kinetic energy of the wind

        1. Rotation types

            a. Cup anemometers

            b. Propeller anemometers

        2. Pressure types

            a. Pressure tube anemometers

            b. Pressure plate anemometers

            c. Bridled anemometers

---

[33] The three components of the Cartesian coordinate system for wind are shown in Figure 3-2.

[34] For details, see Middleton, W.E.K. and A.F. Spilhaus. 1953. The measurement of wind near the surface, In: Meteorological Instruments. The University of Toronto Press. 286 pp.

162

   B.   Thermodynamic forces utilizing the cooling and heating power of the
        wind

        a. Hot-wire anemometers

        b. Kata thermometers

        c. Crystal anemometers

   C.   Energy of Sound Waves

        a. Sonic anemometers

Wind vanes are designed to indicate the direction of the wind. However, the continuous shift in the wind's angle of attack complicates the design of vanes and hence their classification. As a result, only a general classification based on the configuration of vanes[35] is made.

   A.   Wind vanes without a propeller

      1. Single fin (or tail) types

        a. Flat plate (any shape or size, generally rectangular)

        b. Aerofoil

        c. Streamline

      2. Double fin types

        a. Splayed vane (single flat wedge at an angle up to a
            20° spread)

        b. Parallel flat wedge

        c. Parallel curved wedge

        d. Double flat wedge

        e. Double curved wedge

   B.   Wind vanes with a propeller

      1. Bivane types

        a. Vector vane (Meteorology Research, Inc.)

        b. Aerovane (annular fin bivane)

        c. Gill bivane (R.M. Young Company)

---

[35]For details, see Wieringa, J. 1967. Evaluation and design of wind vanes. J. Appl. Meteorol. 6(6): 1114-1122.

      d. Axiometer (Climet Instruments Division, E.G. & G.)

   2. Pressure tube types

      a. Aerofoil

      b. Streamline

To reduce frictional drag between the wind and the earth's surface, an anemometer or vane should be exposed at a height of 10 m (33 ft) or more above the ground. With the exception of a moving ship or ground vehicle, all wind instruments in the Eulerian wind system are installed on such surface fixed platforms as a mast, tower, roof of a building, or mountain top. In the case of a ship, it is the relative velocity instead of the true wind velocity that is measured. This relative velocity is the vector difference of the velocity of the wind and that of the ship. As shown schematically in Figure 6-29, let $\overline{PQ}$ be the vector representing the course and speed of the ship and $\overline{RQ}$ the true wind velocity; then the relative velocity which is observed on board ship is the vector $\overline{RP}$. Thus, the direction of the true wind is given by the angle NQR, where N is oriented northward.

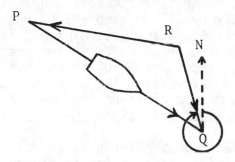

Figure 6-29 Estimation of Wind

A discussion and illustrations of the dynamic performances of both anemometers and wind vanes in conjunction with their operational mechanisms follows shortly.

(2) Lagrangian Wind System. In the Lagrangian system, wind speed and direction are determined mainly by ground or spacecraft observation of a moving platform or tracers. The moving platform is usually a balloon; tracers may be oil fog, fluorescent particles, aerosols, gaseous particles, visible plumes, or even moving cloud elements. Three techniques are available for ground detection of these moving objects.

(a) Optical Technique. A single theodolite, for example, is employed for tracking the speed and direction of a pilot balloon by assuming a constant rate of ascent for the balloon with a horizontal component moving with the wind. This technique has been widely employed for over a century and is now the major worldwide method used for upper-air wind measurement. Other optical techniques are double theodolite and aerial photography.

(b) <u>Radio or Radar Technique</u>. A radio theodolite is used for tracking the radio signal emitted from the transmitter carried by an airborne balloon. Here again, the vector position of the balloon at any instant in time is assumed to be that of the air parcel moving with the wind. This technique makes it possible to measure wind speed and direction above clouds.

Radar detection of balloon and air motion uses a similar technique. A corner reflector or any other reflector which is attached to an airborne balloon reflects a portion of the tracking radar signal. When the balloon is equipped with a transponder, its radar signal from the transponder is detected by a ground radar. This technique extends the range of upper-air wind measurements. Still another method is to use weather radar or sidelooking radar[36] to measure the movement of clouds and, hence, the wind. Doppler radar, in particular, can track three-dimensional motions; the signals are reconstructed to yield air flow patterns in clouds. Weather satellites equipped with a scanning spectrometer or video scanner have been used in detecting cloud movements. Using a cloud element for wind observation is not an accurate procedure; nonetheless, it provides generalized patterns of global wind fields which would not be obtainable otherwise.

(c) <u>Tracer Technique</u>. Various artificial tracer particles are released and detected by radar, lasar, lidar, or time cameras. In addition, these particles also can be collected locally by filter samplers or rotor rods. The tracer technique is unique in its applications for determining diffusion and turbulence of the atmosphere since turbulent currents of the atmosphere are continually in existence. These turbulent forces are $\sim$ 100 times greater than the gravitational settling force on the tracer particles. The drawbacks to this technique are those of its time-consuming method of operation using the sampler method, and its expense when using lasar or lidar detection.

In summary, the first two techniques offer a large scale survey of wind fields, while the last technique provides information on micro- and macroscale wind measurements. Strictly speaking, only the last technique adheres closely to the Lagrangian wind measuring system. Constant-level balloon techniques, discussed in Chapter 7, better illustrate the principles of Lagrangian system tracing than does the free lifting balloon technique.

The above techniques in addition to several not metnioned here are discussed in Chapter 7, Upper-Air Instruments. Surface wind measurements are described in the following paragraphs. This discussion includes wind speed and direction instruments. These instruments, in practice, employ the Eulerian system because of its simplicity and ease of operation.

5.4.2 <u>Measurements of Wind Speed</u>. As mentioned previously, the three major principles applied to wind measurements are aerodynamic, thermodynamic, and

---

[36] Side looking radar (SLAR) refers to pulsed and phased wavelengths just under 1 cm that are transmitted and received to the side of the flight path of the instrument. The electron beam of a CRT (cathode ray tube) display scans vertically, with position of the image proportional to distance from the flight track. Film is translated past the beam of the CRT, recording a succession of single line images. Both real aperture and synthetic aperture SLAR systems are employed to reconstruct images derived from reflected radar signals.

acoustic. The first, which uses the kinetic energy of the wind, is most commonly used in the design of wind instruments. All anemometers employed in worldwide national weather services are based upon aerodynamic principles. Instruments based upon the cooling power and acoustic energy of the wind, respectively, have a more sensitive and precise response to the wind. They are restricted for use by research workers because of their cost and calibration requirements. The instrument applications of the aerodynamic principle are discussed under the heading of Routine Wind Devices, while the other two applications are discussed under Research Wind Devices.

(1) Routine Wind Devices. The application of the wind's kinetic energy in anemometer design is exemplified by three major types: cup, propeller, and pressure tube anemometers. The first two are classified as rotation types and the last as pressure type.

(a) Rotation Type. The modern cup anemometer is more commonly employed at present than the propeller or windmill anemometers for routine applications

(i) Cup Anemometer. The three cup system, after numerous wind tunnel tests, has proven a better arrangement than any system using more than three cups arranged around a single axis. Not only does the three cup system have a larger torque per unit weight, but it also has a more uniform torque through each revolution. Several performance characteristics of the three cup anemometer are inherent to its design.

A cup of semi-conical shape is more rigid than a hemispherical cup. The angle of attack by the wind on the concave side of a 45 deg conical cup produces a maximum torque.

Beaded rather than plain edges reduce turbulence effects.

The size and mass of the anemometer must be kept to a minimum to reduce the moment of inertia and friction. This reduces the minimum wind required to start rotation as well as to increase the sensitivity of the instrument to wind gusts.

The most important characteristic is the establishment of the linear relationship between the actual wind speed and the cup tangential speed. This linearity of response requires further elaboration.

In a steady wind, let V be the wind speed and v be the cup tangential speed, as shown in Figure 6-30. It is possible to express the relationship between V and v by a power series.

$$V = \alpha + \beta v + \gamma v^2 + \delta v^3 + \ldots \ldots \ldots \ldots \ldots \ldots \ldots (14)$$

where α, β, γ, δ, etc. are constants.

When Eq. (14) is divided by v, the ratio, V/v or f, is called the anemometer factor. In his wind tunnel experiment, Brazier[37] found that if the ratio of the cup diameter, d, to the wheel diameter, D, is equal to 0.5, the coefficients, γ, δ,

---

[37] Brazier, C.E. 1914. Recherches expérimentales sur les moulinets anémométriques. Annales du Bureau Central Météorologique de France 157-300.

166

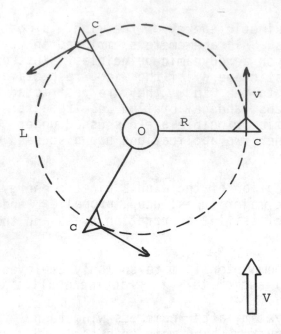

O	Center of rotation
R	Arm length
c	Cups
L	Loci of cup centers
V	Wind speed
v	Tangential speed

Figure 6-30 Cup Anemometer

etc., will become almost zero. Then, Eq. (14) may be reduced to

$$V = \alpha + \beta v \quad \ldots \ldots \ldots \ldots \ldots \ldots \ldots \ldots \ldots \ldots \quad (15)$$

where α is the threshold value of the anemometer (i.e., the maximum value of V for which v is zero) and β is the slope of the regression line, or $\beta = \Delta V/\Delta v$.

For a small, lightweight, and sensitive anemometer having an arm length of 7.5 cm and cup diameter of 5 cm, the value α may be as low as 0.5 mph. Therefore, it is suitable for the measurement of low wind speeds with high linearity. The large anemometer adopted by the US National Weather Service and the Federal Aviation Agency has a wheel diameter of 33 cm to 45 cm and cup diameter of over 10 cm. Its linearity is limited to a range between 1 and 30 m s^{-1} (or approximately 4.5 to 47 mph). In terms of Eq. (15), the α value may be as high as 5 mph.

The distance constant, or response distance, is the length in feet of an air column required for the anemometer to respond to 63.2% of (1-1/e) of a step change in speed exponentially. It has a first order characteristic. The relationship between the distance constant and time response for both steady and varying winds has been discussed. In Mazzarella's large inventory on sensor specifications for wind instruments, the distance constant of the cup anemometer varies from 2.5 to 26.2 ft in more than two dozen selected cup anemometers.[38] For a steady wind of 5 mph, this variation in the distance constant corresponds to time responses of about 0.4 s to 3.7 s.

---

[38] Mazzarella, D.A. 1972. An inventory of specifications for wind measuring instruments. Bull. Amer. Meteorol. Soc. 53(9): 860-871.

The Beckman and Whitley cup anemometer,[39] shown in Figure 6-31, has two separate three-cup units mounted on a common vertical shaft with a 30° displacement of the cup arms between the first and second level. The manufacturer claims that such a two-level arrangement offers the best distribution of torque. This arrangement appears to be superior to a six-cup anemometer with all cups at the same level in that the former minimizes the turbulence effects. The two-level, three-cup anemometer, while responding better to wind changes than the one-level, three-cup anemometer, is expected to have a larger moment of inertia and mechanical friction and have higher production costs than the single level anemometer. This is an example of a technically improved device that is over-designed for routine, accurate wind measurements.

(ii) <u>Propeller Anemometer</u>. There are two simple and popular types of propeller anemometers: the flat blade and the helicoidal blade. The former is represented by the airmeter and the latter by the aerovane. Airmeters are generally employed as portable field instruments and, therefore, are small and of lightweight construction. The aerovane is used extensively as a permanent installation in most of the national weather services. Unlike the cup anemometers with a vertical axis, both the aerovane and airmeter are constructed with a horizontal axis.

In gusty winds, the airmeter is preferable as the blades are made of a lightweight material such as mica. Thus, the moment of inertia of the system is small, enabling the airmeter to give good results in an unsteady wind. The indication is usually made by a counter mechanism including a horizontal spindle, a train of gears and several dials. Therefore, the airmeter must be read with the naked eye and does not give a continuous recording. The axis of the spindle must be parallel to the wind direction in use. In order to do so, the airmeter may be mounted on a wind vane or similar device. When held by hand, the airmeter must be held out at arm's length to avoid the effect of the body on the wind measurement. The wind direction must be measured at the same time.

As shown schematically in Figure 6-32, the wind blows in a direction parallel to the axis of the spindle with a speed, V, and at an angle of θ with the blade, B, when the blade is at rest. As soon as the wind reaches the blades, which begin to rotate with a linear speed of v, the tan θ = v/V, or

$$v = V \tan \theta \dots \dots \dots \dots \dots \dots (16)$$

An airmeter is designed such that when the axis of the horizontal spindle is parallel to the wind, the blades are at an angle of 45° to the wind direction and, hence, v = V. A calibration sheet from the manufacturer must be available, as the error of an airmeter is not constant over the entire speed range. The speed range of a low speed airmeter (eight light blades) is about 1 to 30 mph, while that of a high speed airmeter (four heavy blades) is 10 to 120 mph.

In contrast, an aerovane transmits both wind direction and wind speed at the same time. It consists of a three-bladed plastic propeller, a wind vane, a pair of

---

[39]Beckman and Whitley originally designed and manufactured their cup anemometers. In recent years Teledyne/Geotech has modified the anemometer system and presently manufacturers it.

168

BECKMAN & WHITLEY, INC.   BELFORT INSTRUMENT CO.   R.M. YOUNG CO.

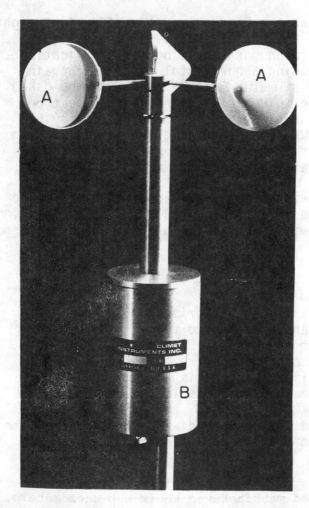

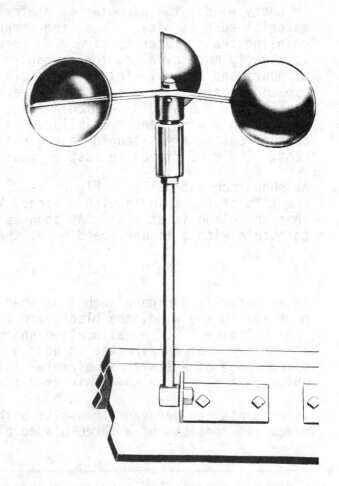

CLIMET INSTRUMENT CO.   HEMISPHERICAL TYPE

Figure 6-31 Cup Anemometers

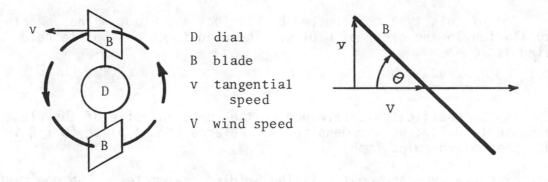

D dial

B blade

v tangential
   speed

V wind speed

Figure 6-32 Dynamics of Hand Anemometer

AC self-synchronous motors, and a DC magneto-generator. The DC magneto-generator is driven by the plastic propeller, converting the mechanical energy of the rotating propeller into electrical energy. The electrical voltage generated is recorded by a potentiometer calibrated in mph. Thus, an aerovane is a self-supplying, recording wind instrument. The power supply is used for the amplifier, the AC self-synchronous motors, and the potentiometer. The wind direction is transmitted to a remote indicator by the synchronous motors. As the propeller and the fin are mounted on the same arm, the axis of the propeller is always parallel to the wind direction.

(b) <u>Pressure Type</u>. There are two major types of anemometers which utilize the pressure force of the wind on the surface of plates, cups, or water. The pendulum anemometer, the normal plate anemometer, and the bridled anemometer use wind pressure plates or cups. Pressure tube anemometers employ wind pressure on a water surface. The latter is widely used in European countries and Japan, while the two former ones are not employed as routine wind instruments. Therefore, our discussion gives more complete details on the pressure tube anemometer and gives only a brief description of pressure plate anemometers.

(i) <u>Pressure Plate Anemometer</u>. The earliest form of the anemometer, based on an idea conceived by Leonardo da Vinci in 1500 and designed by Robert Hooke in 1667, is the pendulum anemometer. It consists of a plate which is free to swing about an horizontal axis when the wind exerts a pressure upon it. The angular deflection of the plate increases with wind pressure, and is thus a function of wind speed. This deflection is read on a vertical scale as a measure of wind speed. When the frequency of gusts and the natural frequency of the swinging plate coincide, the readout becomes very difficult to make. The moment of inertia of the plate is large so that oscillation of the plate counteracting the wind fluctuations causes considerable inaccuracy. Nonetheless, the pendulum anemometer is mentioned here to provide the historical and scientific background leading to the development of the modern anemometer. The pendulum anemometer is considered obsolete and unsuitable for actual practice except for use by amateurs.

Another type of pressure plate anemometer has the plate held perpendicular to the wind by a stiff spring. This instrument, known as the normal plate anemometer, is useful for recording gusty winds. The wind-activated motion of the plate is recorded electrically by an oscillograph. The natural frequency of this instrument system can be made high enough so that resonance magnification does not occur. One of this type constructed by Sherlock and Stout[40] has a natural period of oscillation in the order of 0.01 s. The wind force on a plate held normal to the wind is

$$F = \frac{c\rho A V^2}{2} \quad \ldots \ldots \ldots \ldots \ldots \ldots \ldots \ldots \ldots \ldots \ldots \ldots \ldots \quad (17)$$

where c is the coefficient determined by the shape and size of the plate with a value close to unity; ρ, air density; A, the area of the plate (9×8 in.^{2}); and V, the speed of the wind (mph).

Another pressure type anemometer is the Bridled anemometer. The one manufactured by the Friez Instrument Division, Bendix Aviation Corporation, has a wheel with 32 cups. When the wind exerts a pressure on these cups, the force of angular displacement of the wheel is transmitted to a position motor, which then prevents the wheel from turning, giving a measure of wind speed.

(ii) Pressure Tube Anemometer. As shown in Figure 6-33, the British Meteorological Office anemograph consists of three main units: the head and vane, the mast, and the recording device. The head and vane are similar to the structure of the head of the Dines anemometer.[41] The British pressure tube anemograph (see Figure 6-33$_a$) consists of a 24-in. aerofoil vane V, pressure opening A, the inner tube connected to pressure outlet B, the outer tube connected to the suction outlet C, and suction holes outfit S.

The pressure tube anemometer measures wind speed by determining the difference in wind pressure between the suction force (the dynamic wind pressure) and the static force (the hydrostatic wind pressure). When wind blows into the pressure opening, it develops a pressure greater than the static pressure; when wind blows across the suction holes, it develops a pressure less than the static pressure. A suitable manometer, such as a pitot static tube, is used to measure the difference in these pressures. This difference in pressure is proportional to the square of the wind speed, thus

$$P = \frac{\rho V^2 (1 + c)}{2} \quad \ldots \ldots \ldots \ldots \ldots \ldots \ldots \ldots \ldots \ldots \ldots \quad (18)$$

where c is an instrument constant less than unity, and all the other symbols are the same as in Eq. (17).

The mast of the British pressure tube anemograph is a 40 to 50 ft tower which balances the head and vane at its center of gravity. Usually the mast is mounted vertically atop an existing building or a specially built house. The main features of the mast proper consist of a directional rod for transmitting the wind direction signals and two gas pipes for transmitting the dynamic and static pressures. This is done by connecting the outlets of the inner pressure tube and the outer suction tube to the recording manometer. Other minor features are clearly shown in Figure 6-33$_b$.

As shown in Figure 6-33$_c$, the Dines float manometer is a recording device consist-

---

[40] Sherlock, R.H. and M.B. Stout. 1931. Eng. Res. Bull. Univ. of Michigan, Ann Arbor, Michigan.

[41] Dines, W.H. 1892. Collected Papers.

A. HEAD & VANE

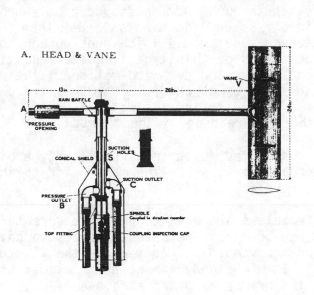

C. RECORDER

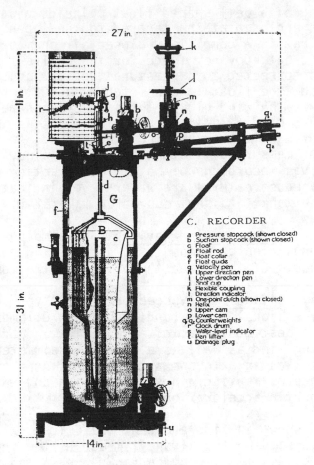

a Pressure stopcock (shown closed)
b Suction stopcock (shown closed)
c Float
d Float rod
e Float collar
f Float guide
g Velocity pen
h Upper direction pen
i Lower direction pen
j Shot cup
k Flexible coupling
l Direction indicator
m One-point clutch (shown closed)
n Helix
o Upper cam
p Lower cam
q q₁ Counterweights
r Clock drum
s Water-level indicator
t Pen lifter
u Drainage plug

PRESSURE TUBE ANEMOGRAPH WITH DIRECTION RECORDER AND 40' MAST OF STANDARD PATTERN

B. MAST

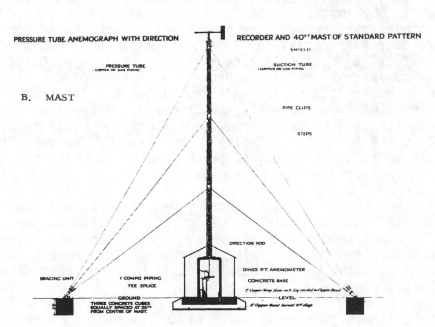

Figure 6-33 Pressure Tube Anemograph

ing of a bell-shaped float c inside a water tank. The inner part of the float is enclosed at the tapered top and open at the bottom. The outer part of the float is a completely closed air chamber surrounding the inner part providing sufficient buoyancy force for the entire float. The air space B inside the float is connected to a pressure tube through pressure stopcock a. Air space G outside the float is connected to a suction valve through suction stopcock b. Whenever the wind blows, the static pressure in air space B is always larger than that of G. To the top of the float is attached float rod d which runs up and down during fluctuating wind conditions. An anemograph of this type is quite sensitive for measuring fluctuations of the wind. The upward and downward motion of the rod gives recordings of both wind direction and speed on clock drum r. Other miscellaneous features are water level indicator s for measuring the water level of the tank, and shot cup j for maintaining the zero reading of the chart.

(2) Research Wind Devices. In atmospheric sciences research, propeller type bivanes, sonic anemometers, and cooling power anemometers are the major wind devices used. Of the three the bivane is the most popular.

(a) Bivanes. Advanced propeller type anemometers are bivanes which measure three-dimensional winds ranging between 0.1 mph and 50 mph. The bivane appears to have the best qualities of an airmeter and/or an aerovane. It can measure the fluctuating wind as well as a sensitive airmeter. It withstands strong gusts almost as well as an aerovane and, yet, has none of the drawbacks of an aerovane (i.e., underestimating winds below 5 mph and exaggerating high and gusty winds by inertia acceleration).

As shown in Figure 6-34$_a$, the Climet axial flow anemometer, which is a bi-directional vane (or bivane), measures the vertical and horizontal wind velocity. It provides the direction (in degrees) and the accompanying speed (in mph) of the natural air flow at any instant. The instrument consists of a four-blade propeller, a vane, a translator, transmitter and several recording devices. Propeller A and vane D are made of polystyrene. The transmitter consists of wind speed transducer B, vertical direction transducer C, and horizontal direction transducer E. The transmitter converts the mechanical signal of air flow into an electrical signal, which is then fed into translator G. The translator amplifies the wind direction signal and converts the frequency of the wind speed signal into electrical currents. It also transforms, rectifies, and regulates the power supply from 110-125 volts AC to the 12 volts DC required by the amplifier. Any recording device, such as a milliammeter, potentiometer, punched-tape digital recorder, or magnetic tape recorder, may be employed.

The recording range for wind speed is about 0.3 to 99 mph, and that for wind direction is 0° to 540° azimuth angle and 0° to ±60° elevation angle. Its accuracy is ~ ±0.2 mph for wind speed, ±1.8° for azimuth angle, and ±0.6° for elevation angle. In contrast to conventional wind instruments that measure only the horizontal wind over 5 mph, the bivane measures horizontal and vertical winds at both high and low speeds.

The MRI vectorvane (see Figure 6-34$_b$), the prototype of the Climet bivane, measures the natural airflow for 0° to 540° azimuth angles and ±60° elevation angles. The capacity range of this instrument for the wind speed measurements is 0.5 to 125 mph, with a 1 mph accuracy. The vertical angle is detectable to ±2° with about 0.25° resolution. It may contain a sigma meter, which is set to within 5, 30, or 180 s for sampling times, and detects fluctuations in azimuth and elevation angles expressed by standard deviations of wind direction. This is a very valuable and unique feature of the vane for both azimuth and elevation angles

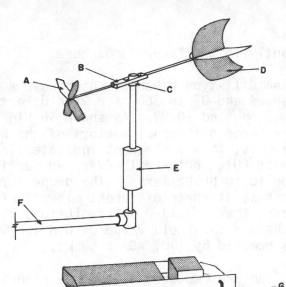

A. Four-blade propeller
B. Wind speed transducer
C. Vertical direction transducer
D. Polystyrene vane
E. Horizontal direct-ion transducer
F. Supporting arm
G. Translator

(a) CLIMET AXIOMETER

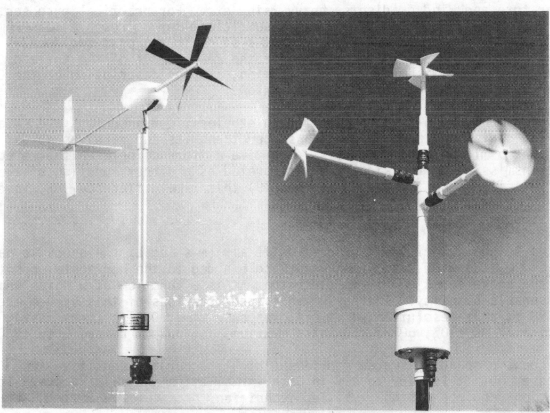

(b) Vectorvane
Meteorology Research, Inc.

(c) Gill Bivane
R. M. Young Co.

Figure 6-34 Bivanes

of the airflow fluctuate constantly in the lower atmosphere.

Another series of bivanes are the Gill-type bivanes, which have a detection range of 0.3 mph to 70 mph for wind speed and 0° to 360° for wind direction, with respective accuracies of about ±0.2 mph and ±0.2°. As shown in Figure 6-34$_c$, the Gill UVW anemometer measures the three orthogonal vectors of the wind speed simultaneously. Unlike the other bivanes, this instrument indicates the wind speed at three fixed directions: east-west (U), north-south (V), and vertical (W), with three separate sensors called helicoid propellers. The propellers are mounted at right angles to each other on a mast at three different elevations. According to the manufacturer's specifications, they rotate 0.96 revolutions per ft for all wind speeds above 2.7 mph, and have a threshold value of 0.8 mph with a maximal range of 50 mph. The device is powered by 110V AC at 60 Hz.

(b) Sonic Anemometer. The aeolian anemometer utilizes the principle that the pitch of the aeolian tones (i.e., sounds produced by winds in the lee of obstacles, such as telephone wires) is a function of the wind speed. This instrument is still in the developmental stage with units now commercially available.

While all bivanes indicate the vertical wind speed at a point in space, sonic anemometers measure the vertical wind along a pre-designed acoustic path. Theoretically, the path could be designed up to a length of 1500 m (about 5000 ft) with little interference from such environmental factors as strong turbulence, temperature and wind gradients, heavy rain, etc. For practical purposes, however, a path length of 15 m (about 50 ft) is sufficient for microscale measurement. Rapid development in acoustic instrument systems in the past two decades has produced a variety of sonic anemometers for the measurement of the three orthogonal components of the wind. As an illustration, a brief description of the Cambridge single axis sonic anemometer is given below.

As shown in Figure 6-35, two miniature piezoelectric transducers are separated by a 15-in. horizontal path. Each transducer transmits and receives pulses of sonic energy at 5 millisecond intervals. The time difference between the reception and the transmission of the signal is interpreted as a direct function of the mean wind velocity along the path. In Eq. (19) this time difference, Δt, is defined as

$$\Delta t = \frac{2L}{c^2} \cdot \frac{w \cos \theta}{1 - (w/c)^2} \cdots\cdots\cdots\cdots\cdots\cdots\cdots\cdots\cdots \quad (19)$$

where c is the velocity of sound in still air (m s^{-1}); L, the acoustic path length (m); and w, the wind speed (m s^{-1}) at angle θ deg to the horizontal position. From the above equation, it is clear that Δt is a function of w which represents a measurable quantity. The performance specifications of this system are: range, 0 to 98 ft s^{-1}; resolution, 1 ft s^{-1}; response time, 0.01 s (very fast); input power, 115V AC ±10%, 60 Hz, 50 watts maximum; and output power, 0 to 1V DC.

The acoustic sounding system provides a highly accurate, sensitive, and linear response, but it is very costly in both installation and maintenance. Its operation requires a highly skilled technician.

(c) Cooling Power Anemometer. Based upon the principle that the time response of a thermometer is a function of its ventilation, the kata thermometer was developed by Hill in the early nineteenth century. It consists of a liquid-in-glass thermometer with a large bulb, and two calibration markers corresponding to 100F and 95F on the stem. The bulb is heated to 102F or higher and kept dry before it is exposed to the wind. The time, t, in seconds required for the liquid column to fall from 100C to 95C is measured by a stop watch and is used to compute the wind

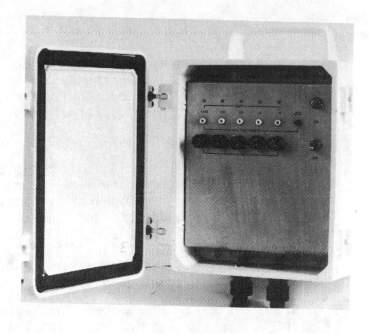

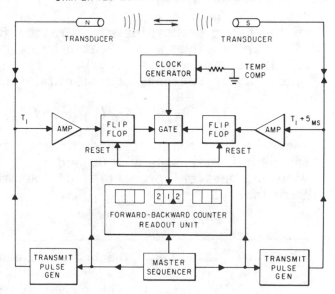

SIMPLIFIED BLOCK DIAGRAM — SINGLE AXIS

Figure 6-35 Sonic Anemometer

176

speed, V, in ft s^{-1} from an empirical formula. For wind speeds below 3 ft s^{-1} (about 2 mph), the formula is

$$K = (0.11 + 0.016V^{1/2}) (97.7 - T)t \quad \ldots \ldots \ldots \ldots \ldots \quad (20)$$

where T is the air temperature in degrees Fahrenheit and K, the kata factor, is determined by the size, thickness, and thermal conductivity of the bulb. Nomograms are available from the manufacturer for the direct readout of wind speed from stop watch recordings. For wind speeds above 3 ft s^{-1}, Eq. (20) becomes

$$K = (0.07 + 0.019V^{1/2}) (97.7 - T)t \quad \ldots \ldots \ldots \ldots \ldots \quad (21)$$

From either Eq. (20) or (21) it is obvious that an increase in V is accompanied by a decrease in t. Thus, a low temperature kata thermometer can be used only to measure a slow moving vertical or horizontal air current. A high temperature kata thermometer is employed when the atmospheric temperature exceeds 99F. In this case, the kata cools from 130F to 125F or from 150F to 145F. Here again, specific nomograms are available to facilitate the direct reading of wind speed. When any kata thermometer is used, several readings (usually three or four) should be taken until a consistent reading is obtained.

Similar to the kata thermometer, the Yaglou thermometer[42] measures low wind speeds of about 0.1 mph to 6 mph. It is constructed by winding a coil of fine wire around a mercury thermometer bulb. This coil is heated at a constant voltage. The excess heat produced to keep it above the air temperature is then measured.

A more sophisticated instrument, the hot wire anemometer, employs an electrically heated fine platinum wire to measure wind speed. Using a well designed sensor and circuitry, anemometers of this type have a time response as short as 0.1 s and a range as wide as 0.1 to 22 mph.

Although many different types have been made and greatly improved upon, the first design of the hot wire anemometer was introduced in 1914[43] by L.V. King, who set up the basic principles of operation for this instrument. He used a 0.003 in. diameter platinum wire, heated by electric current, i, and held perpendicular to an air current with speed, V, at temperature, T_i. When the equilibrium temperature, T_e, is reached, equilibrium electric resistance, R, is obtained. Thus

$$T_e - T_i = R\, i^2/(k_1 + k_2 V^{1/2}) \quad \ldots \ldots \ldots \ldots \ldots \ldots \quad (22)$$

where k_1 is a constant determined by the radiative and convective heat losses from the heated wire when $V = 0$, and k_2 is a constant determined by the diameter of the wire and physical properties of the ambient air. Both k_1 and k_2 may be determined empirically.

---

[42] Yaglou, C.P. 1938. The heated anemometer. J. Ind. Hyg. Toxicol. 20:497.

[43] King, L.V. 1914. On the convection of heat from small cylinders in a stream of fluid: Determination of the convection constants of small platinum wires with applications to hot wire anemometer. Philosophical Transaction A 214.

When the platinum wire is heated to a very high temperature on the order of several hundred degrees, $T_e - T_i \sim T_e$. Also, T_e and R may be kept constant by including the wire in an arm of a Wheatstone bridge. Then, Eq. (22) is reduced to

$$i^2 = i_0^2 + KV^{1/2} \quad \ldots \ldots \ldots \ldots \ldots \ldots \ldots \ldots \ldots \quad (23)$$

where i_0 is the electric current measured in still air (a constant) and K is another constant. Eq. (23) indicates that i is a measure of wind speed, V. Alternately, when i is kept constant, the electric potential drop across the wire is a function of wind speed.

The advantages of the hot wire anemometer over other anemometers are its high accuracy and its great sensitivity at low wind speeds if a very fine wire is employed (e.g., 1 cm long and 2 to 5 μm in diameter). Its disadvantages are that the hot wire design is not suitable for such wet weather conditions as rain or heavy fog and that its suitability is generally lower than conventional anemometers.

With the advent of modern electronic technology, hot wire anemometers have undergone many improvements. Several commercially available hot wire anemometers and their performance specifications are illustrated in Table 6-14.

Table 6-14 Examples of Hot Wire Anemometers

Manufacturer	Disa Elektronik A/S Herlev, Denmark	Toa Gijutsu Center 4-11-5 Kotobashi Sumida-ku, Tokyo Japan	Wallac Oy Turku 5 Finland
Range	up to 150 m s^{-1} (335 mph)	0 to 3 m s^{-1} (0 to 8 mph)	0.1 to 5 m s^{-1} (0.2 to 11 mph) 2 to 30 m s^{-1} (4 to 67 mph)
Response Time	–	1 s for 63% of signal change	0.2 s for 50% of signal change
Accuracy	–	2 cm s^{-1} (0.04 mph)	8% of mid-scale
Form of Output (indicator or recorder)	voltage (voltmeter or potentiometric recorder)	current (meter)	voltage (meter or recorder)

A comparison of four types of anemometers by Acheson[44] is summarized in Table 6-15.

---

[44]Acheson, D.T. 1969. The principles of wind speed sensors and records, NAPCA course on Meteorological Instrumentation in Air Pollution. Research Triangle Park, North Carolina, USA.

Table 6-15 Comparison of Several Types of Anemometers

Type	Speed of Response	Required Electronics	Cost
Rotating	slow	none	very modest
Dynamic pressure	fast	linearizing (AC or DC)	modest
Aerodynamically cooled	very fast	linearizing (AC or DC)	high
Sonic	very fast	complex digital	very high

6.4.2 <u>Measurements of Wind Direction</u>. Criteria for obtaining desirable characteristics for the dynamic performance of the windvane are the following:

1. Reduce the friction at its pivot to a minimum in order to get a
better response to low winds.

2. Obtain a perfect balance in weight of the mass (the counterweight)
and the tail (the fin) so that the vane does not tilt in a prefer-
ential direction.

3. Measure the verticality of the support and the true north orientation of the vane in order to obtain an accurate wind direction.

4. Produce a maximum torque for a given change in wind direction in
relation to its moment of inertia.

5. Reduce the moment of inertia to minimize the aerodynamic effect on
the vane from gusts.

6. Make the vane slightly underdamped in order to reduce the distance
constant and to increase the speed of response.

For a better understanding of the above criteria we shall describe the construction of a vane, its dynamic performance, and its transmitting and recording
systems.

(1) <u>Windvane Design</u>. Windvanes may take many forms. For simplicity, only the
shapes of the tail are shown schematically in Figure 6-36. The design of a
streamline vane is given in Figure 6-37.

The accuracy of a windvane is determined primarily by its design. The overall
design becomes a compromise between the best designs for the fin, the supporting
arm, and the counterweight. The entire load of the vane is balanced at the axis
of rotation. The arm length of the fin must be as long as possible to produce a
strong torque, yet as short as possible on the counterweight side to reduce
inertia. The vane should be built of lightweight material for fast response and
yet it should be rigid enough to resist bending of the arm, deformation of the
fin, and destruction of the entire vane due to high wind speeds. It should rotate
about its axis freely with a minimum of friction. The axis of rotation must be

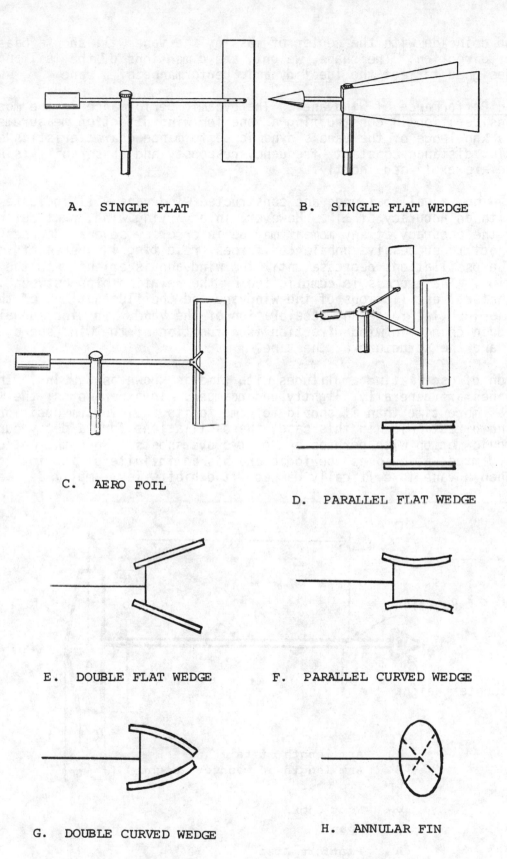

A. SINGLE FLAT

B. SINGLE FLAT WEDGE

C. AERO FOIL

D. PARALLEL FLAT WEDGE

E. DOUBLE FLAT WEDGE

F. PARALLEL CURVED WEDGE

G. DOUBLE CURVED WEDGE

H. ANNULAR FIN

Figure 6-36 Illustration of Windvane Forms

vertical and coincide with the center of mass or the vane will show a bias toward a particular direction. The shape, weight, and dimensions of the tail and arm should be designed to meet the ideal dynamic performance of a vane.

(2) <u>Dynamic Performance of Windvanes</u>. The dynamic performance is the most fundamental consideration in constructing a vane for wind direction measurements. Therefore, a knowledge of the vane's dynamic performance characteristics such as damping ratio, distance constant, frequency response, and phase shift is necessary. These are explained shortly.

In a steady wind, a well balanced and constructed wind vane will indicate the wind direction with an accuracy of ±1°. However, in a varying wind, particularly a gusty wind, the accuracy at any moment may be in error by several degrees. A windvane is subject to successive unbalanced forces, resulting in overshoot and oscillations. The oscillations decrease until the windvane is oriented to the true wind direction. The process is complicated by the relationships between the forced and natural oscillations of the windvane and the fluctuations of the wind involving the inertial force and acceleration of the vane. In wind tunnel experiments, a sudden change in wind direction is a function of the wind speed, deflection angle, and the structure of the vane.

The reduction of oscillating amplitudes with time is known as damping. In practice, windvanes are generally slightly underdamped. In other words, the windvane requires more time than it should to come to its equilibrium position (or the true wind direction). In this case, the oscillations form a decay curve of simple harmonic motion with perhaps one or two overshoots. When the motion of the vane is overdamped, it becomes aperiodic and has an infinite period in its oscillations. When a vane is critically damped, it exhibits less amplitude, less

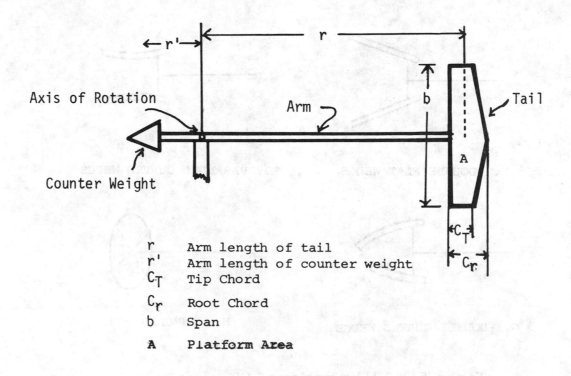

r	Arm length of tail
r'	Arm length of counter weight
C_T	Tip Chord
C_r	Root Chord
b	Span
A	**Platform Area**

Figure 6-37 A Schematic Diagram of Streamline Vane Construction

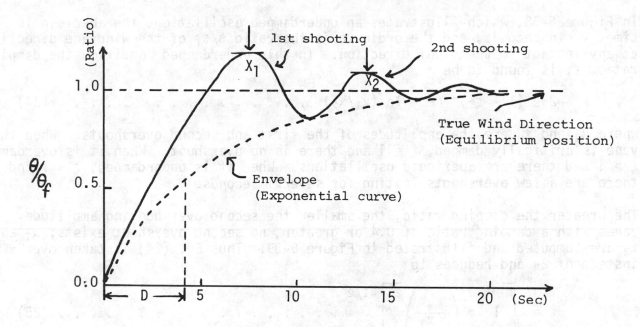

Figure 6-38   An Illustration of an Underdamped Oscillation

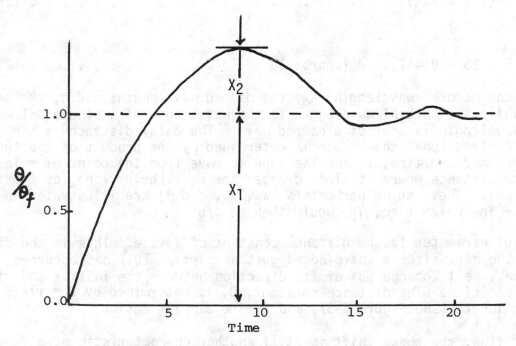

Figure 6-39   Underdamped Oscillation Without the Second Overshoot

oscillation and no overshooting. As a critically damped vane condition is not likely to occur in a sensitive windvane, it is better to have a slightly under-damped rather than overdamped condition.

In Figure 6-38, which illustrates an underdamped oscillation, the abscissa is time, t, in seconds, and the ordinate is the ratio θ/θ_f of the windvane direction at any instant to the final direction. In this underdamped condition the damping ratio, ξ, is found to be

$$2\pi\epsilon \, (1 - \xi^2)^{-1/2} = \ln \, (x_1/x_2) \quad \ldots \ldots \ldots \ldots \ldots \ldots \quad (24)$$

where x_1 and x_2 are the amplitudes of the first and second overshoots. When the vane is critically damped, $\xi = 1$ and there is no overshoot. When it is overdamped, $\xi > 1$ and there are aperiodic oscillations. When it is underdamped, $\xi < 1$ and there are a few overshoots lasting for several seconds.

The greater the damping ratio, the smaller the second overshooting amplitude. For vanes with a damping ratio of 0.4 or greater, no second overshoot exists; x_1 and x_2 are computed and illustrated in Figure 6-39. Thus Eq. (24) is taken over π instead of 2π and reduces to

$$\xi = \left[1 + \left(\frac{\pi}{\ln\frac{x_1}{x_2}} \right)^2 \right]^{-1/2} \quad \ldots \ldots \ldots \ldots \ldots \ldots \quad (25)$$

For vanes with a damping ratio of less than 0.4, there are third and possibly fourth overshoots. The damping ratio is then determined either by

$$\xi = (1 - \lambda_n/\lambda_d)^{1/2} \quad \ldots \ldots \ldots \ldots \ldots \ldots \ldots \ldots \quad (26)$$

or

$$\xi \cong 0.25 - 0.417 \, \lambda_n/d \text{ (approx.)} \quad \ldots \ldots \ldots \ldots \ldots \quad (27)$$

where λ_n is the natural wavelength; λ_d, the damped wavelength; and d, the delay distance. While the natural wavelength is the wavelength of a sinusoidal wave, the damped wavelength is that of a damped wave. The delay distance is the 50% recovery of a directional change and is determined by the product of the tunnel wind speed, u, and the time, t, for the vane to move from the point of release to 50% of the distance where it first crosses the equilibrium line (or the true wind direction). These three parameters (λ_n, λ_d, and d) are illustrated in Figure 6-40. The delay time, T_D, would then be d/u.

Another useful parameter is the distance constant of a vane. This is the distance traveled by the air, after a sharp-edged gust or partial lull has occurred, in order for the vane to change 63% of its direction between the initial and final equilibrium position. The distance constant, D, is determined by $D = r/2\xi^2$, where r is the arm length (see Figure 6-37) and ξ, the damping ratio.

In turbulent flow, the phase shift is still another characteristic of a vane.

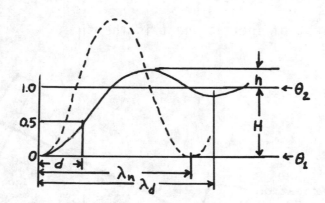

Figure 6-40 The Relationship of λ_n, λ_d, and d

Le Gette[45] has formulated the expression for phase shift as

$$\phi = \tan^{-1} (2\ f/f_n)\ [1 - (f/f_n)^2]^{-1} \dots \dots \dots \dots \dots (28)$$

where ϕ is the phase shift in degrees; f, the frequency of the direction change of the vane; and f_n, the natural frequency of the vane.

As shown in Figure 6-37, A is the tail platform area;[46] C_T, the tip chord; C_r, the root chord; and b, the tail span. The torque, T, may be expressed as

$$T = c\rho R_e V^2 rA \sin \theta \dots \dots \dots \dots \dots \dots \dots \dots (29)$$

where ρ is the air density; R_e, the Reynold's number; θ, the instantaneous angular deflection; V, the wind speed; r, the arm length; and c, the constant at less than 15° (see Figure 6-41).

For a given vane, A and r are constants. For wind speeds of 5 to 30 mph and with an angular deflection of less than 15°, r and R_e essentially become constants. The air density is considered as a constant without introducing appreciable error. Then, Equation (29) becomes

$$T = KV^2 \sin \theta \dots \dots \dots \dots \dots \dots \dots \dots \dots \dots (30)$$

[45]

Le Gette, M.A. 1962. The theory of recording galvanometers, Consolidated Electrodynamics Booklet 1582.

[46]

A = 1/2 $(C_T + C_r)$b.

For a given wind speed, V, and for a small angle, θ, Eq. (30) may be written as

$$T = K\theta \quad \ldots \ldots \ldots \ldots \ldots \ldots \ldots \quad (31)$$

where K is a constant.

According to the definition, torque, T, may also be written as

$$T = I\ d^2\theta/dt^2 \quad \ldots \ldots \ldots \ldots \ldots \ldots \quad (32)$$

where I is the moment of inertia and t is the time.

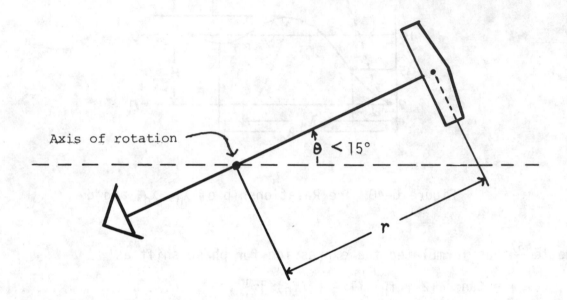

Figure 6-41   Dynamics of Windvane at Deflection Angle Below 15°

As the wind force acts to decrease θ, Eq. (31) is more properly written as T = -Kθ. From this and Eq. (32), without considering damping effects, we have

$$T = -K\theta = I\ \frac{d^2\theta}{dt^2}, \text{ i.e.,}$$

$$I\ \frac{d^2\theta}{dt^2} + K\theta = 0 \quad \ldots \ldots \ldots \ldots \ldots \ldots \quad (33)$$

However, the damping effect or damping torque, T_d, (i.e., the reduction of oscillation of the vane by increasing the speed of the wind) is directly proportional to the product of the angular speed of the rotating vane, dθ/dt, and the arm of the torque, r.   Thus

$$T_d = ar\ \frac{d\theta}{dt}$$

where a is taken as a constant of proportionality. Let the constant, ar, be defined as F, then

$$T_d = F\ \frac{d\theta}{dt} \quad \ldots \ldots \ldots \ldots \ldots \ldots \quad (34)$$

where F is the damping factor. The term F dθ/dt, or the damping effect, should be added to Eq. (33). We then have

$$I \frac{d^2\theta}{dt^2} + F \frac{d\theta}{dt} + K\theta = 0$$

or

$$\frac{d^2\theta}{dt^2} + \frac{F}{I} \frac{d\theta}{dt} + \frac{K}{I}\theta = 0 \quad \dots\dots\dots\dots\dots\dots\dots\dots\dots (35)$$

Eq. (35) is known as the Force Equation, and is identical with the standard differential equation

$$\frac{d^2\theta}{dt^2} + 2\alpha \frac{d\theta}{dt} + \beta^2\theta = 0$$

Eq. (35) has the roots

$$m = -\alpha \pm \sqrt{\alpha^2 - \beta^2}$$

The solutions are $\alpha = D/2I$ (where D = distance constant) and $\beta = \sqrt{K/I}$.

Thus, three damping conditions exist:

a. When $\alpha < \beta$, the roots are imaginary; the motion is underdamped and is a simple harmonic motion.

b. When $\alpha > \beta$, the roots are real and unequal; the motion is overdamped and is exponential.

c. When $\alpha = \beta$, the roots are real and equal; the motion is critically damped and there is no overshooting.

(3) Transmitting and Recording Systems. As wind direction usually fluctuates, most windvanes are connected to continuous recording systems instead of indicating systems such as directional meters. Four transmitting and recording systems are used for windvanes: mechanical, electrical, electronic, and photographic.

The mechanical system connects the revolving spindle of the vane directly to either the recording pen or the drum. The former represents the early type of windvane transmitting system typified by the pressure tube anemograph (see Figure 6-33). The directional rod of the pressure tube anemograph is connected directly to the recording pen, which is in contact with the drum chart. The latter type has greater flexibility in usage and has an improved mechanical system. The pen is controlled by a train of gears and a clock device, allowing recordings to be made horizontally and vertically. The main drawbacks of a mechanical system are the large frictional force in the system and the lack of telemetering capability (i.e., the recorder has to be placed within close proximity to the vane).

The electrical system records wind directions by using a ring-shaped resistance coil or by relying on a self-synchronous motor to vary the electric current. The

186

former system employs a spindle of the vane which rotates a brush inside a resistance ring of eight or more contacts. These contacts are connected in series with a battery. A meter gives a direct readout. The change of resistance, as a result of different contact positions, varies the current in the meter and thus is a measure of the wind direction. For telemetering wind direction signals for distances over a kilometer a self-synchronous motor is employed. The vertical shaft of the windvane is connected to the rotor of one of the position motors. Several of these position motors are interconnected and supplied with single-phase alternating current. A phase shift of one motor is transmitted simultaneously to the other motors. Thus, a torque produced by the wind on one shaft is transmitted to the directional transmitter and then to the recorder. The advantages of an electrical system are its telemetering capability, its accuracy in recording, and its fast response. Readouts of an electrical system can be as accurate as ±1° in direction, compared to ±5° accuracy of a mechanical system.

The electronic system utilizes photoelectric cells as position detectors. The transducer converts the interruptions of light into electrical charges. Nearly all the bivanes shown in Figure 6-34 use an electronic transmitting system. The photographic system, on the other hand, utilizes a suitable camera to photograph the dials at chosen intervals. Some systems use a combination of photographic and electronic procedures.

Some improved recording devices have charts marked to cover 540° instead of 360° This means that the chart covers all directions from N-E-S-W-N-E-S. When the pen reaches the outer edge, a cam allows it to return automatically to an inner position, reducing the need for a full-scale traverse. The benefit of this arrangement is noted especially when the wind direction oscillates around 360°. The swing of the pen across the entire chart produces sufficient inertia to slow down the time response, resulting in inaccurate readings. However, the overlap of 360° towards 540° does not result in full-scale excursions of the pen with oscillation around N, producing a better record.

New instruments for the measurement of wind speed and direction use either the features of vortices generated behind an obstacle or frequency changes of quartz crystals to heat advection. Production and prototypic vortex anemometers using the first principle above, and a prototypic microbalance incorporating wind velocity features using the second principle above, are now available and promising. Each system is described below.

(4) Vortex Anemometer. Wind speed can be measured as a function of the spacing between vortices generated by an obstruction to the wind flow. In 1969, J-Tec Associates[47] began utilizing an ultrasonic method of vortex sensing in their wind flow instruments. The theory of operation is based on the regular pattern of vortices produced downwind of an obstacle once the wind speed, v, reaches a minimum value. Spacing between the vortices is a well defined constant that is ~2.5 times the diameter of the obstruction.[48] The frequency of vortex formation, f_v,

[47] J-Tec Associates. 1979. VA-320 Vortex Anemometer. Cedar Rapids, Iowa. 19 p. Also: Beadle, D.W. 1978. Vortex sensing: An operational review. Preprint Volume: Fourth Symposium on Meteorological Observations and Instruments, April 10-14, 1978. Amer. Meteor. Soc., Boston. p. 563.

[48] Scorer, R.S. 1978. Environmental Aerodynamics. Halstead Press, New York. 488 p.

in Hz is found to be

$$f_\upsilon = \frac{Sv}{d} \quad \cdots\cdots\cdots\cdots\cdots\cdots\cdots\cdots\cdots\cdots \quad (36)$$

where S is the Strouhal number of 0.207; v, fluid speed upstream from the obstructing strut (ft s^{-1}); and d, width of the strut normal to the flow (ft). Frequency of vortex formation is directly proportional to wind speed, and is independent of temperature, pressure, or humidity. The threshold of vortex formation decreases with increasing width of the strut normal to the flow. Frequency of vortex passage is detected by passing an ultrasonic beam (f $\cong$ 150 kHz) perpendicular to the vortex path downwind of the strut through the transceiver mechanism (see Figure 6-42).

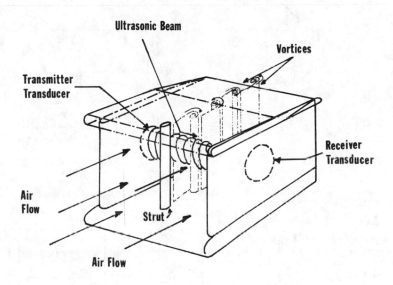

Figure 6-42   Vortex Speed Sensor

Since vortex formation occurs in alternating succession to either side of the strut, each pair of vortices causes one cycle of amplitude modulation of the ultrasonic carrier signal due to the beam scattering effect of the pairs of oppositely rotating vortices. Figure 6-43 depicts the linear correlation of frequency versus wind speed for a VA-320 anemometer. A block diagram of the vortex sensing electronics is shown in Figure 6-44. The vortex sensor can be mounted on a conventional vane to keep the sensor pointed into the wind, or mounted in mutually perpendicular horizontal tubes to derive a mean resultant wind vector.

The VA-300 series of vortex anemometers manufactured by J-Tec Associates employ the vortex sensor in a movable vane, where the latter provides wind direction information (see Figure 6-45). A cutaway view (Figure 6-46) of the VA-320 system highlights the essential details of vortex and directional sensors when mounted

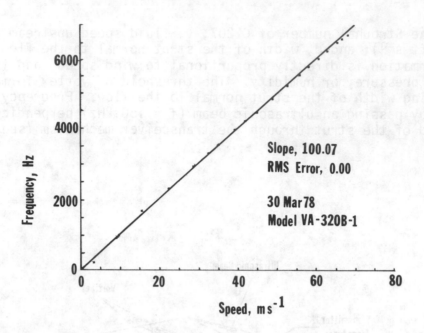

Slope, 100.07
RMS Error, 0.00

30 Mar 78
Model VA-320B-1

Figure 6-43   Representative Full Range Calibration
of VA-320 Vortex Anemometer

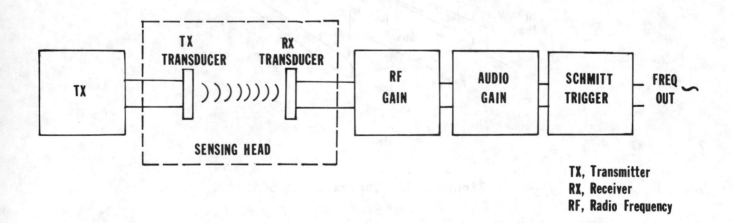

TX, Transmitter
RX, Receiver
RF, Radio Frequency

Figure 6-44   Vortex Anemometer Electronics

on a rotating vane. Output of the speed sensor is transferred through slip rings
in the base of the vane to a sensor (transmitter and receiver) and regulator
boards enclosed in the base. The frequency output of the vortex sensor is converted by the regulator into an analog signal between 0 and 5V DC. Performance
specifications of the VA-320 are given below. (Since this is a prototypic
instrument, specifications are subject to change.)

189

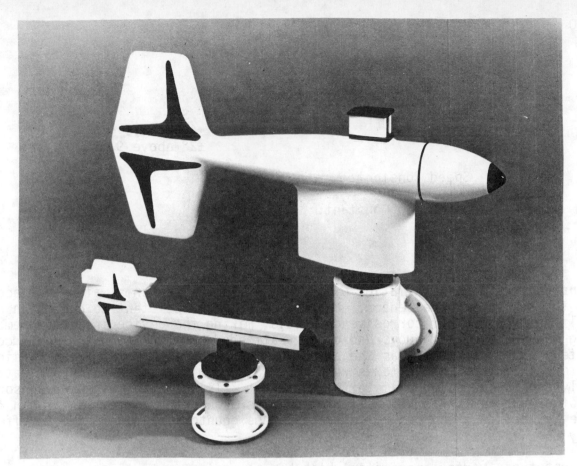

Figure 6-45 VA-300 and VA-320 Anemometers

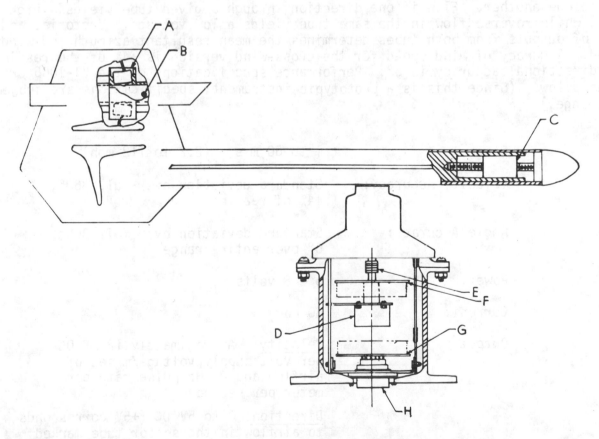

Figure 6-46 Cutaway Drawing of the VA-320

Range:	2-135 mph
Accuracy:	Speed ±1% F.S. Direction ±4° at 4.5 mph ±2° above 9 mph
Speed Constant:	6 mm
Direction Constant:	10
Power:	10-24V DC @ 30 ma

Construction and installation of the vortex sensor prevents water from accumulating on the surface of the sensor. Materials used are resistant to corrosion due to sea salt spray and other corrosive chemicals, making the device ideal for remote sensing on environmental buoys and other unattended platforms. Water that does accumulate on the transducer has the effect of reducing gain of the detected signal; water buildup on the vortex strut appears to result in speed underestimation, probably due to increased diameter of the strut. Icing of the vane and sensor is another problem. A small heater with coils around the vortex sensor can reduce ice buildup on these non-moving parts, thus counteracting the adverse effects in icy conditions.

A prototypic vortex anemometer which has no moving parts is ready for production. It is the cross-wind version of the vortex anemometer (Model VT-1005) also manufactured by J-Tec. Two tubes with vortex sensors in each are mounted perpendicular to one another. Flow in one direction through a given tube yields a high voltage, while reverse flow in the same tube yields a low voltage. Vectorial addition of outputs from both tubes determines the mean resultant azimuth and wind speed. Accuracy of wind speed for the cross-wind version is ±4% of the reading, and directional accuracy is 5°. Performance specifications of the VT-1005 are given below. (Since this is a prototypic instrument, specifications are subject to change.)

Range:	1 to 50 m s^{-1} (2.2 to 112 mph)
Velocity Accuracy:	Standard deviation over full 360°, ±4% of reading
Angle Accuracy:	Standard deviation over full 360°, 6° over entire range
Power:	6 - 8 volts
Current:	50 ma
Output:	Velocity: Approximately 12 mV DC per volt supply voltage/m/sec of airflow and 86 Hz pulse rate per meter per second Direction: 0 to 5V DC (+5V corresponds to airflow in the sensor tube marked North or East)

(5) <u>Quartz Crystal Microbalance</u>. A prototypic instrument, designed to measure wind speed and direction, humidity, solar irradiance, and particulate deposition, was designed and patented by James B. Stephens and Eric G. Laue of the Jet Propulsion Laboratory in 1975.[49] Specially cut quartz crystals (Y cut) are arranged on a flat sensing surface. The frequency changes of the crystals when cooled and heated are used to deduce wind speed and direction. In addition to the crystals and attendant circuitry for the wind velocity functions, the microbalance has other crystal configurations and circuits for measuring air humidity (paired hydrophillic and hydrophobic coated AT cut quartz crystals), solar irradiance (two Y cut crystals, one blackened and exposed to incoming radiation and the other shielded at ambient air temperature), and particulate deposition (paired moisture crystals referenced to the hydrophobic coated crystal). Refer to Figure 6-47 for schematic diagram of the microbalance.

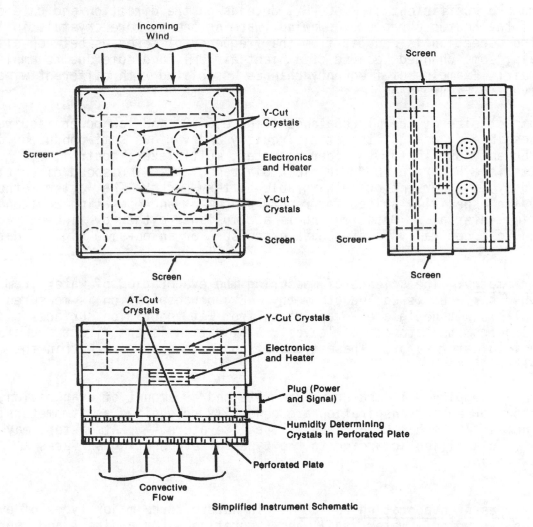

Figure 6-47 Simplified Microbalance Schematic

[49] Stephens, J.B. and E.G. Laue. 1975. Quartz crystal microbalances to measure wind velocity and air humidity. NASA Tech Brief 75-10124. NASA, Pasadena, Calif. 4 p. Stephens, J.B. and E.G. Laue. 1975. Quartz crystal microbalances to measure wind velocity and air humidity. JPL Invention Report 30-3112/ NPO-13462. Jet Propulsion Laboratory, Pasadena, Calif. 14 p. Stephens, J.B. and E.G. Laue. 1975: Specifications of wind sensor. Patent Application, NPO Case No. NPO-13462, TB 75-10124. National Patent Office, Washington, D.C. 21 p.

Wind speed and direction are determined by the frequency changes of a crystal measured with respect to its base frequency (established at wind temperature) and the frequency changes between crystals, respectively. Wind speed is determined by comparing the frequency of a crystal at its base or wind temperature and its frequency change after heating the crystal with a predetermined amount of current. The temperature increase of the crystals produces a frequency shift which is proportional to wind speed. Wind direction is determined by the spatial comparison of frequency changes of four crystals arranged in a square around a centrally located heater. The base frequency of each crystal, each of which differs, is determined by an oscillating circuit activated prior to the use of the heater. The crystal frequencies due to ambient wind flow are summed for all crystals to derive wind temperature. The heater, when activated, generates $\sim$ 0.1 W of energy, sufficient to produce up to a 0.5C temperature change in the quartz crystals. The rate of temperature change in the crystals, each having a positive temperature coefficient of $\sim$ 400 Hz, depends on the direction and rate of movement of the heated air plume downwind, warming one or more crystals in its path. The wind direction is deduced from the frequency differences between all four crystals, some changed because of ambient air only and some due to ambient plus plume air. Examples of frequency changes associated with different wind directions are given in Figure 6-48.

The compact size of the microbalance (Figure 6-49) and low power requirements are made possible by using integrated circuitry and various semi-conductor devices. The frequency oscillator, divider, counter, multiplexer, control, and computing circuits (see Figure 6-50 for a block diagram of the wind speed circuit) can be powered by line current or rechargeable battery packs. The latter method makes the device especially suited for remote placement and operation unattended for long time intervals. Data logging is accomplished either by line or radio transmission to a centrally located data processor, or an on-site digital data logger.

6.5 _Atmometry_. The science of measuring the evaporation of water from solid or liquid surfaces is called atmometry. Evapotranspiration is measured for green plants; it designates evaporation from wet vegetation surfaces and transpiration from the stomata of leaves. Sublimation is the direct transformation of ice into water vapor. These types of phase changes come within the scope of atmometry.

Instruments employed in measuring the rate and/or amount of evaporation, transpiration, or evapotranspiration are generally recognized as atmometers, although other nomenclature has been cited in the literature.[50/] Atmometers may be conveniently classified according to the types of evaporating surfaces and the methods of measurement.

6.5.1 _Types of Evaporating Surfaces_. In nature three major types of evaporating surfaces are noted - water, soil, and vegetation. For a _direct_ and _representative_

---

[50] The various names appearing in the literature are: atmidometer, evaporimeter, evaporometer, evaporation pan, gage or tank, lysimeter, evapotranspirometer, evapotron, phytometer, and potometer. The term atmometer was coined by Sir John Leslie in 1813. It is the oldest name and is generally accepted as the general name for this kind of instrument.

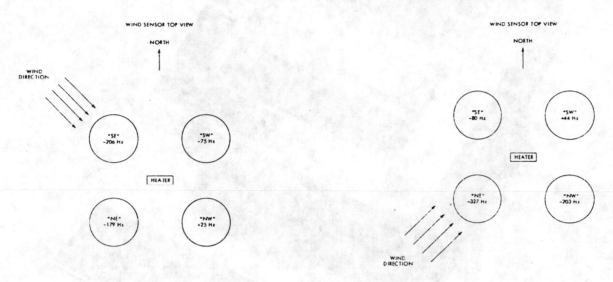

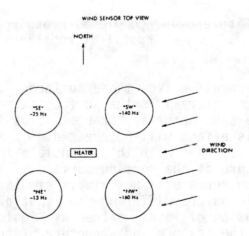

Figure 6-48 Illustrations of Frequency
Dependence on Wind Direction

Figure 6-49   Completed Prototype of Quartz
Crystal Microbalance

measurement of evaporation from these surfaces, the sensor of an atmometer should be made of the same material as the surface it measures. In other words, it should possess the same natural characteristics as the surface in its surroundings; otherwise, an oasis effect will be created.[51]   Most of the presently available sensors have been designed with this principle in mind. However, when the amount of available moisture of the sensing surface is taken into consideration, the natural evaporating surface is <u>not</u> always simulated. This results from the water supply mechanism in nature differing from that of an instrument. For example, the vertical diffusion of water, either as liquid or vapor, through a layer of soil depends upon the texture and structure of the soil profile. As the natural soil profile cannot be perfectly simulated, it is impossible for a designer to replicate it exactly in order to bring the same amount of moisture to the sensing surface.

---

[51] The difference between the physical and biological properties of the sensing surface of the instrument and those of the surrounding field affect the evapotranspirating rate of the sensor. This is known as an island effect or oasis effect.

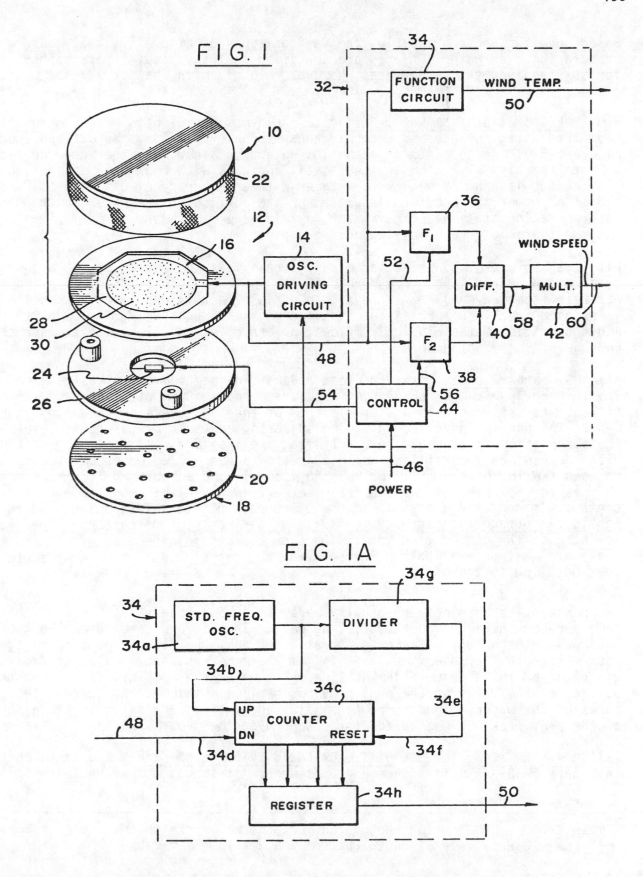

FIG. 1

FIG. 1A

Figure 6-50   Block Diagram of Wind Speed Circuit

196

Based upon the above principle, the various types of atmometers that measure the evaporation over water surfaces are specified in the following paragraphs. Those for soil and vegetative surfaces, however, are given in Chapter 12.

6.5.2 Evaporation from Water Surfaces. An atmometer used in measuring evaporation from free water surfaces is commonly known as an evaporating tank or pan (see Figure 6-51). It consists of a shallow water container with the water surface open to the air as the sensor, and a stilling well with hook gage as the indicator. The readout is usually made by a vernier with an accuracy of 0.1 inches (0.25 mm). A pair of water thermometers is placed on the surface of the water in the pan. In an evaporation station an anemometer, a rain gage, and a sheltered maximum and minimum thermometer are located near the pan.

With the above provisions, a large pan gives a fair indication of the amount of evaporation over a lake or reservoir if the instrument is floating in a body of water (a floating pan).[52] However, an oasis effect is unavoidable when the pan is exposed over land. Brief descriptions of the US Weather Bureau standard pan and the British Meteorological Office tank are given to illustrate the structure and use of such instruments.

(1) US Weather Bureau Standard Class-A Pan. The pan is made of galvanized iron,[53] cylindrical in shape, 10 inches deep and 4 ft in diameter, and is kept filled with water to within 2 or 3 inches of the top. This is shown in Figure 6-51. The pan is installed on wooden supports, and must be carefully leveled. The observation site should be nearly flat, well sodded, and free from obstructions. With the pan are the stilling well to provide an unruffled water surface, a level to keep the top of the stilling well horizontal, and a hook gage to measure the changes in the level of the water in the well and thus the amount of water evaporated. In operation, a small amount of copper sulfate solution is added to the pan to discourage algal growth. Oil films, sediments, and scum should be removed to increase the evaporation rate. There should be an inspection of rust spots and leaks at least once a month. The seams of the pan should be strengthened to prevent buckling to the bottom.

Two consecutive readings are usually made at 24-hour intervals by use of the hook gage vernier installed on the stilling well. The difference between the two readings gives the amount of water evaporated per day. Readings from a rain gage, an anemometer, sheltered maximum and minimum thermometers, and a water thermometer are required in an evaporation station. The water thermometer, floated on the surface of the water in the pan, provides maximum, minimum, and current temperatures of the water. The anemometer installed next to, and slightly above the rim of the pan, provides the speed of the wind over the pan.

(2) British Meteorological Office Standard Evaporation Tank. The tank, as shown in Figure 6-52, is 6 ft square and 2 ft deep. It is made of wrought-iron plates

---

[52] This is done by floating the pan on pontoons to ensure that the temperature of the water is very close to that of the lake.

[53] An alloy similar to monel metal is preferable in areas where the water contains large amounts of dissolved corrosive substances.

Still Well   Hook Gage

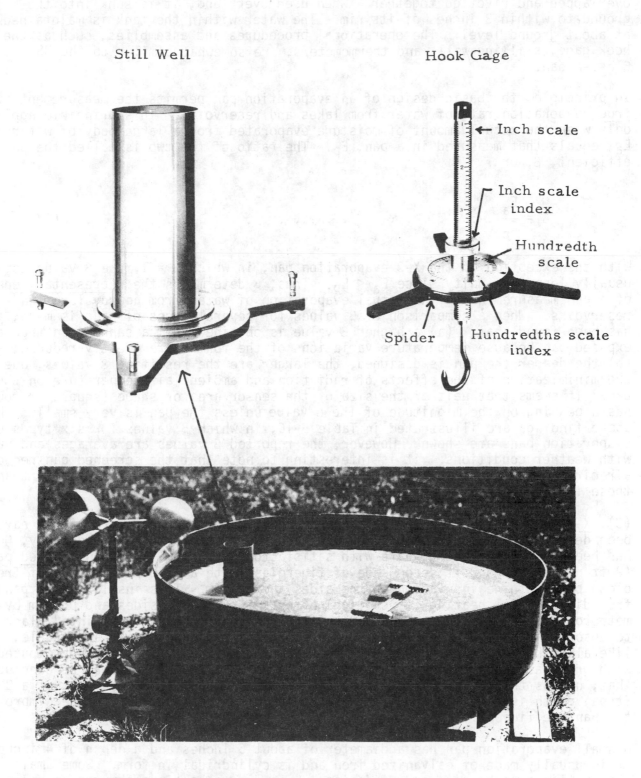

Figure 6-51 US National Weather Bureau Class-A Evaporation Pan

198

overlapped and rivetted together. When used over land, it is sunk into the ground to within 3 inches of its rim. The water within the tank is maintained at about ground level. The operational procedures and assemblies, such as the hook gage, stilling well, and thermometers, are somewhat similar to the US Class-A pan.

In principle, the basic design of an evaporation pan permits the measurement of the true evaporation rate of water from lakes and reservoirs. This principle applies only when the actual amount of moisture evaporated from a large body of water, E_a, equals that measured in a pan, E_p. The ratio of the two is called the pan coefficient, β, or

$$\beta = \frac{E_a}{E_p} \quad \ldots\ldots\ldots\ldots\ldots\ldots\ldots\ldots\ldots\ldots\ldots\ldots\ldots\ldots \quad (37)$$

With the exception of the X-3 evaporation pan, in which $\beta = 1$, the β value is usually less than unity since $E_a < E_p$. Thus, β determines the representativeness of a pan measurement to the actual evaporation of water from nearby lakes and reservoirs. When a comparison of β values for several types of pans is made, it has been found that: (a) a higher β value is obtained from a sunken pan than an exposed pan because temperature variations of the former are greatly reduced; and (b) the deeper the pan is designed, the larger are the resulting β values, due to the minimization of the effects of radiation and ambient air temperature on the pan. It seems that neither the size of the sensor area or shape (square or round) has a bearing on the magnitude of the β value unless the pan is very small. The above findings are illustrated in Table 6-16, in which β values for six types of evaporation pans are shown. However, the reported β values are averages and vary with weather conditions. It is interesting to note that the screened pan reduces air circulation and insulation, making the pan coefficient nearer to unity than those for unscreened pans.

(3) X-3 Evaporation Pan. So far standard pans which are well established have been described. The updated X-3 evaporation pan of the Office of Hydrology, NOAA, has been tested in recent years with satisfactory results. It is an exposed pan (2 ft diameter and 2 ft deep) made of fiberglass with a 3 inch thickness of freon-blown polyurethane insulation on the sides and bottom. This insulates the pan from differential heating (or cooling) effects through its sides and bottom by meteorological variables. In addition, the inside of the pan is painted black to absorb radiation. The resulting pan coefficient is near unity and stable, unlike all other existing pans. With a stable β value, E_a values may be computed with great accuracy. In view of test results of the X-3 pan, it may be concluded that, unless the walls are well insulated, a sunken pan is necessary; that a 2 ft pan depth is required; and that blackening of the inside wall further improves the pan coefficient.

A small evaporation pan has a diameter of about 5 inches and a depth of 4 inches. It is usually made of galvanized iron and is cylindrical in form. Some small evaporation pans have recording devices which draw a continuous curve on paper strip charts in response to the variation of either the weight or the water level with time. Because they are small and exposed, they are usually sensitive to temperature, radiation, and wind variations, and therefore are not as reliable as,

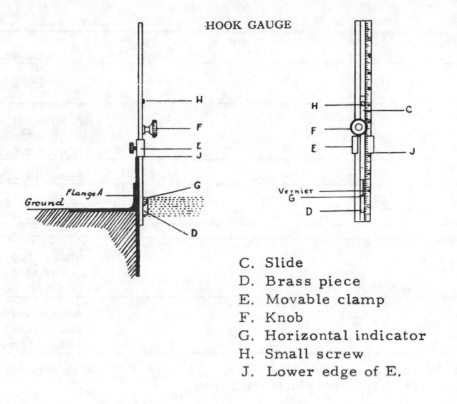

HOOK GAUGE

C. Slide
D. Brass piece
E. Movable clamp
F. Knob
G. Horizontal indicator
H. Small screw
J. Lower edge of E.

EVAPORATION TANK

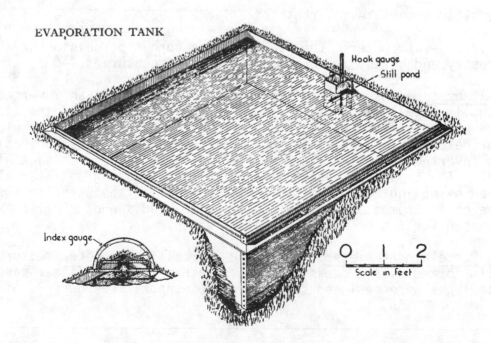

Figure 6-52 British Meteorological Office Evaporation Tank

Table 6-16 A Comparison of Pan Coefficients

Type of Pan	Size (ft)	Depth (ft)	Pan Coefficient (β)	Remarks
U.S. Class A Pan	4 dia.	0.8	0.7	An exposed pan.
Colorado Sunken Pan	3 x 3	1.5	0.8	A sunken pan made of unpainted galvanized iron.
BPI Pan*	6 dia.	2.0	0.9	Same as above.
British M.O. Stn. Pan	6 x 6	2.0	0.9 (?)	A sunken pan.
Screened Pan	2 dia.	3.0	0.9 to 1.0	A sunken pan whose top is covered by 1/4" mesh screening.
X-3 Pan	2 dia.	2.0	~ 1.0	An exposed pan with excellent insulation.

*The initials BPI stand for Bureau of Plant Industry, US Department of Agriculture, which first introduced this pan.

nor have results comparable to, those of a large pan.

Two other types of atmometers, the Piché and the porous porcelain, have been widely used in forestry and botany for over one and a half centuries.[54]

(4) Piché Atmometer. As shown in Figure 6-53, the Piché is an inverted graduated burette, usually labeled in centimeters, with one end closed and the other open. It is filled with distilled water and then covered with a larger circular piece of filter paper held in place by a disc and collar arrangement. In operation the instrument is inverted, allowing the distilled water to come in contact with the filter paper. The amount of evaporation which occurs during an interval of time is determined by noting the change in miniscus level of the water in the burette. As this type of instrument is highly sensitive to wind speed, similar exposures should be chosen for two or more observation sites.

(5) Porous Porcelain Atmometer. The porous porcelain atmometer measures the moisture film formed over a ceramic surface, rather than the filter paper of the Piché. Both types were designed to measure the evaporative power[55] of the

---

[54] Porous procelain spheres, cylinders or plates, for example, have been used by various workers since the time of Sir John Leslie in 1813.

[55] Evaporative power is a measure of the saturation deficiency of water vapor, d, in the atmosphere. It may be expressed by $d = e_s - e$. Since $r = e/e_s$, then $d = e_s (1-r)$, where e_s is the saturated water vapor pressure; e is the actual water vapor pressure; and r is the relative humidity. If the r value is expressed in percent, $d = e_s (100-r)$.

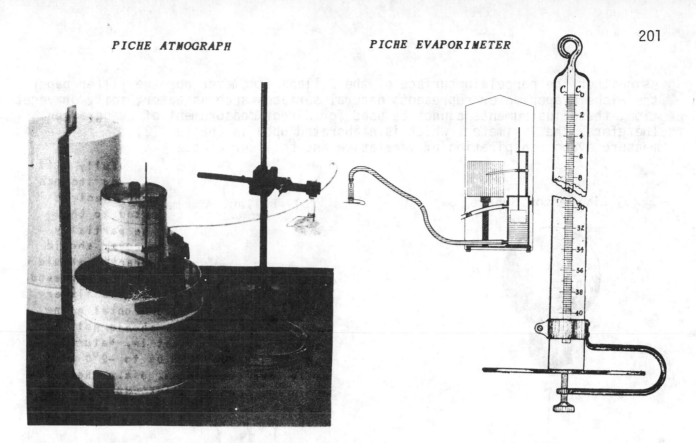

Figure 6-53 Piché Atmometer

atmosphere rather than the evaporation of a natural water surface. They are commonly known as evaporimeters and serve as secondary instruments for measuring the relative value of evaporation.

Porous porcelain atmometers are distinguished by the shape or color of their sensors. Some are spherical or flat-topped, black or white. As shown in Figure 6-54$_a$, the Livingston sphere is a black or white, round-shaped, ceramic, porous cup sensor with a diameter of about 1.5 inches. The neck of the porous cup is inserted through a rubber stopper in the top of a graduated cylinder. A 1/16-inch glass tube extends from inside the sphere, through the rubber stopper, and close to the bottom of the cylinder. The entire apparatus is filled with distilled water which is then siphoned from the cylinder reservoir to the surface of the sphere. As the water evaporates, the difference in two consecutive water-level readings observed in the graduated cylinder is a measure of the evaporative rate in a specific time interval, usually 24 hours.

The Bellani atmometer (see Figure 6-54$_b$) has a 2.5 inch, flat, porous sensor which is usually black, although a white one has been used occasionally. This sensor, known as the Bellani plate, is constructed and operated under the same mechanism as the Livingston sphere. The blackened flat top allows the heating of the sun to be differentiated at different solar angles during the course of a day or seasons of a year. Therefore, this flat surface gives a better representation of the approximately horizontal earth than a spherical sensor to which the sun's rays are always perpendicular. It has been possible to telemeter and record the Bellani readings with the aid of a simple electric device.

As neither the porcelain surface of the Bellani atmometer nor the filter paper of the Piché evaporimeter represents natural surfaces such as water, soil, or vegetation, these instruments cannot be used for direct measurement of evaporation. Therefore, the lysimeter, which is elaborated upon in Chapter 12, is employed to measure evapotranspiration of vegetative and soil surfaces.

(a) Livingston Sphere   (b) Bellani Atmometer

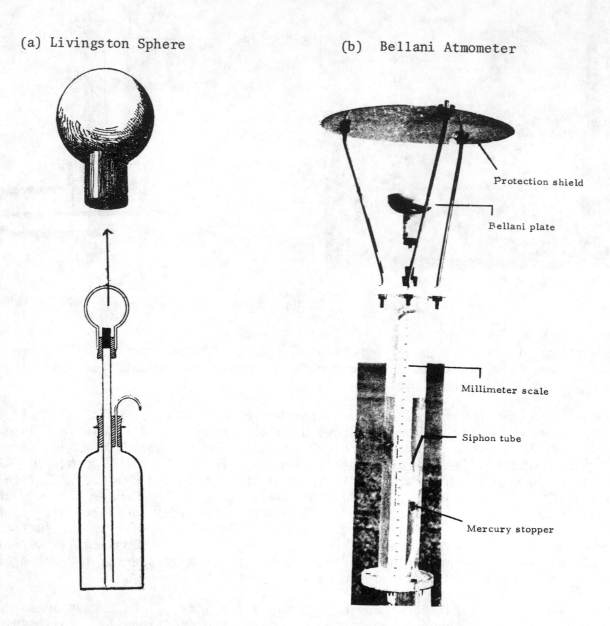

Figure 6-54 Porous Porcelain Atmometers

6.6 <u>Hygrometry</u>. Hygrometry is the science of moisture measurements of evaporation and evapotranspiration (atmometry), condensation and precipitation (hyetometry), and the water vapor content of the atmosphere. Atmometry has been discussed in the previous section and hyetometry is explained in the next section; the discussion here is restricted to the measurement of water vapor content, i.e., humidity, in the free atmosphere near the surface of the earth. Measurements of humidity in the upper air and the soil atmosphere are dealt with in Chapters 7 and 13, respectively.

As is the case with evaporation and precipitation, the techniques for sampling humidity pose the problem of representativeness. Less than 10% of the humidity within an airmass can be detected from a fixed sampling station. An instrument usually measures only its own ambient humidity and not that of the surrounding airmass as a whole. Also, instruments capable of measuring within 1% accuracy are rare. In addition, the great variability of humidity with space and time further complicates measurement. The variability of the saturation mixing ratio ranges from less than 0.001 to about 100 g kg^{-1} at surface atmospheric pressure and temperature conditions. This is to say that the water vapor content could vary more than 5 orders of magnitude in various locales simultaneously over the earth's surface. At the present time, no single instrument is capable of detecting such a wide range of variation. Because of these inherent difficulties, scientists have devoted much effort to hygrometry. As a result, diverse methods, instruments, and procedures have been established.

6.6.1 Classification of Humidity Instruments. Arnold Wexler,[56] of the National Bureau of Standards, has classified the major techniques of measuring humidity into six categories:

1. Removal of water vapor from the ambient air.

2. Addition of water vapor to the test gas.

3. Equilibrium sorption of water vapor by the sensor.

4. Measurement of physical properties of the ambient air.

5. Attainment of a vapor-liquid or vapor-solid equilibrium.

6. Use of chemical procedures.

Using the above six categories, Wexler has specified 24 methods by which 50 types of instruments are identified. A brief summary of these is given in Table 6-17.

Another method of classification is by the type of usage - laboratory, field, or both. The methods and procedures for laboratory instruments are usually slow, tedious, and complicated but their measurements are accurate and reliable, allowing them to be used as primary instruments. Some instruments in the first, second, and sixth categories of Wexler's classification belong to this class. The remaining are primarily field instruments, although some of them have been used in the laboratory. Field instruments are comparatively faster and easier to operate than the laboratory types but are generally less accurate and, therefore, are

---

[56] This is a 4 volume publication including (1) Principles and methods of measuring humidity in gases, (2) Applications, (3) Fundamentals and standards, and (4) Principles and methods of measuring moisture in liquids and solids. Wexler, A., ed. 1965. Humidity and Moisture. Reinhold Publishing Corp., New York.

Table 6-17 Classification of Humidity Instruments

Category	Method	Instrument[1]	Principle of Operation	Primary Measurement(s) Some Major References
1. Water Vapor Removal	a. Gravimetric, volumetric, or pressure	Absorption hygrometer, Chemical absorption hygrometer, Cellophane hygrometer, Paper hygrometer.	Weighing the moisture absorbed by desiccant; or measuring the change in volume of the air at constant pressure and temperature or the change in pressure of the air at constant volume and temperature after the removal of moisture by desiccant[2]	Mass of moisture; volume or pressure of air. (Wexler, 1961; Bongards, 1926a & 1926b; Schweiler, 1933; Mellanby, 1933.)
	b. Coulometric	Electrolytic hygrometer, Mixing ratio indicator.	Measuring electric current from electrolysis of absorbed water vapor.	Electric current. (Keidel, 1956 & 1959; MacCready & Lake, 1965; Goldsmith & Cox, 1967.)
	c. Pneumatic	Pneumatic bridge hygrometer.	Measuring pressure drop in one arm of bridge which contains desiccant to absorb moisture.	Differential air pressure. (Greenspan, 1965; Wildback et al., 1965.)
	d. Diffusion	Diffusion hygrometer.	Measuring pressure drop across semi-permeable membrane[2]	Differential air pressure. (Greinacher, 1944 & 1945.)
2. Water Vapor Addition	a. Gravimetric, volumetric, or pressure		Measuring mass, volume or pressure as in 1(a) by saturating the instrument system with water vapor. It is a direct measure of the saturation deficiency.	Mass, volume or pressure. (Bongards, 1926c.)
	b. Psychrometric	Mercury-in-glass psychrometer(e.g. sling or whirling and aspirated), Resistance psychrometer, Thermocouple psychrometer, Adiabatic psychrometer.	Measuring the change of water temperature of the wet bulb, thermistor or thermocouple due to evaporation.	Temperature, electric or emf.(Bongards, 1926d; Wexler, 1965a; Middleton & Spilhaus, 1953; Greenspan, 1968; Wentzel 1961.)
3. Equilibrium Sorption of Water	a. Electric	Electric hygrometer(e.g. Weaver Water Vapor Indicator, Dunmore Hygrometer, Gregory Hygrometer, Aluminum Oxide Hygrmeter, Cerium Titanate Element, Ceramic R.H. Sensor).	Measuring resistance of aqueous electrolytic solution, impervious solids, porous solids, and dimensionally variable materials.	Electric resistance or reactance. (Weaver & Riley, 1948; Dunmore, 1938; Mathews, 1965; Gregory & Rourke, 1952; White, 1954; Jones, 1967; Jason, 1965; Pope, 1955; Stine, 1965.)
	b. Mechanical	Hair hygrometer (Koppe Hygrometer, Hygroscope, Polymeter), Goldbeater's Skin element; Cellulosed Paper.	Measuring the elongation of rolled and chemically treated human hair; expansion of ox intestine; deformation of cellulosed paper.	Elongation or deformation. (Bongards, 1926e; Kobayashi, 1960a,b; Miller, 1965.)
	c. Weighing	Weighing hygrometer(Cellophane Hygrometer, Mahajan Optical Hygrometer).	Measuring changes in weight due to sorption, or desorption of moisture with increase or decrease of ambient R.H.	Mass or torsion. (Schweitzer, 1933; Mellanby, 1933; Mahajan, 1941, 1944a,b.)
	d. Colorimetric	Solomon's Paper, cobalt salts.	Measuring the change in color of either Cobaltous chloride or bromide impregnated paper.	Color hange. (Swartz, 1933; Solomon, 1945.)
	e. Piezoelectric	Piezoelectric Sorption Hygrometer.	Measuring frequency change of quartz crystal covered with hygroscopic film.	Electric frequency. (King, 1965.)
	f. Heat of Sorption	Thermal Hygrometer.	Measuring heat produced through exothermic or endothermic exchange of heat by hydrophilic material.	Temperature change. (Tyndall & Chattock, 1921; Flumfelt, 1965.)

Table 6-17, continued

4. Physical Properties of Air	a. Absorptive	Spectroscopic Hygrometer(IR Hygrometer), UV Hygrometer).	Measuring attenuation of IR or UV radiation.	IR or UV radiation. (Johns, 1965; Lück, 1964; Randall, et al., 1965; Wood, 1965.)
	b. Refractive	Microwave Hygrometer, Dielectric Hygrometer, Interferometer.	Measuring refractive index at microwave frequency, optical frequency, or radio.	Microwave, optical or radio frequency. (Magee & Crain, 1958; Sargent, 1959; McGavin & Vetter, 1965; Webb & Neugebauer, 1954.)
	c. Thermal Conductive	Thermal Conductivity Bridge, Absolute Humidity Recorder.	Measuring thermal conductivity.	Dew-point or frost-point temperature of sensor. (Daynes, 1933; Cherry, 1948, 1965; Rosecrans, 1930.)
5. Condensation or Solidification of Water	a. Dew- and Frost-point	Dew-point Hygrometer, Frost-point Hygrometer.	Measuring temperature at which dew and frost form.	Temperatures. (Kobayashi, 1960_b; Sonntag, 1967_a; Wexler, 1965_b; Wylie et al., 1965_b.)
	b. Cloud Chamber	Cloud Chamber Hygrometer.	Measuring pressure ratio required to produce fog after adiabatic expansion.	Pressure and temperature. (Griffiths, 1933; Sonntag, 1967_b.)
	c. Saturated Salt Solution	Dewcel Ionic Crystal Element.	Measuring equilibrium temperature of saturated salt solution.	Temperature. (Conover, 1950; Hedin & Trofimenkoff, 1965; Hicks, 1947; Nelson & Amdur, 1965.)
6. Chemical Reaction of Reagent & Water	a. Karl Fischer		Methanol extraction of water vapor and titration with reagent.	Color change. (Mitchell & Smith, 1948; Thuman & Robinson, 1954.)
	b. Water Vapor Conversion		Reacting water vapor with calcium carbide to obtain acetylene.	Temperature. (Wylie, Caw & Bryant, 1965.)

1/ The listing is not exhaustive, but includes instruments that are commonly mentioned in the literature. If this column is blank, those instruments do not have an accepted nomenclature.

2/ A drying agent such as $CaCl_2$, H_2SO_4, P_2O_5, wool, glass or Ca_2C as well as cryogenic fluids such as liquid H_2 and He.

3/ A porous material such as clay, marble, gypsum, cellophane, alabaster or gelatine.

All references listed in the last column of the above table may be found either in the footnotes in this section or the Appendix of this book.

employed as _secondary instruments_.

Similar to other types of instruments, those which measure humidity are grouped according to their method of readout - recording or non-recording (i.e., indicating instrument). Recording humidity instruments are conveniently classified as fast, medium, or slow response types. Aside from the considerations of speed of response and method of readout of an instrument, the following are essential features: performance characteristics, requirements, and responses to various environmental impacts.[57] All of these serve as the bases of discussion for each individual instrument in this section.

---

[57] This designates such impacts as fog, mist, rain, snow, ice, water vapor saturation, ambient temperature and pressure, radiation, winds and air quality (including salt content, dust, oil, and other corrosive gases in the ambient air).

Although there are many types of humidity instruments, this section deals only with psychrometry and hygrometry, particularly hair element and dewpoint temperature measurements. A few instruments are discussed in detail for each of the above methods. For a comprehensive description of other methods and instruments, the reader should consult Humidity and Moisture.[58]

6.6.2 Psychrometer. A psychrometer is an instrument which measures atmospheric humidity by means of a pair of wet and dry temperature sensors. Its simplest form consists of two identical thermometers, one of which (dry bulb) is an ordinary glass thermometer, and the other (wet bulb) is a similar thermometer whose bulb is covered with a jacket of clean muslin which is saturated with distilled water prior to an observation. The temperature indicated by the dry bulb is the air temperature, T; that by the wet bulb is the wet-bulb temperature, T_w. The difference between the two temperatures is known as a wet-bulb depression which is a measure of atmospheric humidity. The humidity values in terms of relative humidity, absolute humidity (i.e., water vapor density), or dewpoint temperature can be obtained from a psychrometric chart or a table.[59] Figure 6-55 is a typical psychrometric chart.

(1) Theory. The wet-bulb depression varies directly with the difference between the atmospheric vapor pressure, e, and the saturation vapor pressure, e_w, at temperature, T_w. The proportionality constant, A, or psychrometric constant, is a function of air ventilation, temperature, and density. In 1886, Ferrel examined a large number of observations and concluded that the psychrometric constant $A = 0.000660 (1 + 0.00115 T_w)$ where T_w is in degrees C. In psychrometry,[60] scientists have determined that the psychrometric constant is

$$A = \frac{e_w - e}{p(T-T_w)} \quad \cdots\cdots\cdots\cdots\cdots\cdots\cdots\cdots \quad (38)$$

where p is the atmospheric pressure in mb and T is the dry-bulb temperature, in degrees C. In construction of psychrometric tables, slide rules, and nomograms,[61] p is taken as 1000 mb.

---

[58] See Footnote 56.

[59] See List, R.J. 1958. Smithsonian Meteorological Tables. Smithsonian Institution, Washington, D.C.; Marvin, C.F. 1941. Psychrometric Tables, No. 235. U.S. Weather Bureau, Washington, D.C.; or British Meteorological Office. 1964. Hygrometric Tables, Part I and II. HMSO, London. For a psychrometric chart, see Albright, J.G. 1939. Physical Meteorology.

[60] The science and techniques associated with psychrometric measurements. Psychrometry is the most popular subject of study in hygrometry.

[61] Among the many available nomograms and slide rules are: Brooks, D.B. 1935. Psychrometric Charts for High and Low Pressures, National Bureau Standards Misc. Publ. M 146; Weather Bureau Psychrometric Calculator, Nos. 1183 and 1184. Bilham, E.G. 1950. New version of the humidity slide rule. The Meteorological Magazine 79(933); and Nottage, H.B. 1950. A proposed psychrometric chart. ASHVE J., Section on Heating, Piping and Air Conditioning.

Spilhaus[62] made a simple derivation of the psychrometric constant by letting m be the mass of air passing the wet bulb in unit time such that m_W is the mass of air which becomes saturated. Thus, the heat given up by m grams of air to the wet bulb is equal to the heat the wet bulb absorbs to evaporate enough water to saturate m_W grams of air. When a steady state is reached

$$m \ c_p \ (T-T_W) \ p = (e_W-e) \ r \ m_W \ L \ . \ . \ . \ . \ . \ . \ . \ . \ . \ . \ . \ . \ . \ . \ (39)$$

where c_p is the specific heat of air at constant pressure (0.24 cal $g^{-1} \ C^{-1}$); r, the ratio of the density of water vapor to that of air (0.622); and L, the latent heat of vaporization of water (597.3 cal g^{-1} at OC).

By comparing Equations (38) and (39), the psychrometric constant is

$$A = \frac{m \ c_p}{m_W \ r \ L} \ . \ . \ . \ . \ . \ . \ . \ . \ . \ . \ . \ . \ . \ . \ . \ . \ . \ . \ . \ (40)$$

In wind tunnel tests, it has been found that the psychrometric constant decreases with increasing ventilation rate. With a rate between 4 and 10 m s^{-1}, the product of A and p will be 0.660 mb C^{-1}. This is the constant used in the calculation of most psychrometric tables.[63]

Somewhat similar to the response time of a dry-bulb thermometer, the wet-bulb response time, λ_W, may be expressed as

$$\lambda_W = \frac{C_W \ p}{m_W \ r \ L(Ap+\beta)} \ . \ . \ . \ . \ . \ . \ . \ . \ . \ . \ . \ . \ . \ . \ . \ (41)$$

where β is approximately equal to $(de_W)/(dT_W)$ at temperature T_W, and C_W is the thermal capacity of the wet bulb including water and muslin. For the same size and ventilation conditions, wet-bulb response time, λ_W, normally is smaller than dry-bulb, λ_d, with the exception of the thermocouple, as shown in Table 6-18.

Table 6-18 Time Response of Some Psychrometers

Thermal Sensor	Dimension	Ventilation	λ_d	λ_W
Mercury-in-glass	Spherical: 1.12 cm in diameter	4.5 m s^{-1}	56 s	52 s
Mercury-in-steel	Cylindrical: 1.90 cm in diameter, 14.5 cm in length	4.5 m s^{-1}	280 s	120 s
Aspirated thermo-couple	4 thermocouple junctions	11 m s^{-1}	1.8 s	2.7 s

[62]Spilhaus, A.F. 1936. Trans. Roy. Soc., S. Africa 24: 185-202.

[63]Resolution 145, International Meteorological Organization, 12th Conference of Directors (Wash., D.C., 1947) recommends the air to be drawn past the bulbs at a rate not less than 4 m s^{-1} and not greater than 10 m s^{-1}, if the thermometers are of the types ordinarily used at meteorological stations.

208

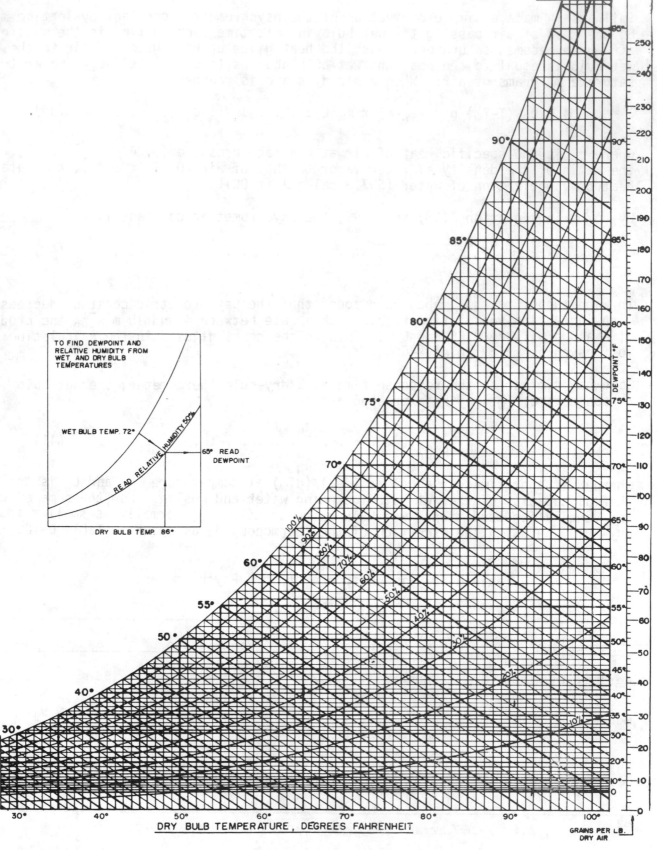

Figure 6-55 Psychrometric Chart

In addition to the common sources of error in a dry-bulb or an ordinary thermometer, wet-bulb thermometers are subject to errors caused by improper ventilation, overly thick muslin, dirty muslin, and encrusted muslin. All of these errors tend to result in a high, erroneous humidity reading. Improper ventilation is the most serious single error affecting the accuracy of psychrometric measurements. While the remaining errors are much easier to overcome, the elimination of ventilation error is a major task in psychrometry. A ventilation rate of 3 to 4 m s^{-1} has been found ideal for most purposes. Psychrometric measurements may suffer from insufficient evaporative cooling of the wet bulb in an unventilated laboratory, and overventilation by a wind with speed greater than 10 m s^{-1} in the field. Therefore, improper ventilation must be corrected or controlled. The influence of ventilation on the accuracy of a psychrometer is considered.

Two types of ventilation, natural and artificial, are available. With natural ventilation, the psychrometer is usually placed in a weather shelter without a ventilating device; ventilation then depends upon the natural air flow to evaporate moisture from the wet bulb. Such a screen psychrometer is apt to give a high humidity reading when the weather is calm. Therefore, a ventilation device such as a fan is required by most national weather services throughout the world. Another natural ventilating type is the sling (whirling) psychrometer shown in Figure 6-56.

U. S. National Weather
Service Type

British Meteorological Office Type

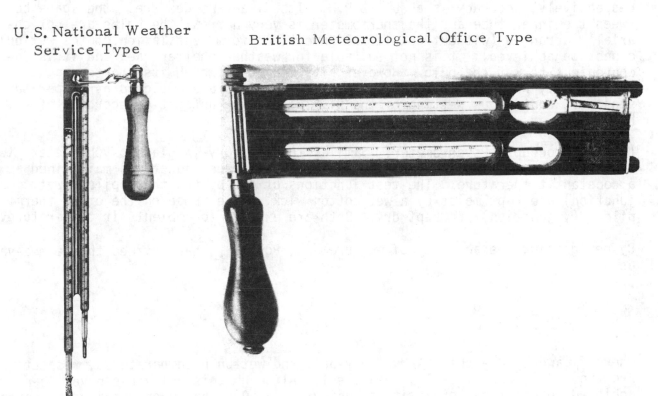

Figure 6-56 Sling Psychrometers

Two thermometers are mounted in a frame which can be rotated by means of a handle. A whirling rate of 2 revolutions per s for a 60 s duration is required to obtain sufficient ventilation. With some practice, a sling psychrometer can be stopped smoothly and in a proper position for reading. Two or more whirlings should be made until two consecutive readings of either dry bulb or wet bulb agree to within 0.2F. For an atmospheric temperature range of 32F to 104F and a relative humidity (R.H.) range of 10 to 100%, the accuracy of a sling psychrometer is about 3 to 4% R.H. In either rainy or freezing weather conditions, accuracy is greatly diminished and an error of 10 to 15% R.H. is to be expected. Accuracy will also be reduced for ambient humidity below 10% R.H.

Artificial ventilation provides greater accuracy. A larger ambient humidity range is achieved by using either a mechanical (clockwork) or an electrical air blower. An instrument equipped with forced ventilation is generally known as an aspirated psychrometer. One such aspirated type, the Assmann psychrometer shown in Figure 6-10, has a ventilation rate of 2.5 to 5.0 m s^{-1}, depending on the manufacturer. It consists of two highly polished, often tubular, concentric metal shields which have forced air flowing in the space between them as well as the space immediately surrounding the thermometer. The advantages of the Assmann psychrometer are the minimization of errors due to radiative or conductive heating by the sun, and the avoidance of the evaporative cooling of the dry bulb due to rain or snow when it is used in the field. Also, readings can be taken within one minute. Within the atmospheric temperature and humidity ranges of 41 to 149F and 5 to 100% R.H., respectively, accuracy is about 1% R.H. Its disadvantages are: the space between the inner tube and the thermometer is very narrow; the bulbs are not coaxially arranged (without coaxial alignment, uniform ventilation of the wet bulb cannot be achieved); it is not suitable for weather shelter use; and it is inconvenient to wet the bulb. However, its merits outweigh its shortcomings, particularly when small diameter thermocouple wires are employed as temperature sensors. Continuous wetting of the wick allows for automatic recording of humidity.

(2) <u>Thermocouple Psychrometer</u>. Similar to a mercury-in-glass psychrometer, two identical thermopiles are employed and their reference junctions maintained at a constant temperature. The test junctions of one of the thermopiles (wet junction) are kept moist by a wet cotton wick, while those of the other thermopile (dry junction) are kept dry. Both are exposed to ambient air temperature.

By designating the accuracy of measurement, Powell[64] has defined the term a-value as

$$a = \frac{T - T_o}{T - T_w} \quad \dots \dots \dots \dots \dots \dots \dots \dots \dots \dots \dots \quad (42)$$

where T_o and T_w are the observed dry-bulb and wet-bulb temperatures, respectively. From Eq. (42), when $T_o = T_w$, then $a = 1$. Although this unity has never been achieved in practice, proximity to unity of $a = 0.99$ has been obtained by several investigators. The following conclusions from their findings can be made:

a. A fine thermocouple wire 0.1 mm in diameter with a ventilation

---

[64] Powell, R.W. 1936. Use of thermocouples for psychrometric purposes. Proc. Phys. Soc. London 48:406.

rate of 0.1 m s^{-1} gives an a-value in excess of 0.99.[65]

b. A wick less than 0.002 mm in thickness together with a long extension of at least 1 cm on either side of the junction is essential for a high a-value.

c. The length of the wick covering is a function of the diameter of the wire and the ventilation rate. In order to make the a-value equal to 0.99, an increase in diameter requires a longer wick, and an increase in ventilation requires a shorter wick.

d. The dry junction must always be kept away from the wet junction either by separating the former above the latter at a distance of at least 1.5 cm or by forcing the airflow from dry to wet junction.

(3) Other Alternatives. During the recent half century, much research has been done in psychrometry, particularly in the design of precise and automated psychrometers. Three major areas of research are delineated: the choice of sensors for better performance; the designation of procedures and methods for humidity measurement in varied climatic conditions; and the determination and reevaluation of the psychrometric constant. Although much research remains to be done, the progress in these three areas is illustrated below.

(a) Sensor Selection. In addition to sensors such as mercury-in-glass thermometers and wet and dry junction thermopiles, others, including platinum and nickel wires, thermistors, mercury-in-steel thermometers, and their multi-combinations and modifications, have been introduced in the field of psychrometry. Responses of these sensors to the fluctuations of several humidity variables have been examined. Specifications on the performances and requirements of these sensors have been studied in depth, and although only a brief review is given here, more detailed information can be obtained from the listed references.[66]

(i) Resistance Psychrometers. Advancements in the use of fine-wire resistance thermometers have been made by several Australian scientists.[67] One of these men, McIllroy, wound 0.001 in. (0.0254 mm) diameter platinum wire in an open helix around a thin strand of cotton wicking and used this for the wet element. Using platinum or nickel as the dry element, he found that the response times for these may be expressed as

$$\lambda_1 = 5.67 \times 10^3 \; r^{1.653} \; u^{-0.347} \quad \text{(for dry Pt element)} \; . \; . \; . \; . \quad (43)$$

---

[65] Kawata, S. and Y. Omori. 1953. An investigation of the thermocouple psychrometer. J. Phys. Soc. Tokyo 8: 768.

[66] Montheith, J.L. 1954. Error and accuracy in thermocouple psychrometry. Proc. Phys. Soc. 67: 217-226.

Trelkeld, J.L. 1962. Thermal Environmental Engineering. Prentice Hall, Inc., New York.

[67] McIllroy, I.C. 1955 and 1961. C.S.I.R.O. Met. Phys. Techn. Papers No. 3 & 11.

$$\lambda_2 = 7.86 \times 10^3\ r^{1.653}\ u^{-0.347} \text{ (for dry Ni element)} \dots \dots \quad (44)$$

$$\lambda_3 = 8.20 \times 10^3\ r^{1.653}\ u^{-0.347}/(1.52\ s+1)$$
$$\text{(for wet Pt element)}. \dots \dots \dots \dots \dots \dots \dots \quad (45)$$

where r (cm) is the radius of the wire; u (cm s^{-1}), the ventilation speed; and s (mb C^{-1}), the slope of the saturation water vapor pressure curve. With the mean wet-bulb temperature of the air, T_w, ranging from 0 to 20C, and the wet-element diameter ranging from 0.005 to 0.02 in., the response time, λ_3, varies from about 0.2 to 1 s. The response time for the dry bulb, λ_1 or λ_2, however, is as small as 0.01 s. Figure 6-57 shows the response time (in seconds) versus the diameter of wire (in inches) at a ventilation speed of 100 cm s^{-1}. The response time of resistance psychrometers is amazingly fast.

Instead of using very fine wires, Collins has experimented with a larger platinum wire 2 mm in diameter and 25 mm in length. He found that a high ventilation speed of 400 cm s^{-1} is required for both the moistened cotton sheath of the wet element and the naked platinum dry element. These elements were arranged in DC differential bridges and fed to a multi-point potentiometer for recording. It has been found that temperature differences up to +2.5C can be recorded with an accuracy of better than 0.05C for the average of 30 minutes duration. This particular psychrometer is known as the Degussa Hartglas resistance thermometer and is used in microclimatic temperature and humidity gradient measurements.

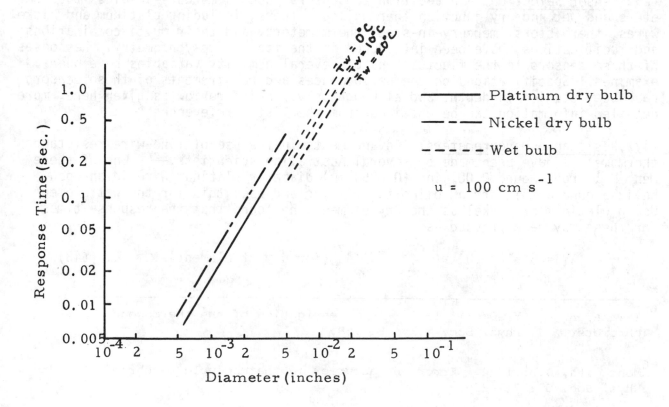

Figure 6-57 Response Time of Resistance Psychrometer

213

Because of corrosion and strain effects, not all resistance elements, such as iron and copper, can be used as wet elements of a resistance psychrometer. Platinum wire is the most suitable because of its high resistance to corrosion and its close approximation to linearity as a function of electric resistance and ambient temperature.

Although we have discussed only positive resistance elements, negative resistance elements such as various types of thermistors have also been employed. For atmospheric temperature ranges, the type F thermistor is often used for both dry and wet elements. However, due to its high sensitivity to fluctuating temperatures, the thermistor has not been used as frequently as the resistance wire in psychrometers. However, portable psychrometers powered by rechargeable or compact battery packs often employ thermistor elements. The low power requirements of these devices, especially for field use where AC power is not available, give thermistor psychrometers a versatility not always found in instruments using platinum elements.

(ii) Mercury-in-steel Psychrometers. The British Meteorological Office has adopted the mercury-in-steel psychrometer. The wet bulb is covered with an open hem made of muslin and the remainder of the wick is extended into a water reservoir whose narrow entrance prevents any effect of the reservoir water on the dry bulb. To prevent conduction of heat from the surroundings to the wet bulb, the wick covers the entire wet bulb and some of the connecting tube. The bulb is approximately 7 inches long and 1 inch in diameter.

Its bulky size necessitates a high ventilation speed to ensure full wet-bulb depression. A large water reservoir is needed for an adequate water supply. Also, because of the high susceptibility of steel to corrosion, the steel bulb is tin coated. Through the connection of Bourdon tubes and capillary tubing, a telemetering capability is possible to a maximum distance of 130 ft, for recording wet-bulb and dry-bulb temperatures in the range of 20 to 120F, or -20 to 100F. The readout on a drum is scaled in increments of 2F and can be estimated to 1F.

Errors encountered in the mercury-in-steel psychrometer, which are greater than those of the mercury-in-glass psychrometer, are caused by the conduction of heat down the thermometer stem and the variability of ventilation. For example, on calm, hot, dry days, the recording of humidity tends to be too high. The only advantages of the mercury-in-steel psychrometer are its capabilities of telemetering and automatic recording through a simple mechanical device.

(b) Modifications of Design. For measuring extreme climatic conditions, several new designs for psychrometers are now available. A thermocouple recording psychrometer[68] for measuring relative humidity in hot arid atmospheres has been studied in Gila Bend, Arizona, where ambient air temperatures may be as high as 110F, relative humidity as low as 10%, and wet-bulb depression as much as 40F.

---

[68] Bellaire, F.R. and L.J. Anderson. 1951. Thermocouple psychrometer for field measurements. Bull. Amer. Meteor. Soc. 32(6): 217-220.

Richardson, R. 1965. A thermocouple recording psychrometer for measurement of relative humidity in hot arid atmosphere, In: Humidity and Moisture, Measurement and Control in Science and Industry, Vol. 1 (see Footnote 56).

214

Copper-constantan thermocouples are employed as both dry and wet elements. To prevent heat conduction from the water reservoir through the wick to the wet bulb, a pre-cooling water reservoir is provided. Cooling is accomplished by passing the aspirated air stream over the surface of the water in a shallow reservoir. As a result, average cooling of the reservoir water is about 70% of the wet-bulb depression. A comparison of this design has been made with the standard sling psychrometer and the aspirated ASTM precision thermometer. It has been found that an average calculated relative humidity is on the order of 0.5% higher than the precision measurement and 1.6% lower than the sling psychrometric reading. The largest variance of the thermocouple from the precision measurements at any observation was 3% R.H., which occurred on days of calm air and rapidly changing humidity. In these tests, humidity measurements ranged from 13 to 36% and temperatures, from 46 to 75F.

For measuring relative humidity at temperatures below freezing, an elevated temperature psychrometer[69] has been designed and tested at Palmer, Alaska. The basic method employed by this psychrometer is to raise both the wet-bulb and dry-bulb temperatures without changing the moisture content of the sample air. This is done by measuring the freezing ambient air at the entrance of the device, and then employing a fan to channel this air through the psychrometer's electric heater. The heated air flows into the psychrometric chamber where both dry-bulb and wet-bulb temperatures are taken. The vapor pressures corresponding to the dew-point and the dry-bulb temperature at saturation are determined using a psychrometric table for conversion. The ratio of these two vapor pressures determines the relative humidity. The accuracy of this instrument is on the order of ±1F. To work effectively the readings should be taken for all three thermometers until the readouts are stabilized.

In addition to the above two extreme climatic conditions, psychrometry has extended its investigations into high humidity and temperature environments. A thermocouple psychrometer utilizing the Peltier[70] effect has been found useful in measuring relative humidities over 95%. In this design, the wet junction of a thermocouple is cooled first by using the Peltier effect. As the moisture of the ambient air condenses on the wet junction, its temperature gradually approaches the dewpoint temperature of the air. When this temperature is lower than the ambient air temperature, the battery current stops, and evaporation at the wet junction takes place. At this instant the instrument acts as a thermocouple psychrometer. According to Box,[71] it takes two minutes to make the recording, and with careful calibration, relative humidities of 98.4 to 100% can be measured

[69] Branton, C.I. 1965. A proposed technique for measuring relative humidity at below freezing temperature. Ibid. p. 95-100.

[70] Spanner, D.C. 1951. The Peltier effect and its use in the measurement of suction pressure. J. Exp. Botany 2: 145-168.

Monteith, J.L. and P.C. Owen. 1958. A thermocouple method for measuring relative humidity in the range of 95-100%. J. Sci. Instr. 34: 443-446.

Korven, H.C. and S.A. Taylor. 1959. The Peltier effect and its use for determining relative activity of soil water. Can. J. Soil Sci. 39:76-85.

[71] Box, J.E. Jr. 1965. Design and calibration of a thermocouple psychrometer which uses the Peltier effect, In: Humidity and Moisture (see Footnote 56).

over a temperature range of 50 to 122F with high accuracy and precision. Box's main interest, however, was the relative activity of water in plant-soil-water relationships, i.e., the extraction of water by plants and suction of moisture by soils.

Measuring relative humidities of high temperature environments, such as environmental chambers, and furnace and brick kiln atmospheres, utilizes self-heating dewpoint sensors capable of detecting dewpoints up to 160F. A ceramic tube psychrometer, however, can measure atmospheric humidity at dry-bulb temperatures up to 325F by using an adequate distilled water supply and placing the dry bulb upstream from the wet bulb. Relative humidity is determined with the use of a high temperature psychrometric chart.

(c) Investigation of the Psychrometric Constant. Although the psychrometric constant was established in August, 1925, many investigations to improve it are still underway today. It is a constant only in a formal sense, as nearly all national weather services have selected or established their own psychrometric constants. Specifications for each psychrometric constant are made in terms of: (a) evaporation of the wet bulb with respect to a water or ice surface, (b) dry- and/or wet-bulb temperature, (c) above and/or below freezing temperatures, and (d) adequate or inadequate ventilation, and/or a specific ventilation rate.

Although the theory of psychrometry has been studied extensively, its application with respect to the above specifications is still problematic. Furthermore, factors influencing the psychrometric constant, such as the specific heat of air at constant pressure, the latent heat of vaporization, and the mass of air passing through the sensors per unit time, are functions of wind speed, air temperature, and a number of other environmental variables. Such complex relationships make it difficult for a psychrometric constant to be a constant value in all weather conditions. Bindon[72] has computed the percentage difference between the selected constants and the thermodynamic constant[73] for several well-known psychrometric tables, monographs, slide rules, and charts. He found that a 1.9% discrepancy

---

[72]Bindon, H.D. 1965. A critical review of tables and charts used in psychrometry, In: Humidity and Moisture (see Footnote 56).

[73]For water, the psychrometric constant $A(T_w) = \frac{c_p}{\epsilon L_v(T_w)} = 6.5205 \times 10^{-4} C^{-1}$, or $3.631 \times 10^{-4} F^{-1}$, where c_p is taken at 0C and L_v at 10C. For ice, $A(T_i) = \frac{c_p}{\epsilon L_s(T_i)} = 5.703 \times 10^{-4} C^{-1}$, or $3.168 \times 10^{-4} F^{-1}$, where c_p is taken at 0C and L_s at -10C. T_w and T_i are the wet-bulb and ice-bulb temperature, respectively; ϵ is the ratio of the molecular weights of the water vapor and dry air; and L_v and L_s are the latent heat of vaporization and sublimation, respectively. The value of $A(T_w)$ changes slowly at sea level (Monteith, 1973). The commonly quoted value of 0.66 mb K^{-1} (mean requires adjustment for elevations above 1000 m (Storr and Hartog, 1975).

existed among them, which would lead to an error of ±2% R.H. for an adequately ventilated psychrometer above the freezing point. The error would be greater for temperatures below freezing or with insufficient ventilation.

In view of the discrepencies in the psychrometric constant and of the recent advances in the knowledge of heat and mass transfer, a new technique utilizing the Lewis Relation[74] has stimulated great interest. The Lewis Relation is essentially the ratio of two coefficients - the heat transfer coefficient and the mass transfer coefficient - from the wet sensing surface to the ambient air. While the psychrometric constant possesses a temperature dimension, the Lewis Relation calculation is dimensionless. For this reason, the Lewis Relation is used in psychrometric calculations. The Lewis Relation is near unity when heat and water vapor transfer occur by perfect turbulent mixing without the effects of radiation and reservoir heat flow. It facilitates the correlation of the ordinary wet-bulb temperature with thermodynamic wet-bulb temperature. Various studies of the physical characteristics of wet evaporating surfaces have associated geometric configurations (flat-plate, cylinder, sphere, or packed beds) and boundary conditions (laminar and turbulent) to water vapor and heat transfer from an evaporating surface. It remains to be seen if the Lewis Relation will be correlated over a wide range of thermodynamic conditions in psychrometry.

The adiabatic psychrometer[75] offers an alternative to the Lewis Relation method for psychrometric calculations. Although the instrument is still in the experimental stage, this psychrometer, which does not require calculation of a psychrometric constant, determines the thermodynamic wet-bulb temperature by saturating the air adiabatically. This instrument, which employs an adiabatic saturator, makes measurements of relative humidity in the range between 0% and 100% with an accuracy of 0.25% R.H. over an ambient air temperature range of +5C to +30C. Both manual and continuous recordings can be made with a medium rate of response. The adiabatic psychrometer would be the most accurate psychrometer on the market if it were commercially available. Its drawbacks are its cost, high maintenance, and non-function at temperatures below freezing.

6.6.3 _Sorption Hygrometry._ As shown in Table 6-17, instruments which are dependent on the equilibrium sorption of water vapor by a sensor include mechanical, colorimetric, electrical, and piezoelectrical hygrometers. The sensors of these hygrometers undergo a change in physical and, sometimes, chemical properties when their surfaces absorb or interact with water vapor. The humidity reading is made at the instant of sorption-equilibrium.

---

[74] See Footnote 62 and see also Eckert, E.R.G. 1959. Heat and Mass Transfer. McGraw-Hill Book Co., New York.

Kusuda, T. 1965. Calculation of the temperature of a flat-plate wet surface under adiabatic conditions with respect to the Lewis Relation, In: Humidity and Moisture (see Footnote 56).

[75] Carrier, W.H. 1911. Rational psychrometric formulas. Trans. ASME 33: 1005.

Harrison, L.P. 1965. Proposal for an experimental triple-tubed adiabatic saturator to measure thermodynamic wet-bulb temperature as a limit approached during the saturation process, In: Humidity and Moisture (see Footnote 56).

(1) <u>Mechanical Method</u>. The criteria for the selection of a suitable substance as a sensor are: fast response to humidity change, high reproducibility or repeatability, durability, and linearity to humidity change. No single substance has been found to meet all of the criteria. However, both human hair and gold-beater's skin meet most of these criteria. Human hair and ox intestines (commonly known as gold-beater's skin) are the materials most frequently employed as sensors in mechanical hygrometers. While the former has been adopted for both surface and upper air humidity measurements, the latter is used almost exclusively in upper air measurements. Other organic substances used in hygrometry include cotton fibers, raw silk, natural sponge, linen, wool, feathers, straw, horn, wood chips, skin of eggs, seed capsules of some plants, and so on. These organic materials indicate humidity changes by their changes in such physical characteristics as length, volume, weight, hardness, and electrical conductivity. Human hair as the humidity sensing element is described below; discussion of gold-beater's skin is found in the following chapter on upper-air measurements.

When human hair is employed as the sensor, the fluctuation in its elongation indicates changing humidity. The fatty materials of the hair first must be removed by such solvents as caustic soda, ethyl ether, or alcohol; then the hair is rolled and flattened[76] to achieve better response rates, a higher coefficient of expansion, and greater linearity than without treatment.

An equilibrium condition must be established to make the humidity readings representative of the actual ambient humidity. The equilibrium condition is a steady state in which the exchange of water vapor and heat between the hair and its surroundings must be equal. A number of physical processes are involved in establishing such a state: (a) transfer of the total amount of water vapor from the ambient air to the hair, (b) diffusion of water within the hair and evaporation of water from the hair surface, and (c) heat gain generated by the sorption process of the hair and heat loss through evaporation. These processes are determined by the number of hairs employed, the expansion coefficient of, as well as the strain and stress of, the individual hair. In addition, atmospheric conditions such as radiation, temperature, wind, moisture (e.g., fog droplets), and atmospheric contaminants can affect hair response.

With a load of 1 gram on each hair, the stress is about 500 psi. Figure 6-58 shows a typical curve of hair elongation under such a stress. The nonlinearity of the natural hair curve can be improved by using Frankenberger hair. Most researchers prefer a lighter load on the order of 0.3 to 0.8 g per hair.

The response time, λ, of a hygrometer is a step function change, i.e., a 63% change between the initial and final humidity indications. If λ is the time required to bring the initial indication, U_0 (at zero time), to the final indication, U_f (at λ seconds), the humidity reading, U_t (after t seconds), may be expressed implicitly as

$$U_f - U_t = \lambda \frac{dU_t}{dt} \quad \ldots \ldots \ldots \ldots \ldots \ldots \ldots \ldots \ldots \ldots \quad (46)$$

---

[76] In 1944, Frankenberger of Germany reported his treatment of hair. He immersed the hair in an oil bath and pressed it with two highly polished steel rollers until its cross-section became elliptical with a 1:4 ratio. The flattened hair was then treated with alcohol. Since then, the term Frankenberger hair has been used widely to refer to rolled and flattened hair, free of fatty materials.

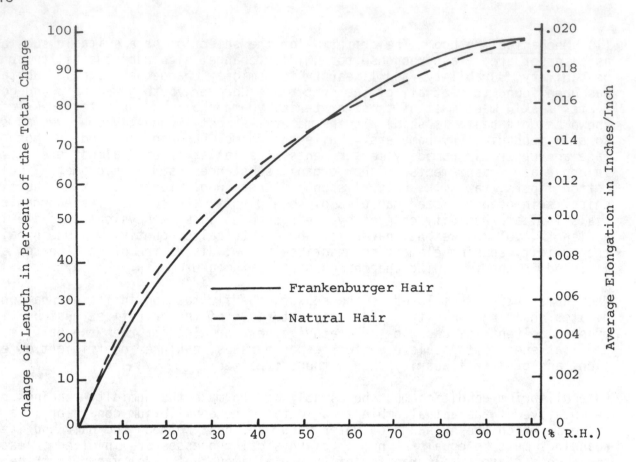

Figure 6-58 Elongation of Hair as a Function of Relative Humidity

where U_f is the humidity of the ambient air after λ s of time, and U_t is the humidity reading at time, t. Let the difference $\Delta U = U_f - U_t$, then

$$\Delta U = \Delta U_0 e^{-\frac{t}{\lambda}} \quad . . . . \quad (47)$$

where ΔU_0 is the increment of the humidity of the immediate preceeding step function.

Frankenberger hair (see Footnote 76) has a reported λ value of 10 s for an air temperature above freezing and absolute humidity above 0.05 g m^{-3}. For air temperatures between 0C and -55C, corrections must be applied for R.H. below 40%. In this case, the λ value is about 60 s. Below -55C, the λ value may be as high as 100 to 200 s, and the usefulness of Frankenberger hair for indicating R.H. becomes questionable.

Numerous problems are associated with the use of a hair hygrometer: (a) the severely reduced sensitivity of the hair to moisture if it is contaminated by dust or oil, (b) the expansion coefficient differing from hair to hair, (c) the increase in response time with decreasing temperature and humidity, and (d) the reduction of reproducibility due to increase of hysteresis as a result of an abrupt change in humidity. In practice, cellophane has been applied to the hair for protection against contamination at the expense of λ values. To avoid diversities in expansion coefficient, either a single hair or a specially mounted, well-balanced group of two or three hairs have been adopted at the expense of breaking strength. The strength of the hair is reduced to 20% or less by the removal of fatty materials and the flattening process, but its sensitivity is greatly improved.

Despite various problems, hair hygrometers have been in use for almost two centuries.[77] Reasons for the continued popularity of the hair hygrometer are its simplicity and very inexpensive cost, its durability, its flexibility for both mechanical and electrical recordings, and its accuracy being comparable, if not superior, to several other types of hygrometers.

(2) Colorimetric Method. In colorimetry, the color humidity indicator is the most inexpensive. When cobaltous chloride is impregnated in blotting paper, the color appears bluish at low humidity, pinkish at high humidity, and a series of lavender-like colors in between. With this device, it is possible to measure R.H. ranges of 40% to 70% with 2% accuracy. The range of a colorimetric hygrometer may extend as low as 8% and as high as 80%, with 5% accuracy. It requires a 30-minute response time in high humidity and as much as 2 hours in low humidity. Cobaltous bromide for the indicator increases its sensitivity about threefold.

In practice, air humidity can be determined by matching the color appearing on the indicator against a color guide. The guide is usually prepared by calibrating it against a standard hygrometer at R.H. 40% (midway between the blue and pink) and at a temperature of 75F. It has been found that a 2.5% R.H. correction is required for each 10F deviation from 75F. At temperatures lower than 75F, the color of the indicator appears bluer than it should and the R.H. correction is added. At temperatures higher than 75F, the color appears pinker than it should and the R.H. correction factor is subtracted. For example, at a temperature of 35F, 10% R.H. should be added to the reading.

The other types of sorption hygrometers, such as the lithium chloride Dewcel (e.g., Dunmore electric hygrometer), piezoelectric sorption hygrometer, and many other electric and electrolytic hygrometers, have a great importance in upper-air sounding, and are discussed in the following chapter.

(3) Electrical Method. Semiconductor devices increasingly are being employed as humidity sensors. These thin film, bulk effect, solid state semiconductors change in electrical conductivity in response to changes in the levels of water vapor in the ambient air. The BR-101B solid state humidity sensor, manufactured by Thunder Scientific Corporation,[78] utilizes such a semiconducting, molecular diffusion barrier as the sensor (Figure 6-59). Noble metal materials are used on the sensor, as opposed to the more commonly used aluminum oxide or other unstable metals. The use of noble metals reduces the hysterisis, drift, aging, and instability of the sensor to negligible levels under a variety of temperature, pressure, and humidity conditions after calibration by the National Bureau of Standards' two pressure system. The dielectric sensor structure is highly porous to water vapor in molecular form; thus, water vapor molecules drift freely in both directions through the sensor surface. Water vapor movement causes electron tunneling, so the sensor varies in complex impedance and resistance as molecular density varies

---

[77] de Saussure, H.B. 1783. Essais sur l'Hygrometrie. Neuchatel.

[78] Thunder Scientific Corp. 1979. BR-101B Solid State Humidity Sensor. Thunder Scientific, Albuquerque, New Mexico. 3 p.

and the water dipoles interact with the lattice. The BR-101B can be operated as either a two-terminal or three-terminal device, similar to either an FET or grating semiconductor.[79] The sensor proper is a die with dimensions of 0.070 in. by 0.010 in., mounted upon a gold TO-5 eight pin transistor header; the entire device is enclosed in a stainless steel protective cover screen. Dirty or wet environments require the use of a sintered filter, 60 μm pore size and replaceable, between the sensor and the outside environment. Complete package size is: diameter, ~ 0.037 in. and height, ~ 0.0375 in. Typical operating temperatures range between OC and 50C for the device, with extension to -60C and +125C for extreme ranges if calibrated for these extremes. Typical instrument specifications for the BR-101B are given in Table 6-19.

Table 6-19 Technical Specifications for the Thin Film Bulk
Effect Humidity Sensor, BR-101B[80]

Specification	Values
Humidity Range	0% to 100% RH
Accuracy	
Guaranteed Accuracy	4% indicated RH
Typical Accuracy	1% indicated RH
Calibrated Accuracy	
(NBS Standard Method)	0.5% indicated RH
Sensitivity	1% RH
Typical Hysterisis	1-2% for 0-100% RH
Response Times	
Typical/Worst Possible Case	300 ms/800-900 ms for 63% step change of ±1-2% RH of reading; 400 ms/600 ms for 2-90% RH range
Drift	3 yrs, 7% RH - exposed to environment; 3 yrs, 10% - with 60 μm filter
Temperature Coefficient	Complex algorithm (function of temperature and vapor pressure)
Temperature Range	OC to 50C
Weight	1 g
Output Signal	Non-linear RH Output

[79] Melen, R. and H. Garland. 1977. Understanding CMOS Integrated Circuits. H.W. Sams and Co., Indianapolis, Indiana. 144 p.; Flynn, G. 1978. Transistor-Transistor Logic. H.W. Sams and Co., Indianapolis, Indiana. 288 p.

[80] Thunder Scientific Corp. 1979. Thin Film Bulk Effect Humidity Sensor, Model BR-101B. Thunder Scientific, Albuquerque, New Mexico. 1 p.

The sensing element can be interfaced with a electronic signal conditioning package Model PC-2000 (see Figure 6-59) for use as a modular component of a humidity or humidity-temperature package (such as the Sierra-Misco Model 2040[81] electronic humidity indicator or the Thunder Scientific Model HS-1CHDT-2A[82] digital humidity-temperature measuring system). Other thin film bulk effect devices using aluminum oxide or other unstable metals, such as that used for the sensor in the WeatherMeasure Model HMI-14 RH indicator,[83] have hysterisis changes between 1% and 2% over full-scale humidity swings, and with accuracy values of ±3% for the RH range of 0-80% and ±5-6% for the RH range of 80-100%.

Calibration of humidity devices, such as the bulk effect and capacitive sensors, does change with time. McKay[84] has found that the vast majority of humidity sensors, whether of the semi-conductor, dewcel, crystal, or hair type, change calibration when exposed to field conditions. It is important for the user to periodically recheck the calibration of these devices against some standard (such as the Assmann psychrometer or the Cambridge cooled mirror hygrometer), and make the necessary adjustments to each instrument's calibration curve. Exposure in the environment to a wide range of atmospheric conditions requires a more frequent check of sensor accuracy since the hygrometer has to function over humidity ranges that vary by a factor of three over the normal range of atmospheric temperatures.

6.6.4 Dewpoint Hygrometer. When moist air at a certain air temperature, T, pressure, P, and mixing ratio, ω, is cooled, it eventually reaches its saturation point with respect to a free water surface and a deposition of dew can be detected on a solid surface. The temperature of this saturation point is called the thermodynamic dewpoint temperature, T_d. A further cooling will cause the formation of ice, and its temperature will then become the thermodynamic frost point temperature, T_f. The saturated vapor pressure with respect to water, e_w, and ice, e_i, is a function of T_d and T_f, respectively, and is designated by

$$e_w(T_d) = \frac{\omega P}{0.622 + \omega} \quad \cdots \cdots \cdots \cdots \cdots \quad (48)$$

and

$$e_i(T_f) = \frac{\omega P}{0.622 + \omega} \quad \cdots \cdots \cdots \cdots \cdots \quad (49)$$

The dewpoint hygrometer is used to measure T_d; or the frostpoint hygrometer, T_f. Despite the great moisture variation in tropospheric air, this instrument is capable of detecting very low or very high water vapor concentrations by means of a thermal sensor without directly measuring water vapor itself.

---

[81] Sierra-Misco. 1979. Weather Instrument Catalog No. 1178. Sierra-Misco, Inc., Berkeley, Calif. 53 p.

[82] Thunder Scientific Corp. 1979. Model HS-1CHDT-2A Digital Humidity/Temperature Measurement System. Thunder Scientific, Albuquerque, New Mexico. 1 p.

[83] WeatherMeasure Corp. 1978. Scientific Instruments and Systems Catalog 1078. Weather Measure, Sacramento, Calif.

[84] McKay, D.J. 1978. A Sad Look at Commercial Humidity Sensors for Meteorological Applications, In: Preprints, Fourth Symposium on Meteorological Observations and Instrumentation. American Meteorological Society, Denver, Colorado.

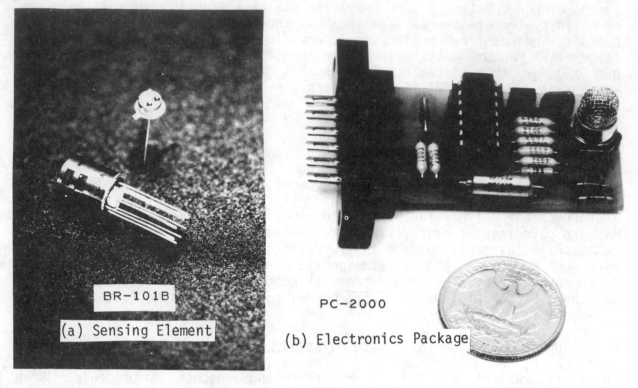

BR-101B

(a) Sensing Element

PC-2000

(b) Electronics Package

Figure 6-59 Thin Film Bulk Effect-Humidity Package with Electronics

It is interesting to note that the dewpoint hygrometer was introduced two centuries ago, but few improvements were achieved before the advancement of modern electronics. The significant improvement achieved in modern dewpoint hygrometry is the great accuracy in the determination of surface temperature, the time of dew (or frost) formation, and the thermal regulation.

(1) <u>Sensor Assembly</u>. The sensor consists of a thin metallic mirror which is 0.06 in. in diameter. It is thermally regulated by a cooling assembly, a temperature measuring device (thermistor, thermocouple, or resistance thermometer) imbedded beneath the mirror. For the surface temperature of the mirror to respond rapidly to the thermal regulation system, the thermal diffusivity of the mirror must be high. The higher the thermal diffusivity, the smaller the temperature gradient that exists between the surface of the mirror and the thermal sensor and, hence, the greater precision of measurement. Thus, such metals as silver, copper, gold, and stainless steel, as well as their alloys, are employed for the mirror. The mirror must be free of atmospheric contaminants, as any water soluble matter on it tends to lower vapor pressure, and hence the dewpoint of the condensate (the Raoult effect). If the condensate forms as very small droplets, then an opposite effect (the Kelvin effect) may occur. Thus, an automatic device should be provided for the detection of liquid or solid contaminants on the mirror surface so that they may be removed. To determine the dewpoint temperature of moist air accurately, an electro-optical system rather than visual observation is employed. In this way the reading of T_d (or T_f) is made at the exact time of deposit of dew (or frost) on the mirror so it is neither dissipating through evaporation nor growing through condensation. This is the point when the partial vapor pressure

of the overlaying air and the saturated vapor pressure of the deposit surface are in equilibrium. The thickness of the deposit is measured by a photodetector.

(2) <u>Optical Detection Assembly</u>. A narrow beam of incandescent light with an incident angle of about 55° is directed at the mirror. The intensity of the reflected light, or reflectance, is detected by a photodetector which, in turn, regulates the cooling assembly through a servo-control. Thus, the output of the photodetector determines the thickness of the deposit. When the deposit is too thick, the effectiveness of the mirror as a reflective surface becomes negligible. Conversely, when it is too thin, the effect of the deposit on reflectance is not detectable. In short, a certain range of effective thickness must be maintained. In an arbitrary scale, the zero value has been assigned to a clear mirror or an undetectable thin deposit, and the 100 value has been designated as the maximum deposit at which the reflectance of the mirror itself is null. By using this percentile scale, Paine and Farrah[85] found that a range of 5 to 40 per cent is most suitable. They claimed that within this range no variations in readings as a function of thickness have been observed. The readouts are sensitive to better than ±0.05C. Other research workers, however, have found that an additional photodetector is required to measure the scattered light of the deposit. The schematic diagram of the optical detection assembly and feedback loop of a basic hygrometer are illustrated in Figure 6-60.

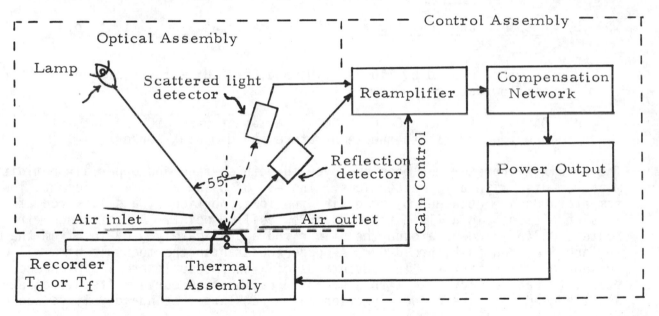

Figure 6-60 Functional Block Diagram for Basic Hygrometer

(3) <u>Thermal Control Assembly</u>. This assembly usually consists of a cooler (heat sink), which aids in the growth of deposits of dew or frost on the mirror, and a heater (heat source) which dissipates the deposit when it either is too thick or heavily contaminated. Low boiling temperature liquids which have been used as

---

[85] Paine, L.C. and H.R. Farrah. 1965. Design and applications of high-performance dewpoint hygrometer, In: Humidity and Moisture (see Footnote 56).

coolants are oxygen, nitrogen, and nitrous oxide. The Peltier element has also been used successfully. The heater is made of an electric resistance wire. The response to either heating or cooling should be within 1 or 2 s in order to maintain its sensitivity and provide a continuous recording.

As shown in Figure 6-61, the Cambridge Systems Hygrometer Model 110 consists of a standardizing circuit and an optical sensing bridge including the Peltier element as a thermoelectric cooler. The sensor is a stainless steel mirror with an imbedded platinum resistance wire. Both reflected and scattered light are detected by a pair of cadmium sulfide photoresistors.

A comparison of the intensity of these two lights regulates the cooler or the heater. For example, a decrease in the electric current supply to the cooler reduces the thickness of deposit, causing an increase of scattered light and a decrease of reflective light. A warning signal appears when the mirror is contaminated. The mirror must be wiped with a cotton swab after unlatching the bottom of the transmitter and rotating the flow shield.

The manufacturer's specificiations on the Model 110 Hygrometer are given below.

a. <u>Range and Accuracy</u>: ±0.5F for -80F to 120F range.

b. <u>Response and Sensitivity</u>: $4F\ s^{-1}$ (Max.) at a sensitivity of ±0.1F.

c. <u>Input</u>: Powered by 115 VAC ±10% and 50-70 Hz with an ambient airflow rate of 60 cfm.

d. <u>Output</u>: A linear 0-50 mV DC, or a linear 0-5 V DC, output is provided from both the temperature and dewpoint sensors.

The Bendix hygrometer has been employed in both surface and upper air humidity measurements, with a few exceptions. Instead of using a Peltier element, low temperatures are obtained by immersing the lower portion of a silver rod (the sensing element) in a fluid that has a low boiling point. A resistance wire heater, which is wound around the silver rod approximately 0.09-in. from the mirror surface is shown in Figure 6-62. A copper-constantan thermocouple of 5 mil wire is buried under the mirror. The advantages of a thermocouple are its high accuracy, stability, and conformance with a standard calibration curve. Its disadvantages are the need for a reference junction and a regulated, standard voltage source.

The specifications of the Bendix hygrometer vary with application. These are the portable field hygrometer, the dropsonde thermoelectric hygrometer, the rocket-sonde hygrometer, the aircraft hygrometer, and a general purpose type thermoelectric hygrometer (Model DHGF-ID of 1963). For details the reader may refer to the article prepared by Paine and Farrah (See Footnote 85).

6.7 <u>Nephometry and Hyetometry</u>. Condensation products suspended in the atmosphere (clouds, fog) and on the ground (dew, frost) are frequently observed. The most commonly observed forms of precipitation are rain, snow, and hail. The measurement of clouds is designated as nephometry and that of precipitation as hyetometry.

PRINCIPLE OF OPERATION

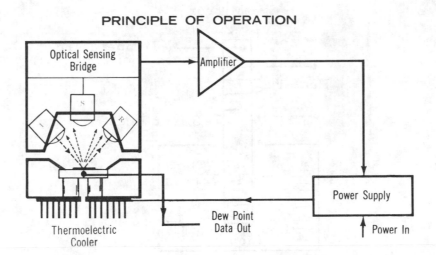

I. = LIGHT SOURCE   S = PHOTOCELL FOR SCATTERED LIGHT   R = PHOTOCELL FOR REFLECTED LIGHT

a)

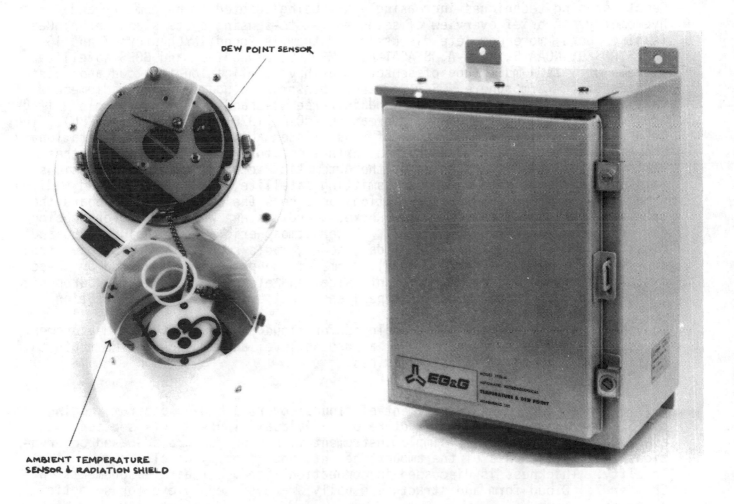

DEW POINT SENSOR

AMBIENT TEMPERATURE
SENSOR & RADIATION SHIELD

b)

Figure 6-61   Cambridge Systems Hygrometer

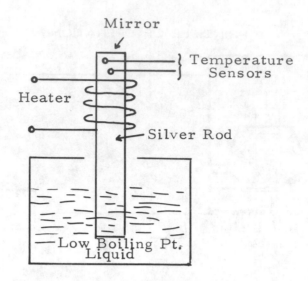

Figure 6-62 Bendix Hygrometer Resistance Heater and Evaporative Heat Sink

Remote sensing techniques increasingly are being adapted to nephometry and hyetometry. A brief overview of selected remote-sensing systems and techniques is given here; more complete discussion of them is found in Chapters 7 and 16. The TIROS-N, NOAA-5, NOAA-A, SEASAT-A, NIMBUS-G, LANDSAT-3, and GOES satellite systems utilize a multitude of sensors that have application to cloud and moisture studies of the atmosphere. Some of these sensors include scanning radiometers of varying resolution in the visible and infrared wavebands that provide cloud imagery and, indirectly, precipitation information (NOAA, 1978; NASA, 1978a, 1978b). The GOES satellites are the only ones of those listed above that remain geostationary above pre-selected points around the earth's equator. Data collected by these various satellites are accessed by the Automatic Picture Transmission stations within the line of site of the transmitting satellite, or by the Advanced Vidicon Camera System where the surface station can access the data stored on board the satellite (Bristor et al., 1977). Rocket soundings are well-suited for probing the upper atmosphere (Quiroz, 1972). Lower atmospheric probes include echo and digitized radars, doppler radar, lidar, sodar, radio sounder, and lidar spectroscopy (Little, 1972; McPherson, 1975). Some of the properties measured by these systems are cloud structure, movement, size, and electrical properties, drop size, charge, and distribution, nuclei types, and areal estimates of precipitation.

Conventional instrumentation is employed for cloud and precipitation measurements because the devices are operational at ground level and provide ground truth calibration for many remote sensing devices.

6.7.1 <u>Nephometry</u>. The measurement of clouds covers a broad spectrum ranging from global coverage to the microstructure of individual clouds. This section, however, describes relatively simple instrumentation used for ground-based observations and measurements of the amount, height, and velocity of clouds at a single locality. Thickness is discussed in connection with upper air measurements in Chapter 7. Cloud form and structure usually are observed by eye for synoptic purposes.

(1) <u>Cloud Amount</u>. The horizontal extent of cloud cover at a station is determined by various means. During the day it can be estimated visually, graphically, or photographically. Visual observation is aided by comparing observed clouds to descriptions and photographs published in manuals and pamphlets. The WMO International Cloud Atlas (1969) details cloud structure, nomenclature, and classification systems used internationally.

The universal practice is to express the extent of the clouds (either that of all the clouds present or of a particular layer) by the fraction of the celestial dome covered. Such an estimation of cloud amount or cloudiness disregards the actual volume or density of clouds present. The World Meteorological Organization and the British Meteorological Office use oktas or eighths of the total sky area to designate cloud cover. A better method than visual estimation employs a conical cardboard tube (Figure 6-63) whose aperture subtends a solid angle equal to one-eighth of an hemisphere. This method improves estimations of scattered cloud cover. This tube has an opening of ℓ inches in diameter and 1.07ℓ inches in length.

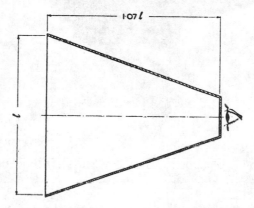

Figure 6-63 Cardboard Tube for Cloud Observation

Graphical methods use sunshine recorders, of which the Marvin and Jordon instruments have been discussed previously. The Campbell-Stokes recorder is discussed here. The sunshine recorder gives a continuous record of hours of sunshine while its converse yields hours of cloudiness. However, this method fails to give an accurate indication of the amount of sky covered should the sun appear in a clear area surrounded by clouds, or shaded by clouds when the sky is only partially covered. Under these conditions, the former is reported as a cloudless day and the latter as an overcast day. Although this method can be biased for any individual day, values derived from it give a fairly good statistical representation of long period cloudiness over a month or season.

The Campbell-Stokes sunshine recorder is a sundial type. As shown in Figure 6-64, it consists of a strip shaped as a spherical metal bowl A with a flange inside the bowl to hold a printed paper card, a spherical solid glass ball B, and a metal frame with two water levels (not shown). The paper cards carry marks indicating the hour and half-hour lines. There are three sets of cards: one for the equinoctial periods, one for the winter, and one for summer. This is necessary in order to cover the entire range of solar declination. The bowl also is adjusted so that the plane containing the mid-line of the equinoctial card is coincident

Figure 6-64   Campbell-Stokes Sunshine Recorder

with the plane of the celestial equator, making the records at the solstice an
equal distance above and below this central line. The glass sphere focuses the
sun's rays on the paper card charring a mark on it. The total length of the
burned trace per day is the total duration of sunshine on the same day. The start
and end of each trace also is computed.

Although the Campbell-Stokes recorder is the recommended standard instrument for
sunshine measurements by the World Meteorological Organization, it can be read only
to an accuracy of 0.1 hr provided errors due to manufacture and observer are elimi-
ated. Those due to manufacture can include such items as the transparency of the
sphere, the accuracy of the card printing, the change of card dimensions due to
moisture, and the size and symmetry of the bowl and the sphere. Most of these
errors can be avoided by obtaining an instrument which has been carefully constructed. Errors due to faulty adjustment of the instrument by an observer can
include such items as levelling of the base, meridional setting of the bowl and
sphere, and latitudinal setting of the framework. In practice, levelling of the
sub-base for E-W and S-N directions with the aid of a set of spirit levels and
levelling screws must be made first. Then the central line of the bowl is levelled
by placing the spirit level on the top of the bowl parallel to the line joining the
pair of horns. The adjustment of the meridian is made on a sunny day at local
apparent noon. The card is turned until the burn comes exactly on the noon line
at this moment. The adjustment of latitude is made merely by turning the screw
of the latitude scale engraved on the bowl adjustment bracket.

During the night the starshine recorder (see Figure 6-65) is the only tool which can be used to measure cloudiness. It consists of a light-tight box with a long focal length lens pointing toward the celestial pole. At the focus of the lens is photographic film which continuously records the image of the Pole star. A shutter mounted in front of the lens is controlled by a clock mechanism which closes off the light during the daytime. The errors involved in such a device are deposit of dew or hoar-frost on the lens, glare light from sources other than the Pole star, and variation in sensitivity of the film. Because the starshine recorder gives indications of what occurs in only a small portion of the sky, it is subject to the same errors in determining cloud coverage as the sunshine recorder.

Cloud photography is not only useful in recording total cloud cover but also records cloud sizes, differentiates high, medium, and low clouds, indicates layering, and reveals details of the structure of breaks in cloud sheets. Fine grain panchromatic films and infrared films are generally employed, while color film transparencies can serve to emphasize cloud details. Appropriate filters are used with black and white films to emphasize specific details: red filters for grey skies or for bright clouds against a dark background; yellow filters for blue or broken skies; and dark yellow or orange for intermediate skies or photography above 6,000 ft.

Cloud photogrammetry is an emerging field in nephometry. By knowing the distance of a cloud and the focal length of the lens, the dimensions of a cloud can be calculated. By using two cameras at the ends of a baseline, a stereoscopic pair of photographs is obtained from which the distance of the cloud is calculated and its contours traced.

Ordinary cameras employ gnomonic projection to form an image. A photograph of the projected image shows rapidly increasing distortion with distance from the center. To avoid this problem, special cameras which employ orthographic projection are used. In orthographical imagery, small circles are represented as straight lines and great circles as ellipses of different eccentricity. If an orthographic image of the whole sky is projected on a flat screen, it gives the impression of a sphere if viewed from a distance.

The Hill's cloud camera, developed in 1923, is capable of taking cloud photographs of the whole sky but it cannot take a clear, undistorted cloud picture with low sky luminosity. This problem was solved in 1957 with the development of the Georgi's portable sky mirror, shown in Figure 6-66. It has a camera mounted below a large spherical mirror from which the cloud image is reflected to a small plane mirror above it. The plane mirror then reflects this same cloud image to the lens of the camera for recording.

The single lens reflex (SLR) camera equipped with an 180° fish-eye lens provides an updated method to collect whole sky negatives without the need for a costly, spherically ground mirror. Pochop and Shanklin (1966) described such a system employing panchromatic black and white film or infrared film. In addition, McArthur and Hay (1978) described an approach using Kodachrome color film that is adaptable to all-sky photography. Pochop and Shanklin used a Nikon-F SLR 35 mm camera with motor drive and timer. (See Figure 6-67.) The distance of the image points from the picture center, when photographed using a Nikon 180° lens, are nearly linear to zenith angle, facilitating exact positioning of every visible

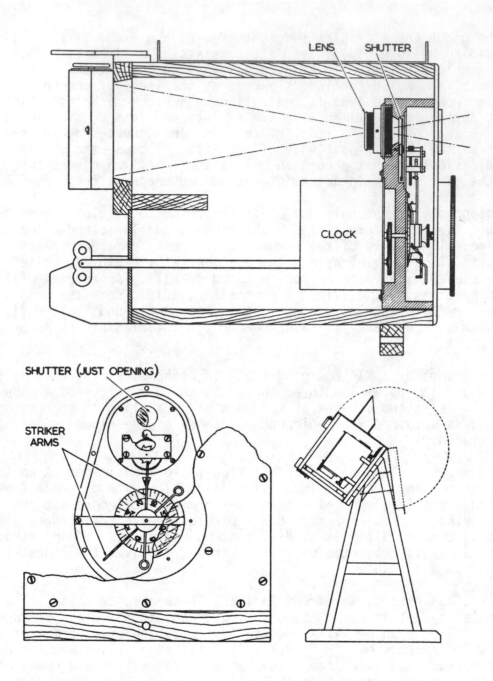

Figure 6-65 Starshine Recorder

detail on the negative. The camera with lens and motor-driven film advance are
protected by a spherically ground 9.5 in. (24.1 cm) diameter glass dome. A blower
system, activated by temperature sensors, ventilates the dome. A circular plate
mounted outside the dome shields the lens from direct solar radiation; the plate
is kept in position with a motor turning at 1 revolution per day. A light sensitive timer activates the system at sunrise and shuts it down at sunset. A special
densitometer with optical scanner views the negative and calculates areas of sky
cover as a function of density differences. The system developed by McArthur and

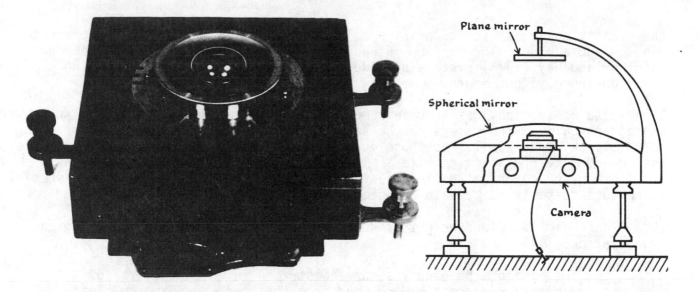

Figure 6-66 Georgi's Portable Sky Mirror

Hay is designed for measuring the distribution of diffuse solar radiation over the sky hemisphere; their approach is readily adaptable to sky cover determinations. They used a Canon F-1 SLR 35 mm camera with a 7.5 mm 1:5.6 fish-eye lens with built-in sky filter. The color negatives are electronically scanned and subdivided into a 32 step density scale (where density is a function of the log of intensity). By employing an electronic counter and by suitable selection of density levels, the negative scanner can yield fraction of cloud cover derived from the negative. Also,

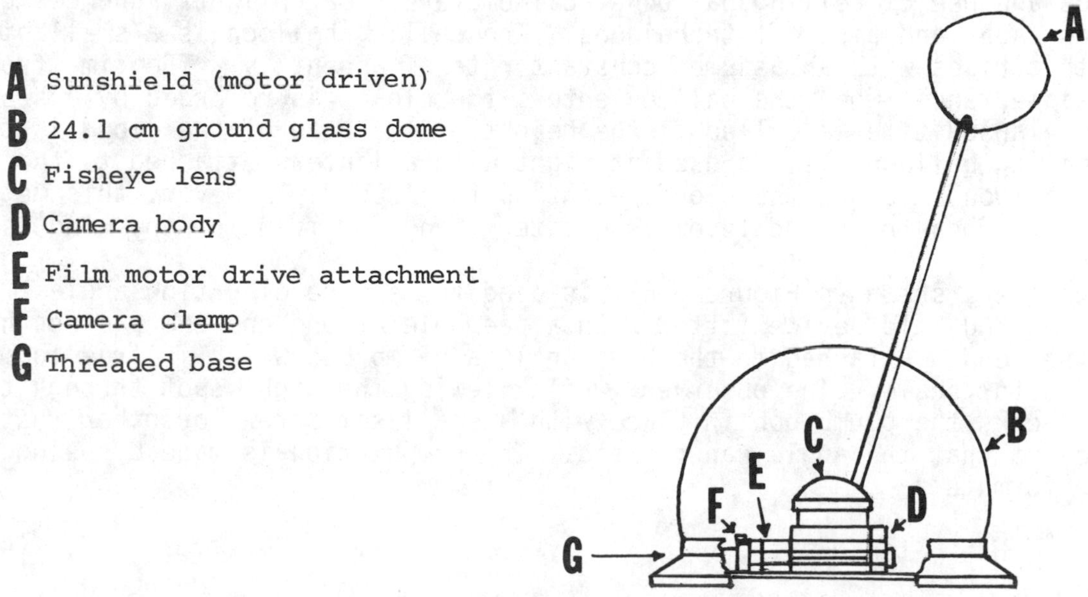

A Sunshield (motor driven)
B 24.1 cm ground glass dome
C Fisheye lens
D Camera body
E Film motor drive attachment
F Camera clamp
G Threaded base

Figure 6-67 Sky Camera with Fish-eye Lens, Motor Driven
Film Advance, and Timer

the density scale is used to print a color image where a yellow color represents minimum and a blue color maximum intensity. The Pochop and Shanklin (1966) and McArthur and Hay (1978) systems provide alternative methods for determining cloud cover during daylight hours.

Time lapse photography using movie cameras is another method of recording sky cover photographically. The camera can be aimed either towards a sky quadrant or used in an all-sky mode similar to those described for still cameras above. Film processing and interpretation require similar methods used for still photographs. Correlating sky coverage to cloud features photographed obliquely requires special techniques suitably adapted to computer handling.

(2) Cloud Height. Cloud height as defined by the International Meteorological Organization (1947) is:

> "The base of the cloud is the lowest zone in which the type of obscuration perceptibly changes from that corresponding to clear air or haze to that corresponding to water droplets or ice crystals. In the air below the cloud those particles which are responsible for obscuration show some spectral selectivity while in the cloud there is virtually no selectivity (due to the different droplet sizes involved)."

The height of the same cloud base is not the same when measured by different techniques. This difficulty in observing the exact location of the base is due to a transition zone existing between the clear air or haze and the actual water droplets or ice crystals, and the fact that the base of the cloud is not perfectly horizontal but may undulate, protrude, or appear convex in form. The variations in the measurements caused by these two effects can be 150 ft or more.

Methods for cloud height measurements (ceiling measurements) include simple visual methods and use of ceiling balloons, clinometers, searchlights, and ceilographs besides radar and aircraft techniques. The ceiling balloon is a small pilot balloon that rises with an assumed constant rate of ascent, v. The time from release to disappearance, when the balloon enters the cloud, is recorded by a stop watch. If the elapsed time is called t, the height of the base of the cloud is then vt. The ceiling balloon also is used at night with a lantern attached to the balloon. These balloons measure the cloud height up to ~1000 m. However, this method is inaccurate when the cloud level is greater than 1000 m and/or the wind is strong.

A clinometer, shown in Figure 6-68, is used to measure elevation angles. It is a simple hand-held device fitted with a peephole at one end and a cross hair at the other end. Attached to the housing is a plumb bob which is free to move over a protractor scale. The observer, while viewing the light spot through the clinometer, locks the plumb bob in place with a small set screw (or other restraining device) so that the angle can be read. The computation is made by using the simple formula

$$H = B \tan \alpha \quad\quad\quad\quad\quad\quad\quad\quad\quad\quad\quad\quad\quad (49)$$

where H is the height of the cloud base; B, the length of the baseline by estimation; and α, the elevation angle from the observer to the observed point as measured by the clinometer.

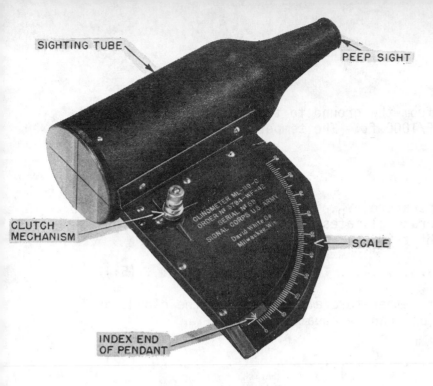

SIGHTING TUBE

PEEP SIGHT

CLUTCH MECHANISM

SCALE

INDEX END OF PENDANT

Figure 6-68   Clinometer

The cloud searchlight shown in Figure 6-69 is used for ceiling measurements at night. This small aperture searchlight, aimed vertically upward, shines a spot of light on the cloud base. It is located at one end of a measured baseline, B. At another end of the same baseline the observer with the clinometer measures the inclination angle, α, between the horizon and spot of light on the cloud base. Usually the average of three consecutive readings of the clinometer is used. The height of the cloud base, H, can be determined by Eq. (49) more accurately with the searchlight method than with a clinometer alone, as the baseline is usually measured precisely. For example, for a baseline of 500 ft., the height of clouds at various angles of elevation are:

B = 500 ft.

α	H
56°	700
57°	700
58°	800
59°	800
60°	800
61°	900

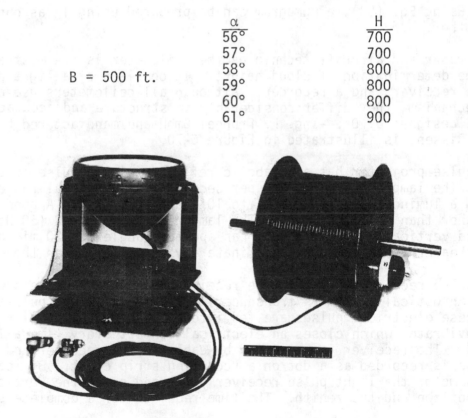

Figure 6-69   Cloud Searchlight

During the day, the base height, H, of a convective cloud can be calculated approximately from surface temperature, T_0, and dewpoint temperature, D_0, observa-

233

tions. The lapse rate of the air from the ground to the cloud base is assumed to be dry adiabatic, where $\gamma = 5.5F/1000$ ft. The temperature at any height z is given by

$$T_z = T_0 - \gamma z \ldots\ldots\ldots\ldots\ldots\ldots\ldots (50)$$

The lapse rate of the dewpoint temperature, γ_D, is a function of D_0 and D_z, but a reasonable approximation in temperate climates is $\gamma_D = 1.1F/1000$ ft. Thus, the dewpoint temperature at a height z is

$$D_z = D_0 - \gamma_D z \ldots\ldots\ldots\ldots\ldots\ldots\ldots (51)$$

The temperature equals the dewpoint temperature at the base of the cloud, or $T_z = D_z$ at a height H so Eqs. (50) and (51) can be equated as

$$H = \frac{1}{\gamma - \gamma_D} (T_0 - D_0) \ldots\ldots\ldots\ldots\ldots\ldots (52)$$

or

$$H = 227 (T_0 - D_0) \ldots\ldots\ldots\ldots\ldots\ldots\ldots (53)$$

Instead of using Eq. (53), a nomogram can be prepared using T_0 as one axis and D_0 as the other.

Apart from radar and aircraft techniques the ceilometer is the most accurate instrument for the determination of cloud height. It consists of a light pulse projector, light pulse receiver, and a recorder. Although all ceilometers use essentially the same mechanism, they differ considerably in structure and accuracy. The ceilograph, designed by Dr. -Ing. F. Früngel GmbH and manufactured by IMPULSPHYSIK of Hamburg-Rissen, is illustrated in Figure 6-70.

The light pulse projector has a parabolic reflector with a pulse lamp centered at its focus. The lamp emits 5 flashes per second with a spectral range of 0.25 to 6.60 μm and a luminous density of 10^7 to 10^8 lumens cm^{-2} (see Appendix 1), which is much higher than daylight level. The lamp requires 220V AC (50 Hz). The light is projected vertically upward through an aperture angle of ±20 min of arc. Usually a built-in heater is installed to eliminate any condensation on the optical system.

The light pulse receiver, or phototube receiver, uses a quartz lens with a 30 min aperture. An optical system (a transducer) converts the light pulses into electrical pulses. These electrical pulses are fed into a broad band amplifier and then fed into a monovibrator which closes an electrical contact every time a light pulse is received. The receiver is connected by cable to the recorder and every electrical pulse is recorded as a dot on a recording strip chart. By means of a synchronous motor the light pulse receiver tilts and scans the vertical plane between the horizon and the zenith. The time required for a complete scanning is one minute.

A base line of 250 ft is set up on a horizontal ground surface, with the projector at one end and the receiver at the other. The recorder is fitted with an altitude scale (100 to 1000 ft) corresponding to the length of the baseline. The higher the cloud base to be measured, the longer the baseline required. The height of

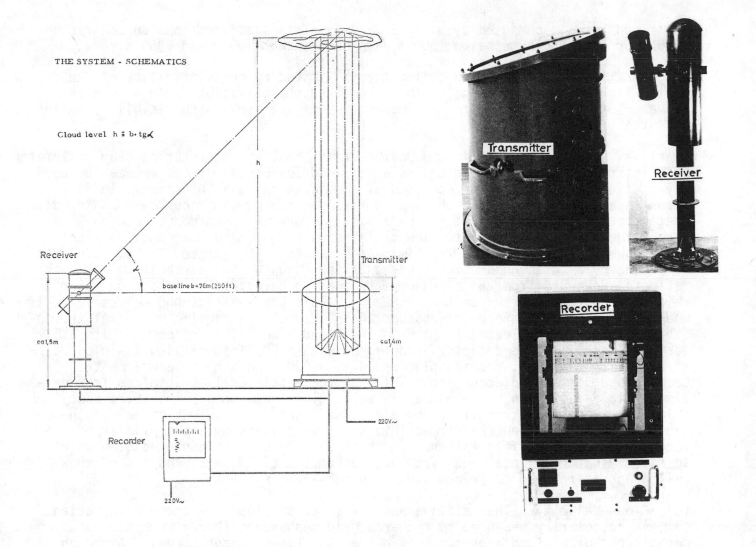

THE SYSTEM - SCHEMATICS

Cloud level h = b·tg∢

h

Receiver

base line b=76m(250ft)

Transmitter

ca1,5m

ca1,4m

Recorder

220V~

220V~

Transmitter

Receiver

Recorder

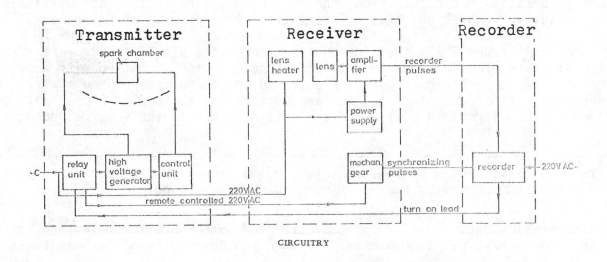

CIRCUITRY

Figure 6-70 Ceilograph

the cloud base is computed by using Eq. (49). This instrument has an accuracy of ±2.5% for a slant range below 1500 m and ±5% between 1500 m and 3000 m.[86/]

In general, the accuracy of a ceilometer is limited by the verticality of the projected light, the slant range of the cloud, and the elevation angle of the receiver. Of course, the system is inoperative if the prevailing visibility is less than the baseline, B.

Nearly every major airport in the United States employs the rotating beam ceilometer. This instrument usually is installed at the middle runway markers and can be used day or night. Its projector consists of two identical 3×10^6 candela iodine-vapor incandescent lamps. Each lamp is fixed at the 10-in. focus of a 24 in. diameter parabolic mirror. The two parabolic mirrors are mounted back-to-back on a rotating motor-driven shaft. The shaft turns at 5 rpm so that emitted light sweeps upward from 0° to 90° ten times each minute. The emitted light, modulated to 120 Hz by a rotating shutter mounted at the focus of the parabolic mirror, is reflected downward from the cloud base to the photoelectric cell of a detector, which is located 400 ft from the projector. This lead-sulfide photoelectric cell mounted at the 10-in. focus of another 24 in. diameter parabolic mirror of the detector senses the reflected light from the cloud base. The projector emits a 5° beam of light, and the detector is designed with a 5° field of view. A cloud base at 2,500 ft elevation, for example, produces a 220 ft diameter beam from the 5° beam spread of the projected light. The photoelectric cell is shielded from daytime visible radiation by either an infrared glass dome or a 5 in. thick honeycomb filter for reducing stray light and direct sunlight. The output signal of the photo cell is amplified and formed into an AC wave train by the amplifier of the recorder. This amplifier is tuned to the frequency of the modulated beam so as to discriminate reflected beam from extraneous light. More discussion of the ceilograph as related to French lidar and ruby lasers appears in Chapter 7.

(3) <u>Cloud Velocity</u>. The relative angular speed of clouds, expressed in radians s^{-1}, is customarily measured by either a grid nephoscope (known as Besson's nephoscope) or a cloud nephoscope (known as the Fineman nephoscope). Although they are different in structure, the two are very similar in principle. For further details on the grid type, the reader should refer to "Instruments for Investigating Clouds" and "Cloud Observation,"[87/] as well as Figure 6-71.

As shown in Figure 6-72, the Fineman nephoscope consists of a mirror A made of a disc of black glass mounted on a tripod stand B, which is fitted with leveling screws. Two concentric circles are engraved on the surface of the mirror. Horizontal azimuth angles of 0° to 360° are engraved on a circular brass plate surrounding the mirror. A millimeter scale is attached to the vertical pointer

---

[86]IMPULSPHYSIK. 1963. Ceilograph (Light Pulse Cloudbase Ceilometer) Service Manual. IMPULSPHYSIK, Hamburg-Rissen, Germany. 109 pp.

[87]Middleton, W.E.K. and A.F. Spilhaus. 1953. Instruments for investigating clouds, In: Meteorological Instruments, 3rd Ed. The University of Toronto Press. 286 pp.

British Meteorological Office. 1956. Cloud observations, In: Handbook of Meteorological Instruments, Part I, Instruments for Surface Observations. Great Britain Meteorological Office. 458 pp.

C, which can be adjusted to various heights by a rack D. Two knobs EE are used to rotate a circular collar around the center of the disc, which has a common vertical axis but is independent of the mirror. Figure 6-72 includes a schematic diagram used in the computation of the relative cloud speed where: b is the center of the mirror; O, the tip of the pointer c; $\overline{cb}$, the image path of the cloud moving from B to C; h, the height of the pointer above the mirror; and H, the height of the cloud above the mirror. Letting $\overline{cb}$ equal d and $\overline{BC}$ equal D, and letting the time interval required for the cloud image to travel from c to b be t seconds as measured by a stop watch, then from the two similar triangles ΔBCO_1 and ΔbcO we have

$$\frac{D}{H} = \frac{d}{h} \quad \ldots \ldots \ldots \ldots \ldots \ldots \ldots \ldots \ldots \ldots \ldots \quad (54)$$

When Eq. (54) is divided by time, t, and rearranged

$$\frac{D}{t} = \frac{d}{ht} H \quad \ldots \ldots \ldots \ldots \ldots \ldots \ldots \ldots \ldots \quad (55)$$

The coefficient of H in Eq. (55) is the angular speed of the cloud or the relative cloud speed. When H is obtained from a cloud height instrument, the true cloud speed can be computed by Eq. (55).

The reverse direction of the mirror image from c to b is the direction of the cloud movement. This is determined by the angle between the compass needle and the path $\overline{cb}$ on the horizontal azimuth scale. As the compass needle is always oriented toward north, the angle is expressed in deg from the north, indicating the direction from which the cloud is moving.

6.7.2 Hyetometry. Hyetology is the study of the origin, structure, and other features of all forms of condensation, sublimation, and precipitation. Condensation and sublimation refer to the changes of water vapor to liquid and solid states, respectively. Examples of condensation are dew, frost, and in some areas, fog drip. Precipitation refers to the falling out of liquid or solid water. The most important examples of precipitation are rain, snow, and hail; they also are called hydrometeors.[88] Although hyetometry is defined as the measurement of precipitation and condensation of all forms, a discussion of condensation products is postponed until Chapter 12 in which hydrologic measurements are emphasized.

(1) Non-recording Precipitation Instruments. The non-recording precipitation gage is the simplest device used to measure the amount of precipitation. Hereafter, the word gage is used to mean precipitation gage, a non-recording device.

As shown in Figure 6-73, the standard 8-in. US National Weather Service raingage consists of three separate parts: the collector A, the inner measuring tube B, and the overflow can C. The collector has a round open surface with an 8.00 in. diameter for receiving rainfall, and a small hole in the apex of its funnel-shaped bottom for conduction of the rainwater to the inner measuring tube, which is 20 in.

---

[88] Buettner, K.J.K. 1958. Sorption of the earth surface and a new classification of kata-hydrometeoric processes. J. Meteorol. 15(2): 155-163.

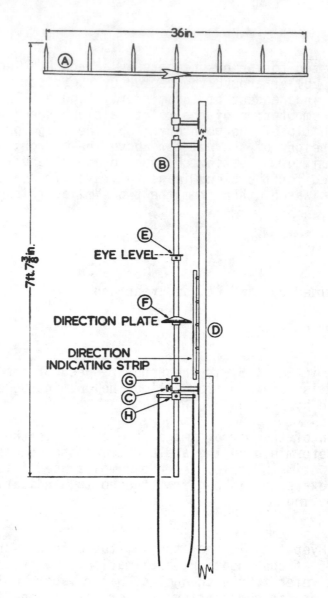

Figure 6-71   Besson Comb Nephoscope

long and has a cross-sectional area one-tenth that of the collector. The inside
diameter of the inner tube is ~ 2.54 in.   The depth of rainwater collected in the
inner tube therefore is 10 times that received by the collector.   The gage is
supported by a metal tripod D, with the rim of the collector more than 30 in.
above the ground.   Water in excess of the capacity of the inner tube flows into
the overflow can.   The amount of rain is measured by a measuring stick E, having
graduations one-tenth of an inch apart.   Each graduation is equivalent to 0.01 in.
of rainfall and is thus labelled.   Similarly, the 5-in. British Meteorological
Office gage consists of a collector, an inner collecting can, and an overflow can.
The collector has an exposed surface with a 5 in. inside diameter, a brass rim
more than 4 in. deep,[89] and a funnel-shaped bottom.   The rain received by the

---

[89] The rim depth of the various approved types of raingage in Great Britain vary
between 4 and 5 3/8 in. For details see Meteorological Office. 1961. Measurement
of precipitation, In: Handbook of Meteorological Instruments, Part I. Instruments
for Surface Observations. Meteorological Office, Air Ministry, London. 458 pp.

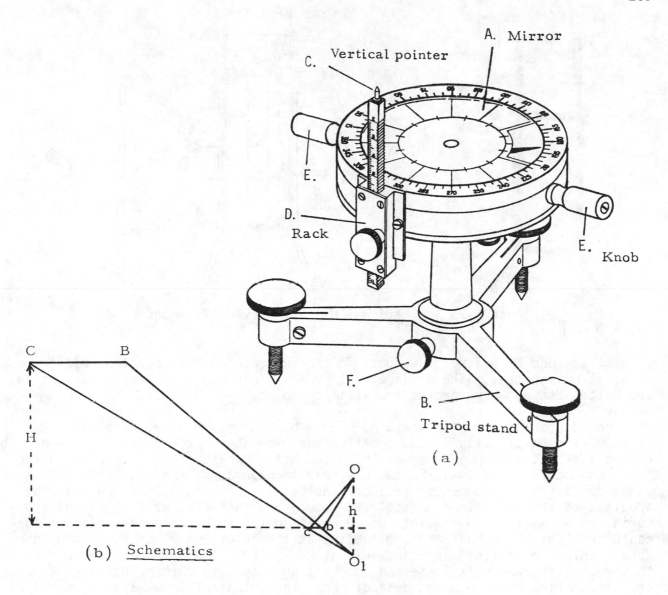

C. Vertical pointer

A. Mirror

E.

D.
Rack

E. Knob

F.

B.
Tripod stand

(a)

C   B

H

O

h

c   b

O₁

(b) Schematics

Figure 6-72 Fineman Nephoscope

collector flows into the inner can and the amount of excess rain is determined
either by pouring it into measuring glass cylinders known as rain measures or by
a diprod of a different type than the measuring stick used in the United States.
Although the overall height of the official British gage varies between 18 and
27 1/8 in., the exposed surface is always set exactly 12 in. above ground level by
placing the gage in a hole of proper depth. The 3.57 in. Canadian Meteorological
Division gage[90] is similar to the British gage except for its extremely small
size. The overall height of the gage is 10 1/2 in. and its base is set 9 in.
below the ground level. The rim depth of the collector is less than the British
collector, but greater than the US National Weather Service type.

[90] Middleton, W.E.K. and A.F. Spilhaus. 1953. The measurement of precipitation
and evaporation, In: Meteorological Instruments. The University of Toronto
Press, Toronto. 286 pp.

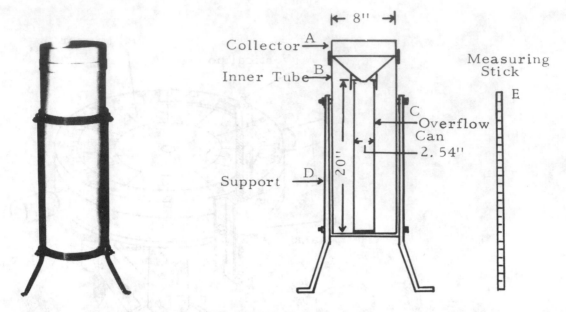

Figure 6-73 US National Weather Service Raingage

Length, instead of volume or weight, is the internationally accepted unit for rainfall measurement. The millimeter is the standard unit of length, except in the United States where inches are still employed.

When solid precipitation is measured with a US National Weather Service gage, it is collected directly in the overflow can with the collector and the inner tube removed; then the snow is completely melted before measurement. In a new design a propane-powered catalytic heater has been incorporated as a heat source in snow gages by Sierra Environmental Products (Berkeley, California). A venturi tube accelerates the gas into the catalyst, regulating fuel flow at 0.5 psi while ensuring even temperature regulation. Heat is ducted from the heater, located beneath the gage, to a heat chamber located at the inlet funnel at the top of the gage. Inlet temperature is maintained at 26C (80F). Thus, the energy is directed to the location where it is needed for melting purposes, rather than used to overheat the entire gage. In the case of snow, the amount of snow water measured with a gage is called the water equivalent. The snow density, ρ_S, is defined as the ratio of the water equivalent, W, as measured with a gage, to the depth of snow, H, as measured with a snowstake. That is

$$\rho_S = W/H \text{ (Dimensionless)} \dots\dots\dots\dots\dots\dots\dots\dots (56)$$

In the case of hailstones, various characteristics such as shape, roughness, structure, size, color, surface temperature, and density are of importance. Hail indicators and other specialized instruments, both electronic and mechanical, have been designed to obtain such measurements. Towery et al. and List[91] have made very

---

[91] List, R. 1961. Physical methods and instruments for characterizing hailstones. Bull. Amer. Meteorol. Soc. 42(7): 452-466.

Towery, N.G., S.A. Changnon, Jr., and G.M. Morgan, Jr. 1976. A review of hail measuring instruments. Bull. Amer. Meteorol. Soc. 59(9): 1133-1140.

comprehensive summaries of various physical methods and instruments for hailstone measurements; Schleusener and Jennings[92] have introduced a simple hail indicator. With an ordinary gage, the water equivalent can be measured only for small and medium-sized hailstones. The measurement procedure is about the same as that for snow.

How well a gage measurement represents the amount of precipitation reaching the ground depends on the following factors.

a. Areal representativeness is related to the gage density (i.e., number of gages per unit area). Usually the sampling area relative to ground surface area is extremely small. For example, with a density of 1 gage per acre, each gage represents less than 8×10^{-6} acre. In practice, there are few large regions in the world that have a density of gages greater than one for every 6400 acres. In most regions the fraction is much smaller. Non-representativeness of measurements is greatly increased where the rainfall is sporadic or the topography is not uniform.

b. The ratio of the diameter of a gage to its rim depth affects the catch appreciably. Small differences in diameter[93] are less significant than rim depth in determining the total amount of rainfall collected. For example, the US National Weather Service type with its shallow rim tends to lose rain that splashes out, while the British type with its deep rim can capture some water that does not belong in it.

c. The amount of rain measured is inversely proportional to the gage's height above the ground. The higher the gage, the greater the wind speed and turbulence above the exposed surface, and the less the catch. Wind shields of various designs often are provided to minimize the effects of wind on the collection of rain, snow, and hail.

An experiment has been conducted by Neff[94] from four locations comparing gages 1 m above ground with gages at the ground surface. For all rainfall events, the catch of the former is between 5 and 15% less than the latter. For single rainfall events the total catch exceeded 12 mm, and the error ranged between 0 and 75% depending upon wind characteristics during the storm.

[92] Schleusener, R.A. and P.C. Jennings. 1960. An energy method for relative estimates of hail intensity. Bull. Amer. Meteorol. Soc. 41(7): 372-376.

[93] A 40 to 50 square foot gage is recommended as the standard gage for calibration of various types of gages, as the diameter to rim depth ratio becomes extremely small.

[94] Neff, E.L. 1977. How much rain does a raingage gage? J. Hydrology 35(3/4): 213-220.

      d. Evaporation of the rain in a gage can be quite large after
          a summer shower if measurement is not made immediately. In-
          tense radiation, high wind speeds, and high temperatures are
          especially conducive to a high rate of evaporation.

      e. Non-uniformity of gages with respect to the diameter of the
          collector, diameter of the inner tube, and graduations of the
          measuring stick reduces the comparability of rain records over
          a geographical region.

In addition to the above sources of error, the effects of gage exposure are of
great importance.

(2) <u>Recording Precipitation Instruments</u>. In the measurement of precipitation
by a recording raingage, intensity, duration, and amount are the three major
considerations. The intensity refers to the amount of precipitation per unit
time, sometimes called the rate of rainfall. The duration signifies not only the
total time of fall but also the beginning and ending times. The amount usually
means either the total catch within a 24-hr period or the total precipitation
during a storm. The ordinary type of raingage described previously collects the
total amount of rainfall and must be read visually with a measuring stick. To
measure the intensity and duration as well as the amount, automatically and in-
stantaneously, recording types of gages are necessary. Many kinds have been designed and are classified according to the nature of measurement: (a) those that
record the rate of rainfall and (b) those that record the total amount and duration.[95] The Jardi rate-of-rainfall recorder (see Figure 6-74) gives a continuous
record of rain intensity as low as 3 to 5 mm per hr and has a capacity of up to
150 mm per hr. The British M.O. rate-of-rainfall recorder and the Sil's rate-of-
rainfall recorder also belong to the first class. The former records at 1 or 3
min intervals, while the latter gives an average rate over a period of 1 min
with a sensitivity of 0.02 in. per hour. The storm gage of Kew Observatory in
England is still another type which records rain or hail every minute. Although
all of these intensity gages record the rate of rainfall over time increments of
one or more minutes, their principles of mechanical operation differ considerably.

For the Jardi recorder, shown in Figure 6-74, the rate of outflow of water through
the circular opening of partition F depends upon the position of spindle H. The
higher the float E rises, the larger the opening, and the float will continue
to rise until the rate of outflow equals the rate of inflow from collector A.
When such equilibrium is established, the float maintains a constant height h
above its zero position.[96] At the zero position, the opening is assumed to be

---

[95]The second class can be further subdivided according to the mechanism employed:
(a) float gages without automatic siphoning arrangements, (b) float gages with
automatic siphoning arrangements, (c) tipping bucket gages, (d) weighing gages,
and (e) tipping bucket and weighing gage combinations. For details, the reader
may refer to Handbook of Meteorological Instruments, Division of Instruments,
British Meteorological Office, HMSO.

[96]It is the position in which the float E is at rest upon the partition F.

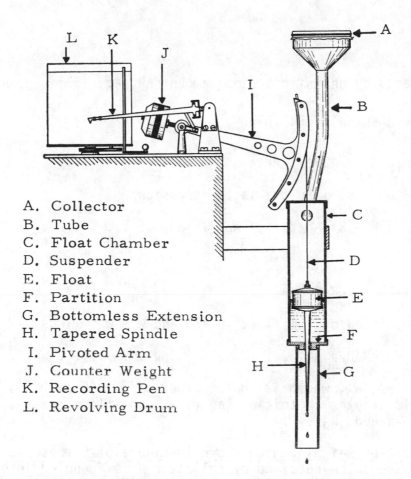

A. Collector
B. Tube
C. Float Chamber
D. Suspender
E. Float
F. Partition
G. Bottomless Extension
H. Tapered Spindle
I. Pivoted Arm
J. Counter Weight
K. Recording Pen
L. Revolving Drum

Figure 6-74 Jardi Rate-of-rainfall Recorder

closed as the diameter of the spindle at this time equals that of the opening.
The movement of the float is magnified by a system of levers and is recorded on
the strip chart of drum L through the angular motion of pivot arm I from which
the float is suspended by a German silver strip D. The rate of outflow through
F is proportional to the area of the opening and to the square root of the depth
of water in chamber C. This rate is directly proportional to the rate of rainfall, w, which is a function of h, i.e. w = bh where b is a constant. Let R be
the radius of the opening in partition F and r, the radius of the spindle at a
distance h from the base of the float. The depth of water in chamber C is (h+a),
where a is the depth of float immersed in the water, a constant value. From the
above statements

$$w = c(R^2 - r^2)\,(h+a)^{1/2} \dots \dots \dots \dots \dots \dots \dots \dots \dots (57)$$

where c is a constant, and

$$r = \left[R^2 - \frac{Ah}{\sqrt{h+a}} \right]^{1/2} \dots \dots \dots \dots \dots \dots \dots (58)$$

where $A = b/c$.

When R increases by an infinitesimal amount ΔR, Eq. (57) becomes

$$w = c[(R+\Delta R)^2-r^2] \sqrt{h+a} \quad\quad \text{.............. (59)}$$

$$w = c(R^2-r^2+2R\Delta R) \sqrt{h+a}$$

Substituting Eq. (58) into the last expression

$$w = c[R^2-(R^2-Ah/\sqrt{h+a})+2R\Delta R]\sqrt{h+a} \quad\quad \text{............ (60)}$$

or

$$w = bh + 2cR\Delta R\sqrt{h+a}$$

At the zero position of the float, $h = 0$ and $w = w_o$. Thus

$$w_o = 2cR\Delta R\sqrt{a} \quad\quad \text{................... (61)}$$

The rate of flow, w_o, which is on the order of 3 to 5 mm hr^{-1}, is not recorded by the pen. This creates an initial lag (i.e., the float remains at the zero position until w exceeds w_o).

The Fergusson type weighing gage shown in Figure 6-75 measures the amount of precipitation. The catch received by collector A is funneled into bucket B mounted on the mechanism and is recorded on drum chart H as in. of rainfall. The vertical column which supports the bucket and platform C is restrained by helical spring G suspended from adjustable arm D. The pen arm movement is controlled by two vertical slotted levers E, and is overbalanced by counterweight F. As the vertical column of the lever system lowers, the pen arm moves upward until the upward traverse is arrested by the limits of the slots in the two vertical levers. The slot limits are adjusted to coincide with the top of the recording chart. When the upward motion of the pen arm has been stopped, further lowering of the vertical column causes the pen arm to move down the chart. This dual traverse feature provides for precipitation measurements up to 12 in., of which 6 in. are represented by the upward stroke and 6 in. by the downward stroke. Oil damper I is provided to minimize the oscillation of the balance by the wind. When precipitation other than rain is measured, it is necessary to remove the collector. The water equivalent of such precipitation can be obtained by melting the snow or ice, done by pouring warm water of known quantity into the bucket and then measuring the total amount of liquid water.

The tipping bucket gage shown in Figure 6-76 consists of a collector, a triangular-shaped bucket, an electric transmitting system, and a revolving drum. The bucket is balanced in unstable equilibrium about a horizontal axis and is divided into two equal compartments by a partition. When there is no rain, the bucket is always tilted with one side resting against a stop. During rainfall, when the weight of the catch reaches a certain small amount (usually equivalent to ~ 0.01 in.), it will tip the bucket and empty it, exposing the other compartment of the bucket to the rain. The bucket makes electrical contact which is recorded from any distance.

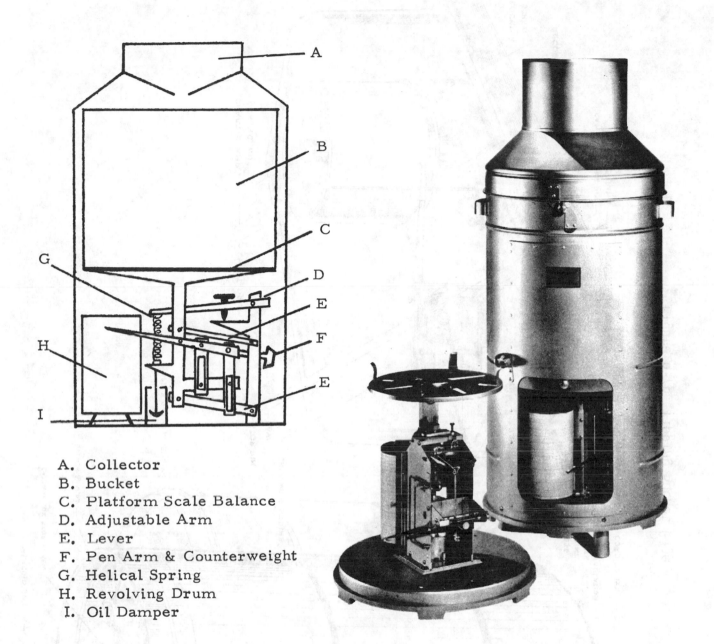

A. Collector
B. Bucket
C. Platform Scale Balance
D. Adjustable Arm
E. Lever
F. Pen Arm & Counterweight
G. Helical Spring
H. Revolving Drum
I. Oil Damper

Figure 6-75 Fergusson Type Weighing Gage

The main advantage of the intensity gage is that it indicates the rate of rainfall with considerable precision. Its disadvantages are: (a) neither the amount nor the duration is recorded, (b) frequent calibration is needed, and (c) the total amount of rain measured is rather limited. The weighing gage measures precipitation of any form and is recommended for use in cold climates. Its disadvantage is that winds of 15 mph or more cause oscillation of the balance. The tipping bucket gage has the advantage of continuous remote recording. Its limitations are: (1) no record is made for less than 0.01 in. catch in the bucket, (2) it takes about 0.2 s to tip the bucket, (3) it records the amount of rain only, and (4) it cannot be used in freezing weather. In short, the tipping bucket gage is neither suitable for use in light rain and drizzle nor is it reliable for

246

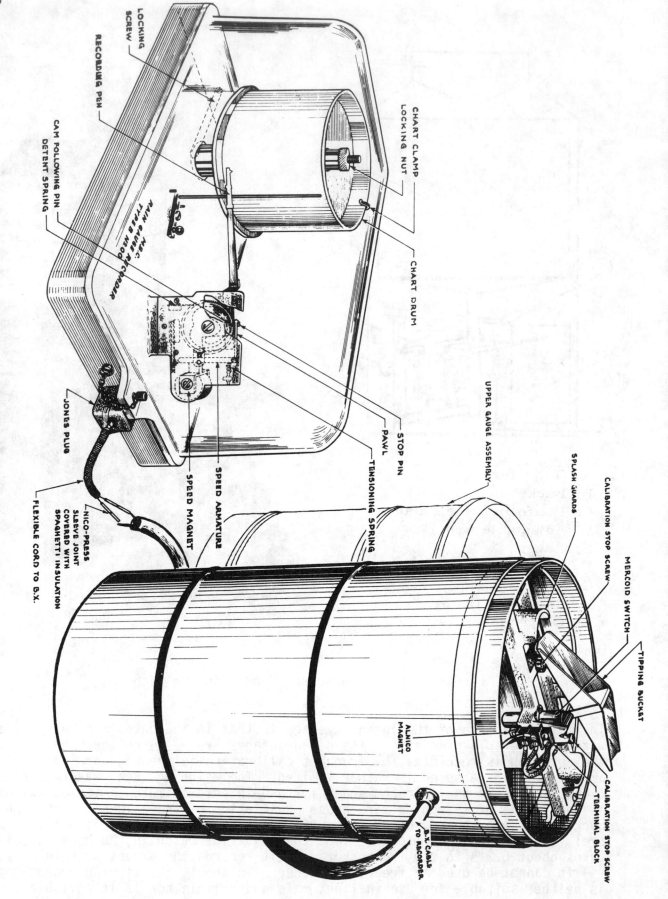

Figure 6-76 Tipping Bucket Gage

recording a shower of about 2 in. hr^{-1} or more. The former is susceptible to evaporation error, and the latter condition causes an inaccurate recording because of splashing.

(3) <u>Innovations Required in Precipitation Measurement</u>. Problems of precipitation measurement can be grouped into two main categories: (a) regional representativeness and (b) accuracy of point source measurements. While the first relies upon sampling techniques, the second is essentially a problem of instrumentation.

(a) <u>Regional representativeness</u>. The amount of rainfall rather than duration or intensity is stressed here to illustrate the representativeness of rainfall distribution.

In macroscale analysis, the distribution of rainfall corresponds closely to the topography and location of the continent. The presently available rainfall station networks seem adequate to describe the seasonal variability of different climatic regions for most developed nations. But when a single storm occurs, the mesoscale distribution of rainfall varies drastically within distances of a few kilometers. This is particularly true during a summer sporadic rain storm. A rainfall station with one gage for an area of 500 to 600 km^2 (193 to 232 mi^2), such as in the United States, would be inadequate. The situation is worse for a microscale study. The representativeness of rainfall distribution becomes a problem of the exposure, location (latitude, longitude, and elevation), and density of gages as associated with the track and type of storm. Among these, gage density is the most crucial problem. The probability of accurate rainfall observation is a function of the total number of stations per unit area for a given period of time. The longer the time, the greater the accuracy or representativeness.

Root mean square per cent errors for the mean in rainfall amount in the USSR have been compiled by Rusin.[97] His data are plotted in Figure 6-77 in which the ordinate is the total number of stations in the area and the abscissa is the area covered by each station. It is obvious that the optimum gage density can only be defined by the degree of accuracy required by the designer. When the data from Figure 6-77 is applied to the United States, the overall representativeness of the USA rainfall sampling technique is found to be 50 to 60 per cent.

For many years the Illinois State Water Survey has conducted research on the representativeness of point source measurement versus radar detection of rain storms.[98] In addition, they made intensive surveys of snow and hail distribution. Their findings indicate that an appropriate increase of gage density according to orographic and geographic features and the application of remote sensing techniques can remedy sampling inaccuracy. As weather satellite observations are now available, further improvements using remote-sensing techniques are in the making.

(b) <u>Accuracy of Point Source Measurements</u>. The point source accuracy as a result

---

[97] Rusin, N.P. 1970. Precipitation observation development. Meteorol. Monographs 11(33):283-286.

[98] Among the many publications the following citation serves as an illustration of their studies. Huff, F.A. and J.C. Neill. 1957. Rainfall relations on small areas in Illinois. Illinois State Water Survey Bulletin 44.

248

of gage exposure, construction, and environmental effects has been thoroughly discussed earlier in this section. A brief summary of the errors in precipitation measurements is given in Table 6-20.

As far as construction is concerned, innovations of precipitation sensors can be made along the following lines.

i. All mechanical gages must be equipped with wind shields (see Figure 6-78) for minimization of turbulence; or the topography of the local site should be modified as recommended by the British Meteorological Office for locations where turbulence is serious.

ii. New prototype sensors should be developed. The measurement of the electric field between two charged plates between which drops are falling is one suggested approach. The use of piezocrystal plates or membranes as well as the application of acoustical methods for measuring the impulses of falling drops is another. Finally, the measurement of light attenuation by precipitation of different intensities provides another angle of attack. Some of the above suggestions are already under investigation by research scientists.

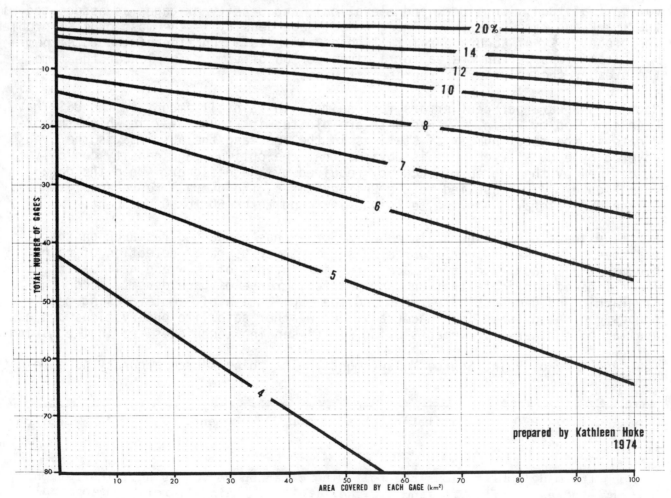

Figure 6-77 RMS Per Cent Errors in Mean Rainfall

(a) Turf Wall for British Meteorological Office Raingage

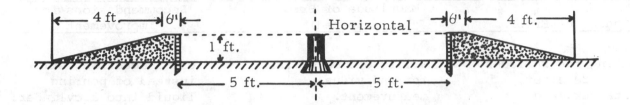

(b) U.S. National Weather Service Raingage Shield

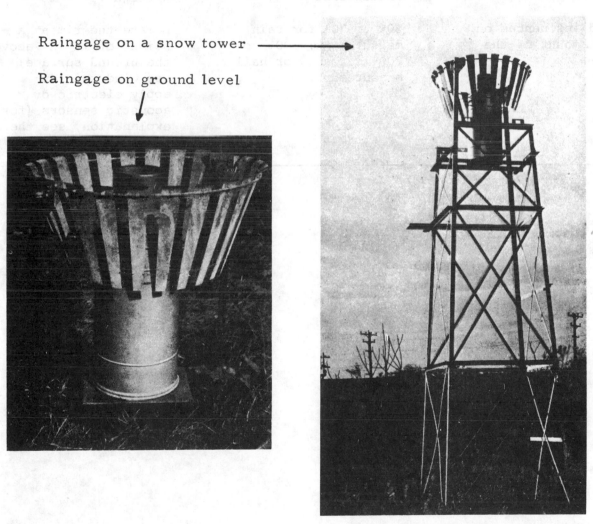

Figure 6-78 Sheltering of Raingage Against Wind Eddies

Table 6-20 Errors in Precipitation Measurement

Causes of Error	Magnitude of Error (Estimated)	Recommendations for Improvement
Wetting the surfaces of the receiver, container, and measuring device.	0.1 to 0.2 mm of rain for individual measurement.	Use a measuring stick instead of pouring liquid into a cylinder.
Evaporation of the catch.	0.3 to 0.5 mm of rain for individual measurement and 3% to 5% for the total annual fall	Use recording instead of non-recording gages; reduce size of aperture and make joints airtight.
Wind influences on the amount of the catch.	30% to 70% for rain measurement; 100% to 200% for snow or hail measurement.	Locate the rim at a height of 12 in. above the ground surface; use wind shields; apply electric or acoustic sensors (for explanation, see the text).

CHAPTER 7   UPPER-AIR INSTRUMENTS

Upper-air instruments are designed for measurements of the free atmosphere up to levels between 30 and 50 km. They are equipped mainly with electronic systems. They utilize essentially airborne platforms such as balloons, aircraft, and space-craft, combine several monitoring systems into one single unit, and include so-phisticated computer programs for data reduction and communication. These char-acteristics, though far from being unique, can serve as a basis for differentiating upper-air from surface instrumentation. Uniqueness becomes more pronounced with innovations in the field. The development of upper-air instrument systems has progressed over the past century; the greatest advances have been made only with-in the last decade or two. A historical overview of free atmospheric measure-ment precedes a discussion of some of the instrument system characteristics.

Historically, upper-air measurements began with temperature observations performed in 1749 by Dr. Alexander Wilson of Glasgow. He used kites as platforms for tem-perature measurements. About one and a half centuries later, routine meteorological kites were launched by the Blue Hill Observatory at Milton, Massachusetts, be-ginning in 1894. For the next few years temperature, pressure, humidity, and wind were measured by meteorographs (i.e., baro-thermo-hygro-anemographs) on large box kites, captive balloons, and free balloons. Systematic observations of tem-perature and humidity aloft using aircraft began in 1925. This same year the ex-ploration of the ionosphere with radio waves also began. Radiosondes were de-veloped in 1928 leading to regular high level measurements of temperature, humidity, pressure, and wind. The first weather radar was introduced in Great Britain in 1935. The use of rockets and aircraft for the exploration of the high altitude atmosphere began in 1946; and the first meteorological satellite, Tiros I, was launched on April 1, 1960. Regular launchings of high altitude rockets began in 1962. During the period between 1970 and 1979 the satellite program of the World Weather Watch made tremendous strides. Technological developments were made in space and ground systems by the United States, which launched and tracked these early weather satellites. Significant achievements included the invention of new sensors and the establishment of Command and Data Acquisition (CDA) stations. It has been less than 20 years since the first global coverage of daily weather con-ditions by meteorological satellites was made. The historical development of upper-air instrumentation is condensed in Table 7-1. Further elaboration is found in Section 17.1.

The first consideration in upper-air measurement is the selection of a platform. Platforms employed include airborne balloons, ground stations, aircraft, and spacecraft. The second consideration is the communication systems for upper-air soundings. Optical, radio, radar, mechanical, and even acoustical systems are utilized. These systems are an integral part of the platforms used and are dis-cussed as part of the sections on balloon, ground, and aircraft techniques. The third consideration is the selection of appropriate sensors facilitating measure-ments of radiation, temperature, moisture, and so forth in the free atmosphere. In this regard solid state electronic equipment and remote-accessed sensors are utilized for purposes of reducing platform payloads and power requirements. The final consideration is the measurement of upper-air variables associated with the type of platform and communications systems. Spacecraft, such as rockets, missiles,

Table 7-1 Historical Development of Upper-Air Measurements

Year	Platform	Variables[1]	Height[2]
1749 to ~ 1905[3]	kite with meteorograph	T, P, U and W	9-7
1784 to ~ 1950[3]	manned balloon with meteorograph	T, P, U and W	30
1892 to ~ 1900[3]	captive balloon with meteorograph	T, P, U and W	2
1873 to present	pilot balloon	W	8-10
1876 to present	radiosonde balloon	T, U, P; sometimes R and O_3	30-35
1912 to present	aircraft (including U-2 aircraft and helicopter)	P, T, ρ, O_3 R, W, U and others	20
1928 to present	rawinsonde balloon	W	30-35
1935 to present	radar: tracking weather	W Almost all weather conditions	30-80
1946 to present	rocketsonde	P, T, O_3, R, W and ρ	300
1960 to present	weather satellite	C, R, A, T, U; W indirectly	1500-35,000

[1] T = temperature, U = humidity, W = wind, P = pressure, ρ = density, R = radiation, O_3 = ozone, A = albedo, and C = clouds.

[2] These are the maximum heights in km. For a balloon, its bursting altitude is usually higher than the maximum heights indicated here.

[3] The final date denotes the approximate year in which the technique was considered obsolete.

balloons, and satellites, measure properties of the surface, and lower and upper atmospheres. The following discussion is restricted to instrument systems for measuring upper-air variables. Systems capable of profile measurements from the surface to upper atmosphere are discussed in Chapters 15 and 16.

7.1 Balloon Techniques.

In descending order of importance, balloons have been utilized in upper-air measurements as carriers of atmospheric sensors, tracers of air motion, and detectors of ceiling height. Balloons carry a variety of sensing packages. One such package, the US radiosonde, has components consisting of a baroswitch, thermistor thermometer, and carbon humidity element. The signals transmitted from the radiosonde to ground provide information on the package's altitude, and environmental temperature and humidity. Ozone concentration and net radiation devices are auxiliary instruments added at times to the basic package carried aloft by the carrier balloon.

Tracer balloons are used in a variety of applications. Common types and their functions are described as follows. The pilot balloon is inflated and released from ground level. It is tracked optically by an observer using a theodolite. The balloon trajectory as a function of time is then determined using the successive azimuth and elevation angles of balloon position, assuming the balloon has a constant rate of ascent. The data are used to graphically or mathematically solve for wind velocity in the free atmosphere. The rawinsonde and its balloon carrier are used to measure wind velocity irrespective of cloud cover, since the package is tracked by ground-based radar. A portion of the signals emitted by the ground radar is returned back to the surface from either a corner reflector on the balloon or by the balloon structure itself. Wind velocity is calculated as a function of balloon trajectory, specified in terms of its slant range, azimuth, and elevation angles. The output of these calculations is displayed by a plan position indicator (PPI) or range-height indicator (RHI).

The balloon method for detecting ceiling or cloud base elevations uses a 10 g ceiling balloon. The time between the balloon's release and its disappearance into the cloud base, assuming a constant rate of ascent for the balloon, is used to calculate ceiling height.

7.1.1 Balloons and Their Construction.

The construction[1] and dynamics of balloons frequently used in atmospheric soundings are discussed in general terms in this section. Details of the performance characteristics of the pilot balloon are given by way of a specific application.

(1) Pilot and Ceiling Balloons. Small expendable balloons, such as those used for ceiling or wind velocity measurements, are made either of natural or synthetic rubber (e.g., neoprene). Nominal balloon weights range between 10 g and 100 g; circumferences after inflation with either hydrogen or helium vary between 50 in. and 250 in. The lifting force, L, (or freelift) of a balloon is the difference between its buoyant force, F, and gross weight, W: $L = F = W$ (g). The gross weight includes the balloon itself, the buoyant gas, and any accessories attached to the balloon. Table 7-2 summarizes the range of characteristics of two common

---

[1] Balloons can be either spherical or elongated in shape. Their texture varies from smooth to rough. Their payloads range between 10 g and 100 kg.

254

balloon types.

Table 7-2 Characteristics of Ceiling and Pilot Balloons

Type	Color*	Weight g	Volume ft^3	Rate of Ascent ft min^{-1}	Lifting Force g	Bursting Altitude ft x 10^3
Ceiling Balloon	B and R	10	1.6	360	40	10
Pilot Balloon	W, B, or R	20+	3.3	500	--	10
		30	5.1	600	132	30
		80+	--	700	--	--
		100	21.2	990	575	45

*W = white, B = black, and R = red; occasionally blue and orange are used.
+Source: British Meteorological Office. 1956. Handbook of Meteorological Instruments, Part 1. HMSO, London. 459 p.

Color selection for a balloon is done to maximize its visibility for optical tracking. Day launchings require balloon colors that provide the maximum visible contrast between balloon and sky. The contrast is either that of brightness, color, or a combination of both. A white or clear balloon is highly visible against a clear or lightly hazy sky. However, should the white balloon be rapidly transported at low elevations where the background is hazy, then the contrast would be poor. When cumulus or cirrus clouds are present a white balloon also has no contrast. Red and orange balloons are compromise colors for use in scattered to broken cloud conditions with blue sky interspersed. Black and blue balloons show best against an overcast sky. The black color, regardless of cloud cover, gives maximum contrast at sunrise and sunset. Generally, a red balloon is the most adaptable color for a range of sky conditions; blue is the least adaptable.

The other performance characteristics of the balloons in Table 7-2 depend on their size, material, and type of buoyant gas used. The ceiling balloon, operated to elevations of 10,000 ft, is produced in one size and lift. Pilot balloons, however, are subdivided into four size and lift categories: 20 g and 30 g balloons for altitudes to 10,000 ft, and 80 g and 100 g balloons up to 45,000 ft. The corresponding rates of ascent are greater in the 80 g and 100 g balloons than in the lighter weights.

The rate of ascent of a balloon is calculated from the relationship between its free lift, L, and the weight, W, of its attachments[2]

$$w = \frac{72L^{0.63}}{(L+W)^{0.42}} \quad \dots \dots \dots \dots \dots \dots \dots \dots \dots \quad (1)$$

[2] The British Meteorological Office uses a different formula: $w = KL^{1/2}/(L+W)^{1/3}$, where K is a numerical factor dependent on the drag coefficient. The numerical value of K is 310 for 80 g balloons (w in ft min^{-1}); 275 for 20 g and 30 g balloons (w in ft min^{-1}); and 84 for 20 g and 30 g balloons (w in m min^{-1}).

Hydrogen or helium is used to inflate pilot balloons. Hydrogen has the lower density of the two gases and provides the greater buoyant force. Various types of hydrogen generators, available commercially, supply gas for fixed base and field operations. When hydrogen is employed, precautions must be taken due to its high flammibility.

Nighttime operation and tracking of a pilot balloon require a lighting unit since balloon color is immaterial under these conditions. The lighting unit consists of a lamp powered by a small 6V water-activated battery. The modified ML-388/AM lighting unit, commercially manufactured in the USA, weighs 31.75 g dry and 42.5 g wet. The battery is activated by immersing the entire lighting unit in water for 3 to 15 min prior to launch. The unit comes equipped with a parachute to lower it slowly to ground once the balloon has burst.

(2) Radiosonde and Rawinsonde Balloons. Two models of balloons with weights greater than 1000 g are used to lift radiosondes and rawinsondes in US operations. The standard balloon, J-350-4A, employed by the NWS to lift radio- and rawinsondes does not have rate of lift specifications. The AWS balloon, ML-537 ( )/UM, lifts rawinsondes at 1000 ft min^{-1}. Payloads carried by the NWS and AWS balloons, respectively, are 2800 ± 100 g and 1300 ± 500 g. The AWS model must pass aging and ozone resistance tests, requirements not enforced for NWS balloons.

Table 7-3 Specifications for Radiosonde and
Rawinsonde Balloons

Type	Color	Weight g	Inflated Diameter ft	Rate of Ascent ft min^{-1}	Bursting Altitude ft x 10^3
Radiosonde Balloon (J-350-4A)	Natural	1200	7.5	---	103.95
Rawinsonde Balloon (ML-537 ( )/UM	Natural	1100	6.0	1000	110.00

(3) Jimsphere. This NASA-developed balloon, shown in Figure 7-2, is a special radar-tracked balloon. Its skin is constructed of 0.5 mil metalized mylar using 12 tailored gores; its surface has 938 protrusions (cones), each having a diameter and height of 3 in. When inflated the Jimsphere is 6.6 ft in diameter. This balloon is tracked by an AN/FPS-16 high precision radar from which are obtained detailed profiles of vertical winds and wind shear. At an altitude of 9 km, the balloon's rate of ascent is accurate to ±1 m s^{-1} under wind speed conditions of 25 m s^{-1} (56 mph). The maximum altitude of the Jimsphere is approximately 18 km.

(4) ROBIN and GHOST Balloons. These two balloons are superpressure devices used for upper-air measurements. The ROBIN (acronym for Rocket Balloon Instrument) is launched by rocket into a vertical path. The GHOST (acronym for Global Horizontal

Figure 7-1 ML-537 ( )/UM
Meteorological Balloon

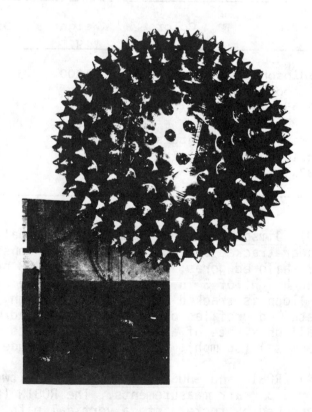

Figure 7-2 Jimsphere Radar
Reflective Balloon

Sounding Technique) balloon travels horizontally; its movements are tracked by weather satellites.

(a) <u>ROBIN</u>. Depicted in Figure 7-3, the Robin balloon is part of a dropsonde package used in the Loki launching system. It is packaged within the nose cone of a rocket vehicle; at rocket apogee (70 km) the balloon is ejected. Liquid isopentane vaporizes inside the balloon and inflates it to a spherical shape 1 m in diameter at a superpressure of 10 to 12 mb. A built-in corner reflector provides a target for the FPS-16 tracking radar. Space and time data on the falling balloon, derived from the tracking radar and computerized data reduction system, are used to derive meteorological information on upper atmospheric density, wind, and temperature. Proper inflation is crucial to the balloon, since it determines the accuracy of the density and wind data based on the balloon motion. Specifications of the Robin system are summarized in Table 7-4.

Table 7-4 Characteristics of Robin Balloon

	Diameter ft	Weight g	Range km	Density %	Accuracy Wind kts
Robin Balloon (PWN-7A)	3.3	122.8	70-30	$\pm 2\%$ at 1.2×10^5 ft	1 kt from 98×10^3 to 198×10^3 ft (inflated balloon)
				$\pm 3\%$ at 2.5×10^5 ft	8 kt for all altitudes (uninflated balloon)

(b) <u>GHOST</u>. This constant level balloon is used in exploration of pre-determined density levels at 30 mb or 50 mb of the southern hemisphere. The balloon is super-pressurized with helium 24 hr prior to its flight; the mylarized skin is pre-stressed at 10,000 psi. The balloon carries an instrument package that relays to the surface values of wind, temperature, and humidity; an orbiting satellite is the communications link between Ghost and the ground. More recently, Ghost measurements have been expanded to include monitoring of radiation, particulates, and gases at altitudes between 24 km and 30 km (30 mb to 10 mb) in the stratosphere above the tropics. GARP Global Atlantic Tropical Experiment (acronym GATE) used modified versions of Ghost balloons to carry payloads up to 100 kg into the upper atmosphere. These superpressurized balloons also act as carrier vehicles[3] for a brood of dropsondes which are released on command from an orbiting satellite at predetermined times. Each dropsonde weighs ~ 40 g and is shaped like a cylinder (50 cm by 7 cm diameter). A lightweight (30 g) cross parachute with two 25 cm by 87.5 cm rectangular panels lowers the dropsonde from 24 km to ground within 40 min.

---

[3] The carrier is called the Mother Ghost and its dropsondes, the daughters. R.S. Quiroz (1972) discussed many of the uses of superpressurized balloons in upper atmospheric research. Specific applications of constant-level, Ghost-type balloons are reported in: TWERLE Team. 1977. The TWERLE experiment. Bull. Am. Meteorol. Soc. 58(9): 936-948; and Morel, P. and W. Bandeen. 1973. The Eole experiment: early results and current objectives. Bull. Am. Meteorol. Soc. 54(3): 298-305.

7.1.2 Balloon Dynamics. The use of
balloons as carriers, tracers, or detectors in upper-air measurements implies a direction of motion that affects the balloon's position in space.
Ascending balloons, exemplified by
ceiling, pilot, radiosonde, rawinsonde, and tracking radar types,
must meet varying standards for rate
of ascent. Rates are assumed to be
constant for ceiling and single
theodolite pilot balloons since both
are tracked optically. Descending
balloons released from dropsondes
(both carried aloft by aircraft or
rocket) must have known rates of
descent. Constant level balloons,
such as the Ghost type, do not
have critical rates of lift; however, the accuracy to which they
are superpressurized determines
their adherence to a specified

Figure 7-3 Robin Balloon

density level. Even captive balloons, wiresonde, and tethered types, require adjustments in order to keep them at constant levels. Hence, balloon dynamics and
the interaction between balloon and environment must be carefully determined. The
basic assumptions and equations used to describe balloon dynamics are detailed in
the following paragraphs.

(1) Basic Assumptions and Equations. A constant rate of ascent for a balloon,
such as ceiling or pilot varieties, is based on these assumptions:

    a. The balloon exerts no compressive force on the contained gas,
       i.e., the internal and external pressures are equal.

    b. The temperatures inside and outside the balloon are equal.

    c. The balloon always remains spherical in shape.

    d. There is no loss of gas from the balloon during the entire
       flight.

Departures from the above assumptions obviously do occur.

As mentioned earlier, the free lift, L, of a balloon is defined as the difference
between the buoyant force, F, and the gross weight, W, of the balloon, i.e., L =
F - W. Rearranging terms, F = L + W. According to Archimedes' Principle, this
force must equal the weight of air displaced by the balloon, or

$$F = \rho V g \ldots \ldots \ldots \ldots \ldots \ldots \ldots \ldots \ldots \ldots \ldots \ldots \quad (2)$$

where ρ is air density; V, volume of the balloon; and g, the acceleration of
gravity.

Should the above assumptions apply, then the buoyant force can be derived in terms of the equation of state: $V = nRT/m_g P$. Density is defined as $\rho = m_a P/RT$, where n is the mass of gas (g) inside the balloon; R, the universal gas constant; m_g molecular weight of the gas; m_a, molecular weight of air; T, internal and external temperature; and P, internal and external pressure of the balloon. Substituting the above expressions for V and ρ in Eq. (2) yields

$$F = \frac{n m_a g}{m_g} \quad\text{. . . . . (3)}$$

In the absence of updrafts and downdrafts, Eq. (3) indicates that the buoyant force, F, is directly proportional to the amount of gas inside the balloon, and inversely proportional to the molecular weight of the same gas. As the balloon rises, however, the inevitable leakage of gas reduces the internal pressure and, hence, the balloon volume. Conversely, the increase in gas temperature by radiation heating increases its volume. These two opposite effects may or may not balance each other.

If a constant buoyant force is maintained throughout the flight, then a uniform upward acceleration results which is counter-balanced by a downward drag force, D, on the balloon surface. From wind tunnel experiments, the drag force is expressed as

$$D = \frac{1}{2} C_D \rho A w^2 \quad\text{. . . . . (4)}$$

where ρ is air density; A, cross-sectional area for a spherical balloon; w, rate of ascent; and C_D, drag coefficient. C_D includes the departure from a perfect square law; drag is a function of the Reynolds number, and, therefore, is an implicit function of w, rate of ascent.

The drag force, D, is a surface force, while the free lift force, L, is a body force acting at the center of the mass. It is permissible, however, to assume that D also acts at the center of mass if the balloon is spherical and well-balanced. Therefore, in equilibrium, the free lift force equals the drag force at all times. By denoting each factor in Eq. (4) with a zero subscript for equilibrium conditions, then

$$(\tfrac{1}{2} C_D \rho_0 A_0 w_0^2) = (\tfrac{1}{2} C_D \rho A w^2) \quad\text{. . . . . (5)}$$
$$\text{equilibrium} \qquad \text{any instant}$$

Eq. (5) is rearranged and simplified as

$$(w/w_0) = \frac{\rho_0 A_0}{\rho A} \quad\text{. . . . . (6)}$$

From the geometry of a sphere

$$A_0/A = (V_0/V)^{2/3} \quad\text{. . . . . (7)}$$

From the equation of state, $V = nRT/m_gP$ and $\rho = m_aP/RT$

$$(V_0/V)^{2/3} = (\rho_0/\rho)^{-2/3} \quad \ldots \ldots \ldots \ldots \ldots \ldots \ldots \ldots \quad (8)$$

Substituting Eqs. (7) and (8) into Eq. (6) yields

$$w/w_0 = (\rho_0/\rho)^{1/6} \quad \ldots \ldots \ldots \ldots \ldots \ldots \ldots \ldots \quad (9)$$

Table 7-5 lists values for the density ratio in Eq. (9) based on the densities of the standard atmosphere at various heights.

Table 7-5 Values for the One-Sixth Root of the Density Ratio
for the First 10 km of a Standard Atmosphere

Height, km	0	2	4	6	8	10
$(\rho_0/\rho)^{1/6}$	1.00	1.04	1.08	1.11	1.15	1.19

From Table 7-5 it is seen that the rate of balloon ascent is ~ 20% greater at 10 km than at sea level.

In practice, Eq. (1)[4] can be used to calculate the rate of ascent, w, with corrections applied for the first five minutes of flight. These corrections, shown in Table 7-6, are necessary to compensate for turbulence or eddy effects on the rate of ascent. This table applies only to pilot balloon flights.

Table 7-6 Corrections for w in the Lower Layer

Per Cent Adjustment to Rate of Ascent	Elapsed Time
20% increase in w	The first minute
10% increase in w	Second and third minutes
5% increase in w	Fourth and fifth minutes

[4]Eq. (1), which is an approximation of w, can be obtained from Eq. (4) when the variation of ρ is ignored. As the balloon cross-section, A, is proportional to $V^{2/3}$, and the volume, V, is proportional to the buoyant force, F=L+W, at constant density; substituting these relationships into Eq. (4) yields $D = L = Kw^2(L+W)^{2/3}$, which is solved explicitly for w to give $w = KL^{1/2}(L+W)^{1/3}$ (see Footnote 2).

For more accurate calculations of the rate of ascent, Luers and MacArthur[5] substituted all the balloon specifications into the equation of motion and solved for the radius, r, of the balloon.

$$r = \frac{- C_D w^2 + 8dg(\rho_s/\rho)}{\frac{8}{3}g(1 - \frac{\rho_g}{\rho})} \quad \ldots \ldots \ldots \ldots \ldots \ldots \ldots \ldots \ldots \ldots (10)$$

where d is balloon thickness; ρ_s, skin density of balloon; ρ_g, gas density; and the other symbols as defined previously. Then they used Eq. (10) to calculate the radii of smooth and roughened spherical balloons needed to achieve various rates of ascent (2 to 20 m s^{-1}) at 10 and 14 km altitudes. Their results are given in Table 7-7 with radii expressed in meters.

Table 7-7 Balloon Radius as a Function of the Rate of Ascent

Rate of Ascent m s^{-1}		Smooth Balloons 10 km	14 km	Rough Balloons 10 km	14 km
	Height:				
2	Radii:	0.14 m	0.20 m	0.17 m	0.25 m
4		0.38	0.44	0.52	0.56
6		0.54	0.84	1.24	1.31
10		0.73	1.20	3.29	3.36
15		1.56	1.61	7.28	7.35
20		2.73	2.78	12.87	12.94

It should be noted that rates of ascent between 10 and 20 m s^{-1} (approximately 2000 to 4000 ft min^{-1}) can be achieved only with smooth balloons less than 3 m in radius. It must also be noted that data presented in Table 7-7 are applicable only to non-expandable 1/4-mil mylar helium balloons inflated to ambient pressure without additional mass, ballast, or an instrument package.

(2) Environmental Effects on Balloon Performance. Ambient pressure, wind, radiation, temperature, and moisture influence both the dynamics of balloons and the length of balloon flights. For a high altitude constant-level balloon, the effects of air quality, particularly ozone concentration, on the balloon must be taken into consideration. Environmental effects on the performance of the pilot balloon and the GHOST balloon are given in order to illustrate.

(a) Excess Pressure of Pilot Balloons. The dynamics of small rubber balloons are affected by whether or not the pressure inside a balloon, P_i, is equal to that outside, P_o. In reality, $P_i > P_o$ by the amount related to the modulus of

[5]Luers, J.K. and C.D. MacArthur. 1972. Ultimate Wind Sensing Capabilities of the Jimsphere and other Rising Balloon Systems, NASA CR-2048. University of Dayton Research Institute.

Luers, J.K. 1974. The limitations of wind measurement accuracy for balloon systems. J. Appl. Meteorol. 13(1): 168-173.

elasticity, M, and the thickness, d_0, of the rubber at any particular degree of stretching. Obviously, the higher the altitude the greater the stretch. The properties of rubber are so complex that M can be treated as independent of d_0 and the degree of stretching. Consequently, the excess pressure (i.e., $\Delta P = P_i - P_0$) and its variations are best investigated experimentally. In his experiments, Väisälä[6] found that the relation between the ΔP and d_0 and the radius of the balloon just before stretching begins, r_0, could be expressed as

$$\Delta P = 2d_0/r_0 P(n) \quad \dots \dots \dots \dots \dots \dots \dots \dots \dots \dots \dots \quad (11)$$

where $n = r/r_0$ (i.e., the ratio of the radius of the balloon at any time during inflation to that immediately before stretching starts). $P(n)$ is a complicated function which depends on the balloon material; it is determined empirically by

$$P(n) = \frac{a}{n^3} e^{[b(n-1)-c/(n-1)]} \quad \dots \dots \dots \dots \dots \dots \dots \quad (12)$$

where a, b, and c are constants for balloons of the same material. For the Finnish balloons used by Väisälä, the relationship between the ratio, n, and the pressure, P (in $kg\ cm^{-2}$), is shown in Figure 7-4. It must be noted that P reaches a maximum at any r-value slightly greater than r_0 (n ~ 1.2) and decreases to a minimum when n = 3.8. P increases monotonically for higher values of n. Using Eq. (11), a 30 g balloon with r_0 = 8 cm and d_0 = 0.04 cm has a first maximum ΔP of 60 mb. The minimum P is 21 mb and occurs at r = 30 cm which is very close to the radius of the fully inflated balloon at sea level.

(b) <u>Response of Pilot Balloons to Varying Winds</u>. Computation of pilot balloon responses to varying winds (see Figure 7-5) is generally based upon the following assumptions, although responses so calculated may deviate appreciably from the real situation:

    i. The balloon is at rest relative to the original wind.

    ii. The rate of ascent is not in any way affected by any horizontal force acting on the balloon.

    iii. The vertical wind always is zero.

Setting the wind speed relative to the balloon as u and the mass of the balloon as m, Eq. (4) can

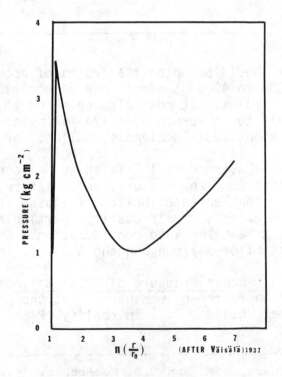

Figure 7-4 Values of P(n)

[6]Väisälä, V. 1937. Ann. Acad. Scient. Fennicae, A 48(1937), No. 8.

Figure 7-5 Launching in High Wind

be used to compute the balloon responses

$$D = - m \frac{du}{dt} = \frac{C_D \rho A u^2}{2} \quad \ldots \ldots \ldots \ldots \ldots \ldots \ldots \ldots \ldots \quad (13)$$

Rearrange the variables in Eq. (13) and integrate with respect to wind speed and time, then

$$\int_{U_0}^{U_1} \frac{1}{C_D u^2} \, du = - \frac{\rho A}{2m} \int_0^\tau dt \quad \ldots \ldots \ldots \ldots \ldots \ldots \ldots \quad (14)$$

where U_0 is the initial value of u, and τ is the time required for the relative speed to fall to any arbitrary value of $U_1 < U_0$.

The relationship between Reynolds number (i.e., $R_e = \frac{d \, u}{\nu}$) and C_D is expressed empirically as

$$C_D = \frac{24.06}{R_e} + 0.376 \quad \ldots \ldots \ldots \ldots \ldots \ldots \ldots \ldots \ldots \quad (15)$$

where R_e is the Reynolds number, and ν is the kinematic viscosity of the air. For air at 1000 mb and 20C, the value of $1/\nu$ is 6.80 s cm^{-2}. If the diameter of a 30 g balloon is about 62 cm and u is expressed in m s^{-1}, $R_e = 911$ u. Substitution of R_e into Eq. (15) gives

$$C_D = \frac{2.64 \times 10^{-3}}{u} + 0.376 \quad \ldots \ldots \ldots \ldots \ldots \ldots \ldots \quad (16)$$

Substituting the value of C_D into Eq. (14)

$$\int_{U_0}^{U_1} \frac{du}{2.64 \times 10^{-3} u + 0.376 u^2} = - \frac{\rho A}{2m} \int_0^{\tau} dt \quad \ldots \ldots \ldots \quad (17)$$

which may be integrated and solved for τ

$$\tau = \frac{-2m}{2.64 \times 10^{-3} \rho A} \left[\frac{U_1 (2.64 \times 10^{-3} + 0.376\, U_0)}{U_0 (2.64 \times 10^{-3} + 0.376\, U_1)} \right] \quad \ldots \quad (18)$$

It should be noted that τ approaches infinity as U_1 approaches zero, so that the balloon never actually achieves equilibrium under the assumed conditions. It can be shown, however, that the integral satisfies $U_1 = 0$, if Eq. (15) is replaced by a relation of the form

$$C_D = \alpha/R_e^2 + \beta/R_e + \gamma \quad \ldots \ldots \ldots \ldots \ldots \ldots \quad (19)$$

A simple solution is to compute the time, τ, for the relative speed, u, to fall to one per cent of its initial value. Taking $\rho = 1250$ g m^{-3}, A = $(0.31)^2$ m^2, and m = 41 g (balloon plus buoyant gas), the value of τ in s is computed as

$$\tau = -29.34 \log_{10} \left[\frac{376\, U_0 + 2.64}{376\, U_0 + 264.0} \right] \quad \ldots \ldots \ldots \ldots \quad (20)$$

Values of τ versus U_0 are listed below:

U_0 (m s^{-1})	1	5	10	50
τ (s)	6.7	1.7	0.86	0.18

The listing above shows that stronger gusts produce faster responses. This is a desirable feature as fluctuations of light winds have no synoptic value. Eq. (20) establishes that the balloon approaches the true wind speed to a greater overall accuracy than the overall accuracy of wind speed derived from a single theodolite observation.

(c) Response of Superpressure Balloons to the Environment. In the case of these types of balloons, exemplified by the Ghost version discussed above, three major changes occur in response to environmental effects with respect to mass, volume, and stability and life span of the balloon.

(i) Mass Losses. The loss of gas occurs either through diffusion or leakage. The increase of free lift force of the balloon caused by the mass loss usually is compensated by the decrease of free lift force resulting from the volume reduction. But a significant mass loss would either shorten the life span of the balloon or cause the balloon to lose its superpressure and descend to the earth's surface. Therefore, balloons should be free from defects, protected from ultraviolet radiation, and kept from damage during launch or ascent. An ultraviolet resistant coating is necessary for balloons floating at 30 mb or higher. The mass loss of typical balloons through diffusion, computed by Lally,[7] is given in Table 7-8.

(ii) <u>Volume Changes</u>. As radiative energy levels change the balloon film heats or cools, producing changes in its gas temperature. A 1C increase in temperature (supertemperature) produces a 0.4% increase in excess pressure (superpressure). For a typical balloon, this creates a 0.04% increase in volume and a 3 m increase in altitude. As shown in Table 7-8, the balloon diameter is about 1 m from ground level to about 700 mb, but it increases to 11 m at 30 mb and 20 m at 10 mb. Balloon gas temperature is less than ambient air temperature at night and vice versa during the day. Lally (see Footnote 7) estimated the day and night supertemperatures of the mylar superpressure balloon, which are shown in Table 7-9.

(iii) <u>Stability and Life Span</u>. The vertical wind in the troposphere displaces a balloon from its floating altitude. A 2.2 mph vertical wind displaces a spherical balloon by about 660 ft. The superpressure balloon, when subjected to perturbations, oscillates in simple harmonic motion in an isothermal region. The period of oscillation is

$$\tau = 11 \, T^{1/2} \quad \ldots \ldots \ldots \ldots \ldots \ldots \ldots \ldots \quad (21)$$

where τ is the period in s, and T is the air temperature in degrees K.

Table 7-8 Gas Loss by Diffusion for Typical Balloons

Pressure Altitude mb	Balloon Diameter m	Film Thickness μm	Assumed Film Temperature K	Gas Loss Per Day %	Six Per Cent Mass Reduction days
900	1*	125	290	0.13	45
700	1*	100	280	0.10	60
500	1.6	75	265	0.05	120
400	2.0	75	255	0.02	300
300	2.3	75	245	0.009	660
200	2.3	50	230	0.005	1,200
100	3.0	37	230	0.0024	2,500
30	11.0	37	235	0.00024	25,000
10	20.0	25	240	0.000086	70,000

*Cylindrical balloon with 8 m length.

[7] Lally, V. E. 1970. Constant-level balloons for sounding systems. Meteorological Monographs 11(33): 392-396. Amer. Meteorol. Soc., Boston, Mass.

Balloon oscillation increases by about 15% in an adiabatic atmosphere. An example of extreme oscillation of a balloon was that of 115 feet on the first overshoot when settling into a predesigned altitude. The oscillation damped out within 3 to 5 seconds.

Table 7-9 Radiation Environment for Mylar Superpressure Balloons

Altitude	Air Temp. Range (K)	$T_g - T_o$ (C)	T_s (C)	Predicted T_s (C)	Predicted P_s (%)
1 km (900 mb)	250 - 300	-6 to 1	5	-6 to 6	-3 to 3
3 km (700 mb)	250 - 290	-8 to 2	6	-8 to 10	-3 to 4
6 km (500 mb)	230 - 270	-10 to 5	7	-10 to 15	-4 to 7
9 km (300 mb)	220 - 250	-15 to 5	8	-15 to 15	-7 to 7
12 km (200 mb)	210 - 230	-10 to 10	10	-10 to 20	-5 to 10
16 km (100 mb)	190 - 230	-20 to 15	12	-20 to 25	-10 to 12
24 km (30 mb)	210 - 235	-20 to 5	12	-20 to 20	-9 to 9
30 km (10 mb)	215 - 240	-20 to 0	12	-20 to 15	-9 to 7

Key $T_g - T_o$: Nighttime supertemperature extreme (i.e. balloon gas temperature minus air temperature).

T_s: Daytime supertemperature increase produced by sun.

Predicted T_s: Predicted supertemperature extreme.

Predicted P_s: Predicted maximum percentage change in superpressure due to radiation changes.

In general, the horizontal wind speed associated with turbulence increases with height. The drag coefficient, C_D, of the balloon decreases with an increase in wind speed aloft and, hence, speeds up the rate of ascent, w. The w value remains approximately constant from 18 to 22 mph, but increases at higher wind speeds, and decreases at lower speeds. If the wind speed is V and the deviation in rate of ascent is Δw, then:

V (mph)	0 to 5	18 to 22	40 to 44	44.1 to 48.4
Δw (m min^{-1})	-33	0	+36	+67.2

A negative sign indicates a reduction in w, and a positive sign, an increase.

Ice and frost formation remain the major problems in achieving long-duration flights between 1 km and 10 km (about 900 to 250 mb), particularly in the cold tropopause over the tropics. Various altitude flights (10, 30, 100, 200, 300, 500, and 700 to 900 mb) have been conducted with balloons of sizes ranging from 1 m to 20 m in diameter. The balloons had UV resistant coatings and thermal massages to protect them against damage. Balloon flights conducted at heights below the freezing level usually were terminated by collisions with solid objects such as mountains, or rupturing caused by severe storms. The cylindrical balloon was specifically designed for these low elevation flights, with some improvement in performance and duration.

Lally's estimation of the leakage of Ghost balloons (see Footnote 7) and their expected life span (expressed as days per hole) is given in Table 7-10.

Table 7-10 Estimated Gas Loss Per Day for Balloons with Various Hole Sizes

Leak	Flight Level: Hole Size:	Gas loss day^{-1} (% per hole)*			Expected life for one hole (days)*		
		500 mb 1.6 m	200 mb 2.3 m	30 mb 11.0 m	500 mb 1.6 m	200 mb 2.3 m	30 mb 11.0 m
very small	10 μm (0.4 mil)	5×10^{-2}	8×10^{-3}	2×10^{-5}	120	750	300,000**
pin-hole	100 μm (4 mils)	5.0	0.8	2×10^{-3}	2	7	3,000
large	1000 μm (1 mm)	-	100	0.5	-	1	12
huge	1 cm	-	-	50	-	1	1

*Assumes 75 mb overpressure for 500 mb balloon, 30 mb overpressure for 200 mb balloon, and 5 mb overpressure for 30 mb balloon (life estimate for 6% gas loss).
**Diffusion limits life to 25,000 days.

7.1.3 <u>Pilot Balloon Observations</u>. The pilot balloon technique for upper wind measurements involves the optical determination of the position of an airborne balloon using a theodolite (see Figures 7-6a and 7-6b). Basically, there are three techniques of observation: (a) single theodolite technique, (b) double theodolite technique, and (c) stadia technique.

In the single theodolite technique, the elevation and azimuth angles of the balloon are measured by one instrument, usually at one minute intervals. These measurements along with the balloon's rate of ascent determine the position of the balloon. The double theodolite technique utilizes two theodolites (see Figure 7-7) located at a known distance (the baseline) and the same elevation for simultaneous observations of one balloon. This technique enables accurate measurements of the balloon position without relying upon the assumed rate of ascent. It requires a systematic synchronization of readings between the two stations through radio or wire communications. The baseline should be normal to the prevailing wind at the stations, or several baselines should be available if the local wind is highly variable. A baseline not less than one-fifth of the maximum range to be observed is desirable. Thus, for a height below 1 km, a line of at least 2 km or more should be used. Despite its high accuracy, however, the double-theodolite technique is unsuitable for routine use because of baseline requirements and installation costs.

A third technique, the stadia method, requires only one theodolite for the observation. Determination of the balloon position involves measurements of the subtended angle of a tail attached to the balloon as well as the elevation and azimuthal angles of the balloon (see Figure 7-8). Although the stadia method does not provide accurate readings for a long distance flight, it gives accurate measurements for the first few minutes of a low level flight, something neither the single theodolite nor the double theodolite can do. Also, the stadia method does not require a baseline, so the operation is rather simple. The three techniques are summarized in Table 7-11.

Table 7-11 Pilot Balloon Techniques

Method	Measurements	Information
Single theodolite	Elevation angle, azimuth angle and elapsed time	Balloon trajectory and velocity
Double theodolite	Elevation angle, azimuth angle and a known length of baseline	Same as above, with addition of balloon height
Stadia theodolite	Elevation angle, azimuth angle and a known length of tail	Same as above

A brief description of the general mechanism of the instrument is in order, followed by a discussion of the geometry of three theodolite techniques.

(1) <u>The Theodolite</u>. During the past 60 some years, a large number of theodolite designs have been introduced but their overall mechanisms essentially are the same. As shown in Figures 7-6$_a$ and 7-6$_b$, the instrument set consists of a theodolite and a tripod. Although similar in many ways to a surveyor's transit set, the theodolite comes equipped with a more adaptable right angle telescope rather than the direct line scope of transit sets.

The theodolite has two telescopes, the primary and secondary, equipped with one eyepiece. The secondary telescope has a lower optical power but a larger range of view than the primary telescope. It is used to locate the balloon during the first minutes of flight at which time the motion of the balloon is likely to be irregular and rapid. The primary telescope is used for the rest of the flight. Some theodolites are equipped with an alidade mounted atop the primary telescope. Separate vertical and horizontal scales are engraved on the instrument's body. Both these scales are controlled by tangent screws which drive the worm gears. With the aid of micrometer drums the scales are read to the nearest tenth of a degree. The rotation of the horizontal scale about the zenith axis (plumb line) is referred to as the azimuthal angle, and that of the vertical scale about the horizontal axis, the elevation angle. The zero elevation angle is obtained when the central axis of the telescope lies in the plane tangent to the earth. The zero azimuth angle is set with the same axis of the telescope pointed directly toward the true north. For pilot observations the theodolite is set in such a way that true north appears as an azimuth angle of 180°. The angle so read gives the direction <u>from which</u> the observer is looking, and is necessary as the wind direction is <u>designated</u> as the direction from which the wind is blowing.

It should be noted that most theodolites are built with an optical system which gives an inverted image; thus the balloon image appears to move in a contrary sense when the screws are turned. A reticle fitted in the eyepiece has a pair of perpendicular cross hairs. These cross hairs are made of spider webs or

A. Dust cap on object lens
B. Battery box
C. Bubble levels
D. Elevation scale tangent screw
E. Bumper
F. Slow motion screw
G. Lower clamp
H. Baseboard
I. Leveling plate
J. Leveling screws
K. Leveling head
L. Block containing capstan, oil, wrenches & spare bulbs
M. Azimuth
N. Azimuth
O. Rheostat
P. Eyepiece
Q. Crosshairs reticule screws
R. Finder telescope
S. Guard
T. Elevation scale
U. Telescope
V. Front sight

Figure 7-6$_a$ Pilot Balloon Theodolite

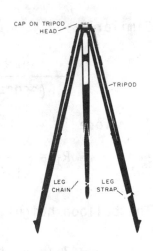

Figure 7-6$_b$
Theodolite Tripod

very fine metal wire. The observer keeps the balloon centered on the cross hairs, giving the exact position of the airborne balloon. For night flights the cross hairs are illuminated from the side by a battery powered flashbulb. Similar illumination is provided for the horizontal and vertical scales. Also attached to the airborne balloon are either electrical bulbs or a Chinese style lantern for illuminating a long distance flight and a small parachute.

Procedures for making pilot balloon observations, maintenance, and calibration of the theodolite, as well as the use of plotting equipment and determination of true north, may be found in USWB Circular O or other technical manuals provided by either the National Weather Service or WMO.

(2) Geometry of Theodolite Techniques. The following section deals with single and double theodolite and stadia techniques.

(a) <u>Single Theodolite</u>. As shown in Figure 7-7$_a$, the line of sight of the airborne balloon OP, its horizontal range OB, and its height BP are related by

$$R = Z \cot \phi \dots \dots \dots \dots \dots \dots \dots \dots (22)$$

where R is the horizontal range; Z, the height; and ϕ, the elevation angle. The height, Z, may be computed by the rate of ascent from Eq. (1) or obtained from a special manual provided by the National Weather Service. The R value also can be obtained from the same manual without using Eq. (22). It must be noted that all optical and electronic balloon techniques for upper-air measurement are grossly inaccurate at elevation angles of 5 deg or less. The main drawback of the single theodolite technique is its dependence on the assumed constant rate of ascent for the determination of the Z value.

(b) <u>Double Theodolite</u>. As shown in Figure 7-7$_{b_1}$, the line of sight of the first observer at station A is AP and that of the second observer at station B is BP. The baseline between the two stations AB is R_0. If ϕ_1 and θ_1 are the elevation and azimuth angles, respectively, at station A, and ϕ_2 and θ_2, the angles read <u>simultaneously</u> at station B, then

$$\angle ABC = 180° - [(360° - \theta_1) + (\theta_2 - 180°)] = \theta_1 - \theta_2 \dots \dots (23)$$

If $R_1 = AC$ and $R_2 = BC$, by the law of sines

$$\frac{R_1}{\sin(\theta_2 - 180°)} = \frac{R_1}{\sin \theta_2} = \frac{R_0}{\sin(\theta_1 - \theta_2)} \dots \dots \dots \dots (24)$$

Similarly

$$\frac{R_2}{\sin(360° - \theta_1)} = \frac{R_2}{\sin \theta_1} = \frac{R_0}{\sin(\theta_1 - \theta_2)} \dots \dots \dots \dots (25)$$

Therefore

$$R_1 = \frac{R_0 \sin \theta_2}{\sin(\theta_1 - \theta_2)} \text{, and } R_2 = \frac{R_0 \sin \theta_1}{\sin(\theta_1 - \theta_2)} \dots \dots \dots \dots (26)$$

The balloon height, Z, is given by either

$$Z_1 = R_1 \tan \phi_1 = \frac{R_0 \sin \theta_2 \tan \phi_1}{\sin(\theta_1 - \theta_2)} \dots \dots \dots \dots (27)$$

or

$$Z_2 = R_2 \tan \phi_2 = \frac{R_0 \sin \theta_1 \tan \phi_2}{\sin(\theta_1 - \theta_2)} \quad \ldots \ldots \ldots \ldots (28)$$

If $(\theta_1 - \theta_2)$ is near zero or 180 deg, or greater than 177 deg, then the method becomes inapplicable as $\sin(\theta_1 - \theta_2)$ in the denominators of Eqs. (26) through (28) is very sensitive to small errors in the azimuth angles. This situation exists when the balloon lies in or near the plane formed by the baseline and the zenith axis, whether or not the balloon is between the stations. Under these conditions, the balloon is assumed to be on the plane and the single plane triangle ABP shown in Figures 7-7$_{b2}$ and 7-7$_{b3}$ is solved.

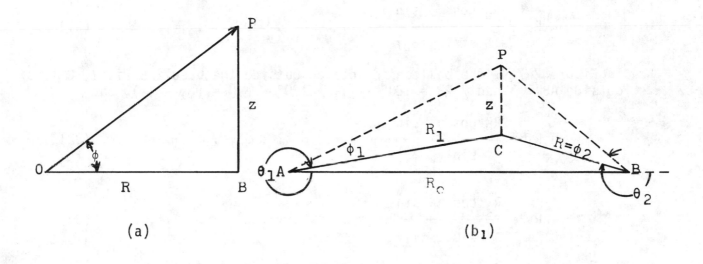

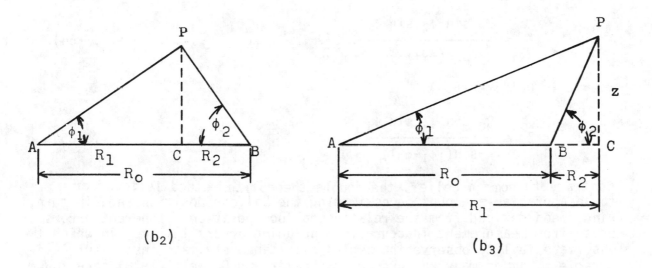

Figure 7-7 Geometry of the Pilot Balloon in Flight

272

In Figure 7-7$_{b2}$, where the balloon is over the baseline, or C is inside AB, then

$$\frac{R_0}{\sin(\phi_1+\phi_2)} = \frac{R_1/\cos\phi_1}{\sin\phi_2} = \frac{R_2/\cos\phi_2}{\sin\phi_1} \quad \ldots \ldots \ldots \ldots (29)$$

Hence

$$R_1 = \frac{R_0 \cos\phi_1 \sin\phi_2}{\sin(\phi_1+\phi_2)} \quad \ldots \ldots \ldots \ldots \ldots \ldots \ldots (30)$$

and

$$R_2 = \frac{R_0 \cos\phi_2 \sin\phi_1}{\sin(\phi_1+\phi_2)} \quad \ldots \ldots \ldots \ldots \ldots \ldots (31)$$

In Figure 7-7$_{b3}$, the balloon is located outside the baseline (i.e., C lies outside AB). Then $\angle APB = 180° - [\phi_1 + 180° - \phi_2] = (\phi_1 - \phi_2)$, and

$$R_1 = \frac{R_0 \cos\phi_1 \sin\phi_2}{\sin(\phi_2-\phi_1)} \quad \ldots \ldots \ldots \ldots \ldots \ldots (32)$$

$$R_2 = \frac{R_0 \cos\phi_2 \sin\phi_1}{\sin(\phi_2-\phi_1)} \quad \ldots \ldots \ldots \ldots \ldots \ldots (33)$$

If C lies on the other side, the denominator is $\sin(\phi_1 - \phi_2)$.

When C is inside AB

$$Z = \frac{R_0 \sin\phi_2 \sin\phi_1}{\sin(\phi_1+\phi_2)} \quad \ldots \ldots \ldots \ldots \ldots \ldots (34)$$

When C is outside AB

$$Z = \frac{R_0 \sin\phi_2 \sin\phi_1}{\sin(|\phi_1-\phi_2|)} \quad \ldots \ldots \ldots \ldots \ldots \ldots (35)$$

Contrary to common belief, the double-theodolite method is in error to an appreciable amount in estimating the balloon position and, in turn, wind speed derived from the relative balloon position. Inherent errors arise from measurement inaccuracies including errors in time, in which two observers fail to observe the angles simultaneously, and instrument inaccuracies. Among many researchers, Schaefer and Doswell (1978) have found errors in wind speed on the order of 5 mph. See Section 7.4 for complete error analysis discussion.

273

(c) <u>The Stadia Method</u>. As shown in Figure 7-8, if ϕ_2 is the elevation angle of the balloon and ϕ_1 is the elevation angle of the tip of the tail which is made of fine string or thread with a flag at the lower end, $\alpha = \phi_2 - \phi_1$. As α is small, it can be expressed in radians as MT/OM. Because triangles MPT and OPQ are similar, MTP = ϕ_2. Thus

$$MT = \ell \cos \phi_2 \quad \ldots \ldots \quad (36)$$

where ℓ is the length of the tail, and

$$OM = MT/\alpha = (\ell \cos \phi_2)/\alpha \ldots \quad (37)$$

and because OM >> MP

$$OP \approx OM = (\ell \cos \phi_2)/\alpha \;. \; (38)$$

But

$$PQ = Z = OP \sin \phi_2 = (\ell \cos \phi_2 \sin \phi_2)/\alpha \ldots \ldots \ldots \quad (39)$$

That is

$$Z = (\ell \sin 2\phi_2)/2\alpha \quad \ldots \ldots \ldots \ldots \ldots \ldots \quad (40)$$

Likewise

$$R_O = OP \cos \phi_2 = (\ell \cos^2 \phi_2)/\alpha \quad \ldots \ldots \ldots \ldots \ldots \quad (41)$$

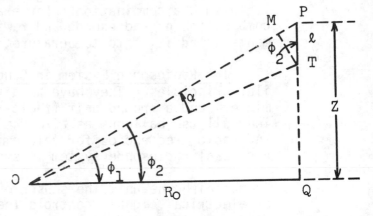

Figure 7-8 Stadia Method

In the stadia technique, a micrometer eyepiece is required to read the very small angle, α. Errors in the application of this method are due to the non-perpendicularity of the tail in a changing windfield, the values of α becoming extremely small at very high altitudes, and the inherent inaccuracy of readings taken in gusts and strong winds.

7.2 <u>Ground Platform Techniques</u>. The major upper-air sounding systems utilize ground platforms as their receiving stations. These are: radiosonde, radio direction finder, radar, lidar, laser, wiresonde, and perhaps diffusion measurement systems. Of these seven systems, the first three are the most important for routine use over the globe.

7.2.1. <u>Radiosonde System</u>. As shown in Table 7-1, the classical radiosonde system has been employed for the measurement of upper-air temperature, pressure, and humidity since 1876. This system uses a balloon-borne instrument for simultaneous measurement and transmission of upper-air data to the ground station. In recent years, research studies have incorporated measurements of other parameters such as ozone content, solar and sky radiation, electrical potential gradient, and even wind direction and speed. Most radiosondes operate on a frequency band between 25 and 100 MHz, with a few at 400 MHz. However, combining two or more instrument systems is now becoming a trend in upper-air instrumentation. Most US radiosondes already use a high frequency band (1670 to 1690 MHz) to facilitate such a combination. For example, the AN/GMD-1 Rawin System is a combination of radiosonde and rawinsonde systems. It gives wind measurements in addition to temperature, pressure, and humidity.

(1) The Radiosonde System in General. All radiosondes operate under similar principles. They have an airborne unit (sensor, converter, and transmitter) and a ground unit (radio receiver, pulse controller, and recorder). They all use balloons as airborne platforms, in situ elements as sensors, and radio frequencies for telecommunicating signals to the ground station. A general radiosonde system is shown diagrammatically in Figure 7-9.

In the airborne unit the sensor output, in the form of either a mechanical or electrical signal, controls the frequency of pulsation in the converter which, in turn, modulates the radio frequency of the transmitter's carrier wave to the ground. Many radiosondes have low transmitter power on the order of a few tenths of a milliwatt. For a high signal to noise ratio, the ground receiver sensitivity must be on the order of one microvolt. This is essential for radiosondes transmitting up to 200 km. In addition to sensitivity, the selectivity of the ground receiver is important for a good signal to noise rejection factor. A superheterodyne receiver, for example, has the ability to separate signals on adjacent wavebands and to eliminate spurious signals and whistles by using a high intermediate frequency.

As shown in Figure 7-9, the pulse controller of the ground unit can be operated manually or automatically. In manual operation, the output signals are presented in the form of audio or visual displays, or both, and interpretation is required to translate the electrical units into meteorological units. Several electronic components are involved in automatic recording equipment. For example, the ground antenna of the British Mark 2B Radiosonde intercepts the radio frequency signal which is then fed to the receiver. The receiver extracts the audio frequency signal and passes it through a filter amplifier with a pass band of 690 to 1020 Hz. The signal then is fed into a pulse shaper unit where one positive and one negative going pulse for each input cycle are produced. These pulses are then fed into a 1/100 scalar unit which produces one output negative pulse for every 100 input pulses. These negative pulses pass into a second pulse unit which feeds a start pulse to a counter chronometer on the arrival of the first pulse and a stop pulse on the receipt of the second. Finally, the translator digitizes the pulses obtained from the counter chronometer, and the recorder further registers the pulses on a paper strip chart. In this system, the recorder output is presented as a fine line structure and the actual meteorological values must be further evaluated. In other systems, direct printout of meteorological readings is made, such as the US ATR (angle, time, range) printer.

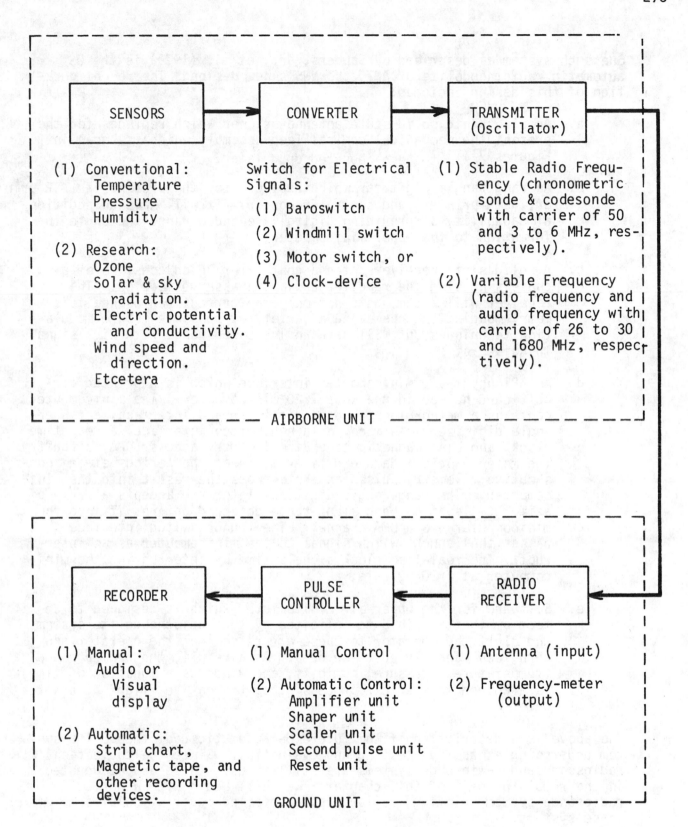

Figure 7-9   A Generalized Radiosonde System

One such system as described by Sanders, Jr., _et al_. (1979) is the US automatic radiotheodolite or ART, the NWS's new design. The reconfiguration of this design includes:

a. An electronic solid state antenna scanner which replaces for the present the mechanical conical feed assembly which is shown in Figures 7-11, to be elaborated upon later;

b. An range/angle digitizer which replaces for the present the ATR printer, data printer, and control recorder. It will have an additional capability of providing digitally encoded range and angle information to the upper minicomputer;

c. A solid state receiver/antenna controller which consists of a chassis housing the receiver and antenna servo-modules. The receiver will accommodate not only the current radiosonde telemetry modulation scheme and a variety of possible future modulation techniques but will have an increased range with high signal to noise ratio;

d. An ART upper-air minicomputer interface which is a printed circuit card housed in the Nova 1220 minicomputer. The minicomputer contains a meteorological data digitizer and interface, a range/angle digitizer interface, a launch switch interface, a real-time clock, and Linc magnetic tape direct memory access (DMA) circuitry. The meteorological data digitizer measures the period between consecutive telemetry pulses and interrupts the result into the minicomputer. The range/angle digitizer interface accepts a frame of azimuth, elevation, and slant-range data and enteres it into the minicomputer via a DMA channel. The launch switch interface passes the launch switch signal to the minicomputer as an interrupt. The real-time clock signals time by interrupting the minicomputer at a 10 Hz rate; and

e. Software for the upper-air minicomputer which is designed to detect ambiguities in data solicited by the radiotheodolite system and alert the operator to these errors so that the operator, in turn, can manually edit the data. As a result, the raw data pool (temperature, pressure, humidity, and winds as a function of time) is archived to enable reprocessing of the entire flight at a later date.

The above is a description of the future US automatic radiotheodolite system undergoing exhaustive testing and evaluation. Since the conventional radiosonde and rawinsonde systems are still in use, these are elaborated in the remaining part of this chapter. We shall begin with a worldwide radiosonde system briefing, followed by a detailed description of some of these systems.

The ten or so different sounding systems vary chiefly in the choice of their sensors and converters, particularly the latter. Four major radiosonde types are classified on the basis of converter systems: chronometric, codesonde, variable radio frequency, and variable audio frequency radiosondes. Table 7-12 summarizes these four types in terms of the mechanisms and users. Radiosonde systems also can be differentiated on the basis of the balloon platform steering method. These radiosonde systems are identified as ordinary radiosonde, dropsonde, and constant-level balloonsonde as described in Table 7-13.

(2) _Performance of Radiosondes_. The performance of some radiosondes are described in terms of their operating principles, mechanisms, and sensors, with particular emphasis on their accuracy and reproducibility. A comparison of the performance quality of each type with the reference radiosonde used as a standard is included. Although several models[8] are available, the following discussion includes only the US, UK, and Finnish models.

(a) _The US Model_. In the audio modulated or variable audio frequency radiosonde, the carrier wave is modulated by an audio frequency signal whose frequency is controlled by sensing elements of the airborne instrument. This type of radiosonde has been employed extensively in the United States, Great Britain, and the Netherlands (see Table 7-12). More nations of the world, such as Finland and USSR, have adopted the audio modulated frequency in their newest models.

As explained earlier, the latest version of the American radiosonde is a combination of both the radiosonde (for measuring temperature, pressure, and humidity) and the rawinsonde (for measuring wind speed and direction), known as AN/GMD-1 Rawin System. Several models of airborne units and ground units have been designed, differing by only minor modifications. Only the ones shown in Figures 7-10 and 7-11 will be described. Models and specifications of their airborne and ground units are listed below.

Airborne Units:

Figure 7-10a   J005 Radiosonde set, VIZ Manufacturing Company

Figure 7-10b   AN/AMT-18 Radiosonde, Bendix Corporation

Figure 7-10c   AN/TMQ-5( ) Radiosonde Recorder, Leeds and Northrup Manufacturing Company

Ground Units:

Figure 7-11a   AS-462 ( )/GMD-1 Antenna, Bendix Corporation

Figure 7-11b   C-577D/GMD-1 Control Recorder, Bendix Corporation

The J005 radiosonde is used with the AN/GMD-1 rawin set for obtaining vertical profile measurements of pressure, temperature, and humidity. The component parts of J005 are illustrated in Figure 7-10a in which the modulator assembly

---

[8] Examples of other models are RKZ and A-22 (USSR), GRAW M6 (Federal Republic of Germany), Mesural (France), and Sangamo (Canada, Portugal, some USA).

Table 7-12 Classification of Radiosondes
(in terms of converter types)

Type	Mechanism	Adopted By
Chronometric Radiosonde (Olland principle)[1]	Electric or windmill driven scanning disc transmits temperature, pressure, and humidity signals one at a time. This is a time interval scanning device.	Canada, Switzerland, France, and India
Codesonde (Digital System)[2]	Electric or windmill driven insulated plate, consisting of a series of contact studs arranged in an arc representing Morse telegraph codes, which transmits meteorological signals one at a time. No special ground equipment is required beyond an ordinary communication receiver. Also, only a mechanical sensor is employed. It is used also in all constant-level balloon soundings.	West Germany, Japan, USSR, and sometimes USA
Variable Radio-Frequency Radiosonde[3]	Variations in capacitance are controlled by meteorological sensors. These capacitors are connected to a windmill switch of an electric transmitter.	Finland (Vaisala type), Germany, and USSR
Variable Audio-Frequency Radiosonde[3]	Meteorological signals are converted into electric signals which, in turn, are modulated and transmitted as audio-frequency signals to a ground receiving set.	USA, United Kingdom, and Netherlands

[1] Lange, K.O. 1937. Bull. Amer. Meteorol. Soc. 18: 107-126.

[2] Molchanov, P. 1928. Zur Technik der Erforschung den Atmosphäre. Beitr. Phys. frei. Atmos., Leipzig. 14:45.

[3] These will be elaborated in the text.

(sensor and converter) and the transmitter assembly (blocking oscillator, cavity triode, and antenna) are clearly shown. A brief description of the components of these two assemblies follows.

There are three sensors in the modulator assembly of the J005 radiosonde set: an aneroid capsule for measuring atmospheric pressure, ceramic rod thermistor for temperature, and carbon hygristor for relative humidity. A contact arm connects the aneroid and commutator by means of a lever system. The commutator, which consists of alternate silver contact strips and insulating segments, is used to switch the humidity and temperature sensors into the circuit by means of a relay. There is a total of over 150 silver strips in the commutator. The contact arm connected to the aneroid barometer capsule moves across the commutator as the pressure decreases with height. The temperature is transmitted whenever the contact arm touches an insulating segment, and humidity is transmitted when the arm is in contact with certain silver strips. Temperature and humidity changes in the atmosphere produce resistance variations in the respective circuits and cause the transmitted audio frequency to vary. The variations in the resistance of a sensor controls the blocking oscillator whose pulse modulates the 1680 MHz cavity triode. This, in turn, transmits a variable audio frequency in the range of 1680 MHz ± 10 MHz to the ground. A low reference signal, which indicates the altitude or its pressure equivalent, replaces each fifth humidity contact up to and including segment

Table 7-13 Classification of Radiosondes (in terms of
method of balloon platform steering)

Type	Flight Pattern	Maximum Distance	Maximum Duration	Remarks
Ordinary Radiosonde	Mainly upward	40 km to 50 km (vertical)	2 hours	An 0.5 to 2 kg balloon with an average ascent rate of about 1200 to 1600 ft min^{-1}; transmits mainly temperature, pressure, and humidity signals
Constant-Level	Horizontal along constant pressure or density surface	10,000 km (horizontal)	Few days	Inextensible balloon with ballast release mechanism for the control of flight level. Meteorological signals are transmitted at intervals of few hours on the 1 to 20 Hz frequency band
Dropsonde	Mainly downward	30 km (vertical)	Few hours	Ground station: 30 to 30 MHz Aircraft station: 13.6 KHz to 300 MHz

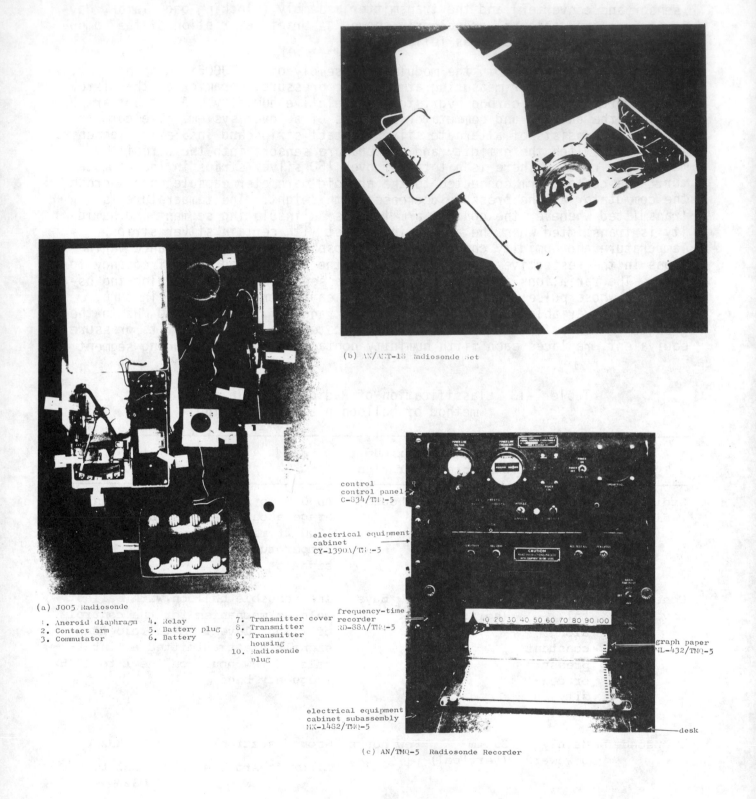

(a) J005 Radiosonde

1. Aneroid diaphragm
2. Contact arm
3. Commutator
4. Relay
5. Battery plug
6. Battery
7. Transmitter cover
8. Transmitter
9. Transmitter housing
10. Radiosonde plug

(b) AN/AMT-13 Radiosonde Set

control
control panel
C-834/TMQ-5

electrical equipment
cabinet
CY-1390A/TMQ-5

frequency-time
recorder
RD-88A/TMQ-5

graph paper
MIL-432/TMQ-5

electrical equipment
cabinet subassembly
MX-1482/TMQ-5

desk

(c) AN/TMQ-5 Radiosonde Recorder

Figure 7-10 The US Radiosonde Airborne Unit
and Radiosonde Recorder

281

(a) ANTENNA AS-467 ( )GMD-1.

(b) CONTROL RECORDER, C-577D/GMD-1.

Figure 7-11 AN/GMD-1 Ground Equipment

number 125 (with the exception that every 15th segment beginning with segment number 30 is replaced by a high reference signal). The low reference signal refers to a low level beginning at about 1010 mb, and the high reference signal, to a level starting at a pressure approximately equivalent to 875 mb. When the sounding reaches 124.4 contacts, the J005 discontinues transmitting humidity. Each segment of the commutator bar is numbered and represents an atmospheric pressure value which is indicated on a calibration chart furnished by the manufacturer for each radiosonde. It must be noted that the numbering systems for the high and low reference values differ from one mode to another. For example, the termination of humidity transmission is at the 124.4 segment for the J005, and at the 104.4 segment for the AN/AMT-4. There are differences in the frequency ranges of the transmitter as well as the range and accuracy of the sensors, even though the same types of sensors are employed for all the US radiosondes.

In addition to the aneroid capsule, the J005B uses a hypsometer to provide accurate measurements of pressure below 50 mb. Here again differences in the millibar threshold value to initiate the hypsometer circuit are used for different models. For example, threshold values are:

AN/AMT-12	J005B
At segment 106, ~ between 40 to 30 mb	At segment 126, ~ below 50 mb

The construction of the hypsometer, however, is essentially the same for both. It consists of a Dewar flask and cap, beadthermistor, wicking, support, and insulation. The non-flammable liquid, Freon 11, is maintained at its boiling point. The temperature of the boiling fluid is converted to a measurement of atmospheric pressure. A special circuit is used to connect the hypsometer to certain segments or contacts on the commutator, beginning at segment 126. Measurements are transmitted when the contact arm is an intermediate conducting segment immediately preceding or succeeding a fifth conducting segment. Each hypsometer is calibrated for pressure in millibars versus recorder divisions, and a hypsometer calibration chart is used to evaluate the radiosonde flight data.

The radiosonde AM/AMT-18, shown in Figure 7-10$_b$, is similar to the J005 and J005B with respect to sensors but differs in other electronic components and circuitry. The sensor resistance values plus three fixed resistance values are sequentially switched into the data oscillator circuits by the solid state commutator. The contact arm of the commutator scans these resistance values once per second and transmits to the modulator (MD-852/U) which consists of two relaxation oscillators. One oscillator supplies data to the transmitter while the other selects the sequence of data supplied. The transmitter (T-1194/U) operates within a frequency range of 1660 to 1700 MHz. Variation in the meteorological resistors modulate the FM pulse which allows transmission of signals up to an altitude of 75 miles.

The AS-462( )/GMD-1 ground unit receiver detects audio frequency signals from the airborne radiosonde. The antenna shown in Figure 7-11$_a$ is a 7 ft parabolic reflector with a conical scan pattern emission with various options of tracking modes such as manual, near automatic, and far automatic. The

C-577D/GMD-1 control recorder shown in Figure 7-11$_b$ controls and records the azimuth and elevation angles of the balloon in flight as well as the meteorological data of the radiosonde. This control recorder is equipped with a special time print circuit switch. (A detailed description of its tracking performance is given in Section 7.2.2.) The AN/TMQ-5 radiosonde recorder, on the other hand, records the data on a paper strip chart by converting the audio frequency pulse to DC voltages. These voltages excite a servo system which then controls a recording pen that traces the signals on a calibrated chart within 2.5 s, including the pen lifting operation. The chart speed is 1/2 in. per min. A pre-flight calibration establishes the relationship between the audio frequency and the resistance values of temperature and relative humidity. Pressure data are obtained by referring to the individual pressure calibration chart furnished by each radiosonde. An elaboration of the ground unit of a rawinsonde system, including its geometry, is given later. This system includes both radiosonde and rawinsonde operation.

During the early part of 1979, the Engineering Division of NWS (NOAA)[9] had completed the specifications for the US National Weather Service Automatic Radiotheodolite System (ART), as mentioned earlier in this chapter. For quality assurance provisions, four phases of tests and evaluations will be made in the near future with publication of the test results following.[10] The system specifications are given below:

(i) Signal Characteristics

(a)	Carrier frequency band	1670-1690 MHz
(b)	Minimum transmitted power	200 milliwatts
(c)	Modulation scheme	PAM
(d)	Pulse repetition frequency	10-1000 Hz
(e)	Pulse off time	45-125 microseconds
(f)	Typical pulse off time	90 microseconds
(g)	Pulse modulation	100% AM-complete cutoff of carrier
(h)	Antenna	
	(1) polarization	Vertical, linear
	(2) electrical length	1/4 wave radiator
	(3) ground plane	Conical skirt
	(4) feed	Transmission line
	(5) SWR as measured at input of transmission line	Not to exceed 1.25 when referenced to 50 ohms impedance level
	(6) radiation pattern	Uniform above vertical axis and does not depart appreciably from that of a vertical half-wave antenna

---

[9] NWS. 1979. Specification for Automatic Radiotheodolite System. Spec. No. J150A-SP001. National Weather Service, Silver Springs, Md. 78 p.

[10] Personal communication with Melvin J. Sanders, Jr., of National Weather Service Equipment Development Laboratory, Silver Springs, Md.

284

(ii) Transponder Characteristics

(a)	Frequency modulation (FM deviation) of 1680 MHz carrier	±125 KHz minimum but not in excess of ±500 KHz
(b)	Modulation rate	74.94 KHz
(c)	Typical deviation	±250 KHz1

(b) <u>The UK Model</u>. The British Mark 2B radiosonde sensors include an aneroid capsule for pressure, a bimetallic strip for temperature, and gold-beater's skin for relative humidity. These sensors control the armatures of the variable inductors. As the balloon ascends, a small windmill-driven switch connects each inductor sequentially to a vacuum tube oscillator. As shown in Figure 7-12, the completed circuit generates a signal whose frequency, determined by the variable gap between the inductance and armature, modulates the carrier frequency generated in the output circuit. The transmitted signal is detected by the ground radio receiver. The operational principle of the Mark 2B system is described in the Handbook of Meteorological Instrumentation as follows:

"The transmitter of the Mark 2B radiosonde is designed to operate within the frequency band 27.0 to 28.5 MHz. The carrier wave is modulated through a buffer circuit by an audio frequency oscillator, the frequency of which is controlled by the three meteorological units measuring pressure, temperature, and humidity. A buffer stage is used in order to prevent reaction between the radio frequency and audio frequency circuits. The oscillator-tuned circuit has a fixed capacitor and is connected, by means of a double-pole three-way switch, in turn to three variable inductors operated by the meteorological elements. The three inductors each consist of two coils, with a mumetal core and a movable mumetal armature, the position of which is controlled by an aneroid capsule, a bimetal element, and a strip of gold-beater's skin respectively. As the armature approaches the core, the air gap in the magnetic circuit is reduced, the inductance rises, and the frequency of the oscillating circuit falls. The audio frequency variation is within the range 700 to 1000 Hz.

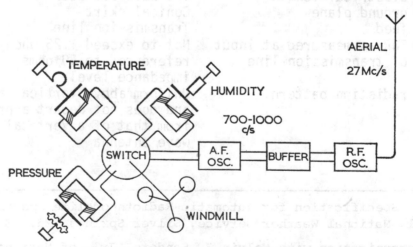

Figure 7-12 Mark 2B Radiosonde

Electrical power is provided by a special light-weight battery, while the power to drive the switch is provided by a three cup windmill but this may be replaced in the near future by a gravity motor.

Reception of the signals is made with a sensitive short-wave receiver, the pitch of the audible modulations being measured by matching it with the note produced by a calibrated oscillator of the resistance-capacity type, the matching being done on a cathode ray oscilloscope by obtaining a one-to-one Lissajous pattern."[11]

The Mark 3 Radiosonde replaced the Mark 2B during 1972. The new design uses a tungsten wire of 13.5 μm diameter with 950 ohms resistance at 20C to measure temperature, an improved aneroid capsule for pressure, and continued to use gold-beater's skin for relative humidity. The change in resistance is used to control an audio oscillator directly, while the deflection of the capsule and gold-beater's skin are used to vary the penetration of a ferrite rod within a ferrite pot. Each of these three devices is connected to the oscillator, one at a time, to generate a note in the frequency range 3500-6800 Hz. The frequency generated by the audio oscillator is used, via an isolating amplifier, to amplitude modulate a small crystal-controlled radio transmitter operating between 27.5 and 28.0 MHz. The instrument uses discrete components together with six transistors. The effect of environmental changes upon this circuit is monitored by three reference signals which together with the three meteorological data signals are selected by an electric motor-driven switch. An interlaced signal pattern is transmitted which provides a temperature signal every 4 s, a relative humidity signal every 8 s, and a pressure signal every 16 s. With the normal rate of ascent of 6.25 m s^{-1}, the corresponding height intervals are 25, 50, and 100 m respectively. Each reference signal occurs at intervals of 48 s. Additional brief signal pulses of two fixed frequencies are sent prior to the temperature and pressure signals to identify these signals for automatic ground-equipment processing. The transmitted notes are picked up by a radio receiver and the periodicity of each is measured. The readings may be converted to the values of air pressure, temperature, and relative humidity by human computations, using calibration curves and tables. An automatic ground-recording system together with a small digital computer to calculate the results, however, are now available.

In practice, wind velocities may also be obtained in the course of a radiosonde ascent, either by tracking the balloon by means of a radio-theodolite or by radar reflections from a radar target carried by the balloon; the latter method is used in the Meteorological Office.

The height attained by the sonde is about 21 km; a large balloon used in favorable circumstances will give observations to 30 km or above.

(c) The Finnish Model: The Vaisala Radiosonde (the variable radio frequency type). The radiosonde system which operates on variable radio frequency was

---

[11] Air Ministry Meteorological Office. 1961. Handbook of Meteorological Instruments, Part II, Instruments for Upper Air Observations. HMSO, London. 114 p.

first introduced by Dr. Vilho Väisälä of Finland in 1934.[12] As shown in
Figure 7-13, each of the three movable plate condensers is connected to one of
the sensors (the bimetallic, the
aneroid, and the hair hygrometer).
One plate of the condenser remains
fixed while the mechanical motion
of the sensor displaces the other
one. The value of capacitance between two metal plates is inversely proportional to the distance
between the two plates as $f = 1/(2\pi\sqrt{LC}$,
where L is the inductance; C, the
capacitance; and $\lambda \propto \sqrt{LC}$. Two fixed
condensers are connected to the
circuit serving as the reference.

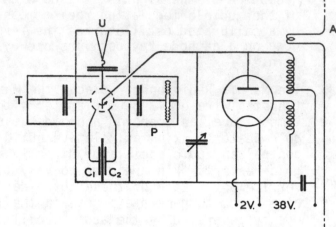

Figure 7-13 Vaisala Radiosonde

When C_1 and C_2 are the capacitances
of the two fixed condensers; C_0,
the capacitance of the vacuum tube;
and C, that of movable condenser
representing the meteorological
variables, the corresponding wave-
lengths will be λ_1, λ_2, and λ,
respectively. Thus

$$\lambda^2 = 4\pi^2L(C + C_0)$$

$$\lambda_1{}^2 = 4\pi^2L(C_1 + C_0) \quad \dots\dots\dots\dots\dots\dots\dots\dots \quad (42)$$

$$\lambda_2{}^2 = 4\pi^2L(C_2 + C_0)$$

where λ is the wavelength of the meteorological condenser. Eliminating L and
C_0

$$\frac{\lambda_1{}^2 - \lambda^2}{\lambda_1{}^2 - \lambda_2{}^2} = \frac{C_1 - C}{C_1 - C_2} = x \quad \dots\dots\dots\dots\dots\dots\dots\dots \quad (43)$$

As C_1 and C_2 are constants and C is a function of one of the meteorological
variables, x must be a function of this variable only. If the scale reading
of the wavelength λ^2 is n and linear, then

$$x = \frac{n_1 - n}{n_1 - n_2} \quad \dots\dots\dots\dots\dots\dots\dots\dots\dots \quad (44)$$

12
    Rossi, V. 1973. On the development of experimental aerology in Finland.
    Vaisala News 58.

With the use of a similar basic circuitry system, Väisälä developed the Finnish high altitude RS 16 radiosonde. Small capacitors are used as sensors. In addition, new sensors were introduced to measure high altitude (at and above 10 mb) temperature, humidity, and pressure. The construction principles of RS 16 radiosonde components are shown in Figure 7-14. The ground unit of this radiosonde is equipped with an automatic receiver which at the same time is used for recording upper winds with a built-in computer. The construction principles of RS 16 sensors as described by Väisälä are quoted below.

"The RS 16 is fitted with a wire thermometer, which has a negligible solar radiation error and small lag. For accurate pressure readings at high levels the radiosonde is provided with two separate barometers, both of the aneroid type-- a conventional barometer for the pressure range from 1050 to 0 mb, together with a special low pressure barometer, which is operative from ~ 100 mb to 3 mb and is nearly independent of temperature. For humidity records the radiosonde has a Frankenberger hair hygrometer of conventional construction.

The wire thermometer is made of aluminum alloy, 0.15 mm in diameter and 130 mm long. One end is fastened on a quartz frame by means of a suitable hanger and an invar wire, while the other end is fastened by means of an invar wire on the movable capacitor plate of the measuring condenser, the insulated plate of which is fastened on the frame. A spring keeps the wire straight and allows the movement of the plate. The hanger has been designed so that it is possible to adjust the meter.

The low pressure barometer is so constructed in order to achieve the best possible temperature compensation effect. Its metal screws determining the fixed condenser plate have been chosen as to length and material with an appropriate thermal expansion coefficient so that the distance between condenser plates is independent of temperature. The thermal coefficient of Young's modulus of the aneroid capsule is nearly zero so that the compensation achieved by the metal screws is mainly for the thermal expansion of the capsule. The compensation screws and the aneroid capsule are in good thermal contact through the base plate, on which both screws and the capsule have been fastened."[13]

In the RS 16 system, the average dispersion of the low pressure barometer (±0.50 mb) is only one-seventh of that of the standard barometer (±3.5 mb). It is read with an accuracy of ~0.1 mb for pressure, and ±0.2 to ±0.3C for temperature. The time response of a ventilated wire thermometer is about 1.6 s at 1 mb level, and 0.2 s at 1000 mb. The radiation errors of this thermometer are small at lower solar elevations (0 to 10 deg), but large at higher elevations (80 to 90 deg). At the 10 mb level, the radiation error ranges from 0.55C to 0.22C for all elevation angles.

After the Vaisala Company put the RS 16 radiosonde into production in 1969, many improvements were made. Further reduction of solar radiation error, par-

13
Väisälä, V. 1970. Instrument problems of radiosonde measurements at 10 mb and above. Meteorological Monographs 11(33): 378-382.

ticularly by accurate methods to correct residual radiation error, and increased accuracy for pressure measurements at high altitudes, were achieved. By 1972, these improvements resulted in a series of three new Vaisala radiosondes: The RS 18 (23.8 to 26.2 MHZ), RS 21-12C (403 MHz), and RS 21-13C (1680 MHz). Transducer components used in all these radiosondes are essentially the same as the RS 18 model with the following exceptions: modifications in the humidity sensor and changes in transmitter frequency band that incorporate solid state devices. A summary of the design and performance of the RS 21 meteorological sensors is given in Table 7-14. Figure 7-15 illustrates the RS 21-12C radiosonde package.

Construction Principle of the Wire Thermometer.

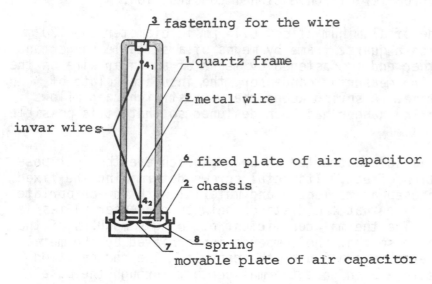

3 fastening for the wire

1 quartz frame

5 metal wire

invar wires

6 fixed plate of air capacitor

2 chassis

8 spring

7

movable plate of air capacitor

RS 16

Construction Principle of the Low-pressure Barometer.

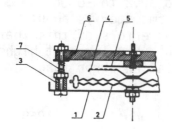

1. Base plate, 2. Aneroid capsule, 3. Supporting (compensating) screw, 4. Movable plate of air capacitor, 5. Fixed plate of air capacitor, 6. Insulating plate, 7. Spring.

Figure 7-14 RS 16 Radiosonde Structure

Table 7-14 Sensor Performance for RS 21 Radiosondes

Variable	Design	Performance
Pressure	Constant modulus alloy used for both upper range (P1)* and lower range (P2)*aneroid barometers; produces minimal hysteresis and reduces thermal and pressure dependence. Range of P1: 1060 to 10 mb. Range of P2: 150 to 3 mb.	Capsule P1*: Accuracy, ±1 mb (s.d.)**; Resolution, (manual data reduction), 1 mb for 1060-500 mb range, 0.5 mb for 500-50 mb range. Capsule P2*: Accuracy, ±0.2 mb (s.d.)**; Resolution, (manual data reduction), 0.1 mb.
Temperature	Circular bimetal (NiFe) alloy, 0.15 mm in diameter, used to minimize heat transfer from radiosonde body to sensor, reduce frictional forces, and decrease radiation errors at various solar elevation angles (protection provided by two concentric radiation tubes).	Accuracy, ±0.15C (s.d.)**; Resolution, (manual data reduction), 0.1C; Response time with 6 m s^{-1} flow, 2.9 s; Solar radiation correction: at 60 deg, 100 mb, 0.2C; at 60 deg, 10 mb, 3.8C.
Humidity	Humicap thin film capacitive device absorbs water molecules on polymer film, with consequent frequency change in oscillator circuit; output approximately linear over 0-100% RH range.	Accuracy, ±1% RH (s.d.)**; Resolution, 1% RH; Time response for 90% of total humidity change at 1000 mb, 20C, 1 s; Linearity on the order of ±1%, 0-80% RH range, slight drift (function of RH change and temperature), 80-100% RH range.

*P1 and P2 refer to Barometer 1 and Barometer 2, respectively, in Figure 7-15.

** s.d. refers to standard deviation.

The humidity transducer used in the RS 21 series is the Vaisala-developed Humicap, a thin film capacitive device.[14] The sensor consists of two small

[14] Salsmaa, E. and P. Kostamo. 1975. New thin film humidity sensor, In: Third Symposium on Meteorological Observations and Instrumentation. Amer. Meteorol. Soc., Washington, D.C. 33-38.
Kostamo, P. 1978. HUMICAP® stability test of 1978. Vaisala News 80: 3-7.

capacitors connected in series, each comprised of an organic polymer dielectric (~1 μm in thickness) deposited on a solid glass substrate and electrodes of a noble metal (such as gold or palladium). Water molecules when absorbed by the capacitive device form bonds with the polymer molecules. The mass absorption of water by the polymer film, via appropriate circuitry, alters the output of a high frequency oscillator. The device is approximately linear in response for relative humidities between 0 and 80%; whereas, slight drift is found for relative humidities ranging between 80 and 100%, varying as a function of the magnitude of the humidity change and time. Appropriate signal conditioning of the output frequency yields a rectified DC output voltage that is directly proportional to relative humidity. Typical specifications for the Humicap device exposed to the environment for short periods of time (less than one year) are, as follows:

Sensor type	HUMICAP thin film capacitive device
Measuring range	0 to 100% RH
Resolution	1% RH
Lag	1 s at 1000 mb, 20C
Accuracy	±1% RH (standard deviation) using repeated calibration method
Temperature range	-5C to 75C
Temperature dependence	~0.05% RH/deg C

The Vaisala ME 11 UHF-radiotheodolite system was developed in 1977, operating at a carrier frequency of 403 MHz. However, any frequencies from 395 to 410 MHz can be tracked by this system. This is a fully automatic radiosonde and upper air wind observing system. It consists of a UHF radiosonde receiver, display unit, control unit, antenna assembly of four 2-element half-wave dipoles, and antenna switch, cables, and other accessories.

Four groups of half-wave dipoles are mounted on the same front plane. A pair of groups at a time are switched to the UHF receiver by means of an automatic solid state commutator. By switching the antenna pairs consecutively, the signal strengths of the four antenna pairs (two for azimuth and two for elevation) can be detected and manually compared. Changes in signal strength of the paired dipoles indicate a deviation of the antenna from its target; this is indicated on a milliammeter located on the display unit. An antenna deviation of 0.1 deg can be detected on the indicators. The angular position of the antenna, piloted by photoelectric transducers, is digitally indicated on the display unit. Resolution of this indicated value is the same, 0.1 deg.

The Vaisala ME 12 radiotheodolite incorporates many of the features of the ME 11 with the exception of the following changes. The antenna is a square array of 32 half-wave elements spaced one-fourth wavelength apart and grouped into four quadrants (see Figure 7-16). Directional characteristics are enhanced by the addition of thin rod reflectors behind the antenna. Tracking is done in much the same manner as the ME 11 except that the ME 12 automatically tracks and activates directional changes in antenna direction in response to

1. Barometer 1 for normal pressure range [Barometer 2 for high altitudes on opposite side
2. Thermometer
3. Hygrometer [HUMICAP]
4. Inner radiation shield
5. Outer radiation shield
6. Rotating switch
7. Switch reel
8. 403 MHz transmitter
9. 403 MHz antenna
10. Battery
11. Balloon suspension string

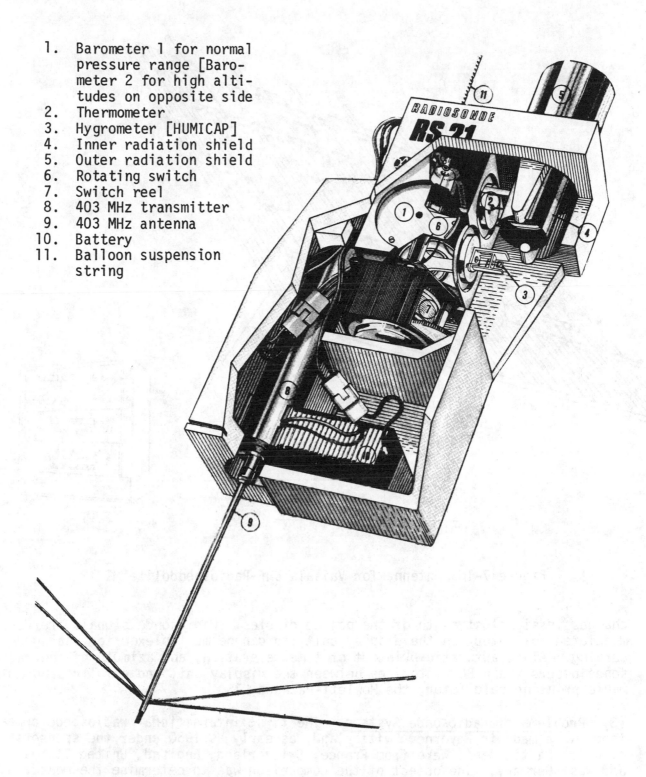

Figure 7-15 The Airborne Unit of Vaisala RS 21-12C Radiosonde

292

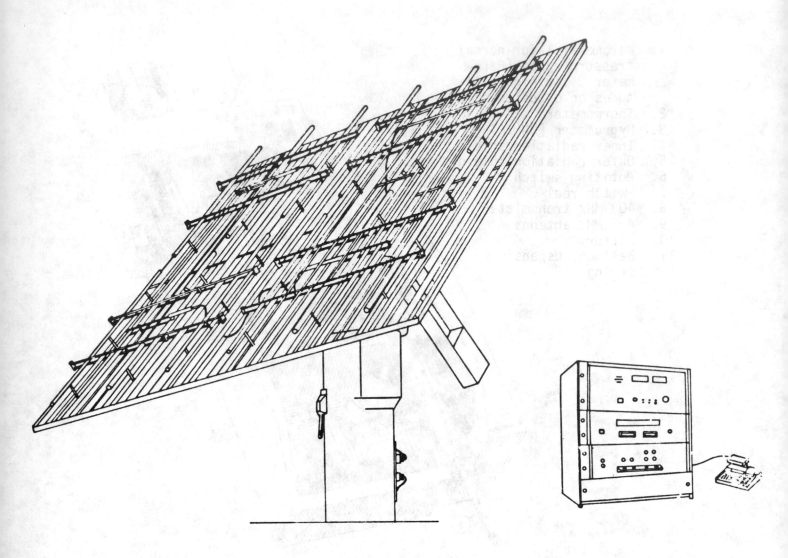

Figure 7-16 Antenna for Vaisala UHF-Radiotheodolite ME 12

changes in signal strength of the paired dipoles. Radiosonde signals are demodulated for output on the display unit and can be multiplexed for analog recording. Also, automatic printout of time, elevation, and azimuth of the radiosonde occurs via a BCD interface between the display unit and a fully program-
mable printing calculator, the Hewlett-Packard 97S.

(3) Problems of Radiosonde Systems. The first international radiosonde comparisons were made in Payerne, Switzerland, as early as 1950 under the sponsorship
of WMO. Participants were from France, Switzerland, England, United States,
and West Germany. The object of the comparison was to determine the measuring
accuracy of radiosonde sensors as well as systematic and incidental accuracy
errors. Special attention was paid to the calibration of sensors and the stability of calibration over long time intervals. In particular, the accuracy
and reproducibility of the radiosonde, especially for high altitude soundings
of 10 mb and above, were of prime concern. In 1956, the second international
radiosonde comparisons were held at Payerne where a total of 13 radiosondes

(Finnish, Belgian, French, Japanese, East German, West German, Dutch, Swiss, English, Soviet, American, Polish, and Indian) were studied. This comparison, far more exhaustive than the previous one, provided data for a thorough study of different types of errors in radiosondes. Five different types of radiosondes (American, Finnish, German, Soviet, and Japanese) have subsequently been approved as reference radiosondes with temperature measurement systems that meet the prescribed requirements. Comparisons were made on dual soundings with sensors carried aloft by a single balloon in daytime and at night. It was assumed that all sensors were exposed to the same solar elevation angle, the same rate of ascent, and identical atmospheric conditions. For a high altitude flight, the increase of solar radiation and the decrease of atmospheric temperature and pressure affect the sensor accuracy greatly. The radiation error especially affects bimetal sensors which are still being employed in nearly 60% of all radiosonde flights. In the comparisons performed, sample mean values of temperature difference were: Finland vs. Germany, 0.11C; Japan vs. USSR, 0.09C; and Finland vs. USSR, 0.01C. It is interesting to note that the German type H50 radiosondes (bimetal sensor) were in good agreement at all levels with the US AN/GMD system (thermistors), even though the thermal sensors were quite dissimilar. Humidity data for all radiosondes, in general, displayed considerable disagreement.

Table 7-15 compares the accuracy of radiosonde sensors of the American AN/GMD system, the British Mark 2B, and the Finnish RS 16. It must be pointed out that within the AN/GMD system alone, the different types of radiosondes exhibit variations in accuracy and reproducibility. Aside from the use of dual sounding techniques, the evaluation of performance can be made by comparison of the radiosonde's soundings against those of aircraft, satellites, and rockets, as done in numerous studies. Although this approach is useful, only a broad evaluation is possible from such a comparision.

In their studies of the compatibility of upper-air data, Finger, McInturff, and Spackman[15] have included data obtained from radiosondes, rocketsondes, and satellites between 1972 and 1976. However, only a summary of the findings is outlined here:

1. In general, the true diurnal variation in temperature is one order of magnitude smaller than the variability in measurement by different types of radiosondes. The variability is essentially caused by radiation heating at great solar elevations and at high altitudes.

2. The diurnal humidity variation between different radiosondes is far greater than those of atmospheric pressure or temperature. This is not only due to the great variety of humidity sensors but also because humidity is always difficult to measure accurately.

3. When the temperatures of radiosondes are compared with those of rocketsondes, it is found that at 25 km the former are on the average 2C to 3C higher than the latter during the winter, with

---

[15] Finger, F.G., R.M. McInturff, and E.A. Spackman. 1978. The Compatibility of Upper-Air Data. WMO Techn. Note No. 163 (WMO No. 512). The World Meteorological Organization, Geneva, Switzerland. 103 pp.

Table 7-15 Summary of Estimated Accuracy of Radiosonde Sensors

RAWIN SYSTEM	AN/GMD-( ) (USA)	M.O. MARK 2B (UK)	RS 16 (FINLAND)*

Atmospheric Pressure (P, mb)	Baroswitch (sfc to 50 mb)
	  3 to 4 mb at the
	    sfc
	  1.5 to 2 mb at
	    50 mb
	
	Hypsometer (100 to 10 mb)
	  ±3 mb at 100 mb
	  ±1.3 mb at 300 mb
	  ±0.4 mb at 10 mb
		Aneroid Capsule
		±1 mb at 1000 mb
			Std Barometer
			0.2 mb (day)
			
			Low pressure Barometer
			0.21 mb (day)

Air Temperature (T, F)	Ceramic Rod Thermistor
	  0.2F to 2.0F
	  (sfc to 10 mb)
	
	  5 to 17 s (sfc to 100 mb)
		Bimetal Element
		  0.20F (day) to
		  0.24F (night)
		  (sfc to 600 mb)
		
		  0.49F (day) to
		  0.32F (night)
		  (20 to 10 mb)
			Wire Thermometer
			  0.2F to 0.8F
			  (100 to 20 mb)**
			
			  0.8F to 1.5F
			  (21 to 3 mb)**
			
			  0.2 s to 1.8 s
			  (1000 to 2 mb)

Air Humidity (U, %)	Carbon Hygristor
	  ±5% (T>32F)
	  ±10% (32≥T≥-40F)
	  (10% to 100%)
		Gold-beater's Skin Strip
		  ±5% (T>32F)
		
		  ±5% to ±30%
		  (32≥T≥-40F)
		
		  6 s (T = 20C)
		
		  200 s (T = -30C)
			Frankenberger Hair Hygrometer
			  ±3% (T>32F)
			  ±10% (32>T>-40F)

* Finnish Vaisala High-Altitude Radiosonde
** For 0 deg to 90 deg solar elevation angle

the difference being less in the summer months.

4. Two methods have been developed for the comparison of temperature profiles obtained from satellites and those from rocketsonde/radiosondes: (a) direct comparision of temperature records and (b) indirect comparison from radiance values.

5. In the compatibility studies of geopotential heights measured in
    the lower stratosphere, there are systematic differences and day-
    night differences. At 100 mb there may be an average difference
    of 80 meters between different types of sondes when measuring
    geopotential heights in darkness, while in daytime the difference
    is usually larger.

The radiotheodolites, discussed in the preceding section, did not include a
detailed discussion of position-finding capabilities. The components and
operation of the radio-detection aspects of the rawinsonde and thunderstorm
detection systems, respectively, are discussed in some detail below.

7.2.2 <u>Radio Detection Finding Techniques</u>. The upper-air instrumentation
techniques which provide measurements of wind speed or thunderstorm locations
by means of radio signals may be identified as the radio-detection finding
techniques. The radio theodolite is employed for upper wind measurements. The
the radio theodolite offers certain advantages over the pilot balloon technique for upper wind observations: it is capable of measuring winds above the
clouds at a greater height and in high winds usually associated with low elevation angles. A radio theodolite is commonly known as a rawinsonde. The atmospherics direction finder is used to determine thunderstorm location.

(1) <u>Rawinsonde</u>. With the aid of the radiosonde transmitter and special ground
rawin equipment, the rawinsonde determines the position of an airborne balloon
from its slant range and azimuth and elevation angles. Three major rawin systems now in use are the French Metrox Radio-Theodolite System, the Finnish
Vaisala Radio-Theodolite System, and the United States Rawin Set AN/GMD System.

The United States Rawin Set AN/GMD-1 is a transportable 1680 MHz radio direction
finder used to measure upper winds to altitudes in excess of 30 km and to horizontal distances of about 200 km, depending upon the surrounding terrain. It
tracks the radio signals of the balloon-borne radiosonde to obtain and record
the azimuth and elevation angles. Manual positioning is used first to align
the antenna of the rawin set with the radiosonde transmitter before releasing
the balloon. When the antenna is aligned, it is switched to automatic tracking. The azimuth and elevation angles of the radiosonde are then plotted
against the height, which is computed from the temperature and pressure data
obtained from the same radiosonde, to determine wind direction and speed. When
the balloon is at an elevation angle of less than 6 deg above the horizon or
any prominent object on it, and/or when the azimuth angle is within 6 deg of
any object on the horizon, the rawin system fails to pick up the signal.

The ground unit of the Rawin System AN/GMD-1 (see Figure 7-11) consists of an
antenna mounted on an adjustable tripod pedestal, a receiver, and a recorder.
In the antenna, a stationary half-wave dipole is mounted at the focus of a 7 ft
parabolic reflector with a lobe of about 8 deg (beam width). A small hemispherical reflector mounted eccentrically in front of the dipole deflects the axis
of the lobe from that of the parabolic reflector by about 3 deg. The sinusoidal
signal from the receiver is amplified, rectified, and compared with two reference voltages from the voltage generator driven by the scanning motor. These
voltages correspond to azimuth and elevation components of the direction of the
axis of the antenna system. A comparison of the reference voltages with the

signal produced voltage gives two DC error voltages which are proportional to the angular magnitude in azimuth and elevation by which the antenna is off target. The error voltages are applied through amplifiers to drive motors automatically controlling the movement of the antenna. Synchronous repeaters transmit the azimuth and elevation angles to a remote recorder which registers these angles at constant time intervals.

As shown in Figure 7-17, the US radio-theodolite system GMD-1, including wind measurement, uses the principle of conical scanning. The parabolic antenna, used to measure the direction of the incoming signal, is offset or squinted from the incoming signal from the radiosonde. The squinted lobe then is rotated about the electrical or boresight axis of the reflector, generating a cone whose vertex is twice the squint angle. The incoming radio wave is modulated sinusoidally

$$A(t) = A_O \left[1 + ke \cos(2\pi f_c t - \phi)\right] \quad \ldots \ldots \ldots \ldots \ldots (45)$$

where $A(t)$ is the instantaneous value of the modulation at time t; A_O, the average value of the received signal over a conscan (conical scan) cycle; k, the lobe slope at the electrical axis; e, the error magnitude or angular distance between the incoming wave direction and boresight; ϕ, the phase of the error with respect to a reference; and f_c, the conscan frequency (from Sanders, Jr., et al., 1979). The modulation of the incoming radiowave remains meaningful as long as the error is contained in the cone. The relative accuracy for the radio-theodolite in azimuth and elevation, given as root-mean-square values of the differences between 1 min increments in angle, is 0.05 deg. Typical errors for various levels of mean wind speed for radio-theodolite systems are given in Table 7-16.

Table 7-16 Typical Radio-theodolite Windfinding Accuracy*

| Altitude | Mean Wind Speed from Surface | | |
	Less than 30 kt	33-60 kt	60-90 kt
10,000 ft	1 kt	3 kt	6 kt
20,000 ft	2 kt	5 kt	11 kt
40,000 ft	3 kt	10 kt	21 kt
60,000 ft	2 kt	7 kt	15 kt
80,000 ft	3 kt	10 kt	21 kt
100,000 ft	4 kt	12 kt	26 kt

*Assumptions: Rise rate = 10 kt; Pressure error = 1 mb; Azimuth error = 0.05 deg; Elevation error = 0.05 deg; Averaging interval = 2 min below 14 km, 4 min above 14 km
From Sanders, Jr., et al., 1979.

With the aid of a transponder from the balloon-borne unit, the United States AN/GMD-2A Rawin System receives and records the slant range of the balloon, in addition to azimuth and elevation angles. The slant range covers 0 to 250 km

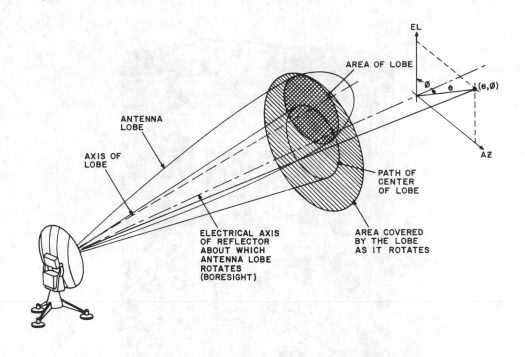

Figure 7-17   Conical Scanning Geometry of GMD-1

in 5 m increments, the elevation angle ranges from -6 to ±96 deg in 0.05 deg
increments, and the azimuth angle extends from 0 to 360 deg in 0.05 deg increments. Improved accuracy for upper wind speed measurement is achieved by this
system for signal detection at altitudes up to 120 km. Still another new
rawin set is the AN/GMD-4, consisting of a modified AN/GMD-2 Rawin Set and an
analog-to-digital converter. Data are reduced on a digital computer which is
not part of the AN/GMD-4 system. The modifications include the use of para-
metric amplifiers to improve the operation of the coarse range system and the
addition of real time digital displays of the azimuth, elevation, and range
data. The meteorological data processor provides azimuth and elevation tracking angles, slant range, temperature and humidity data, and elapsed time on
punched paper tape.

The Finnish rawinsonde, known as the Vaisala Radio-Theodolite, utilizes the
radio signals of the Finnish radiosonde transmitter. It differs from all
others in having pairs of fixed antennas instead of a single rotating antenna.
The French Metox Radio-Theodolite, on the other hand, is similar to the United
States Rawin System. It is a transportable radio-wind-direction finder primarily designed for tracking a balloon-borne transmitter but also capable, if
the latter is frequency modulated, of receiving radiosonde information. The
ground equipment consists principally of a directional receiving aerial system
with a lobe switching assembly, a receiver incorporating a CRT (cathode ray
tube) indicator, and a power supply unit. The whole apparatus, apart from the
power unit, is mounted on an adjustable tripod pedestal.

Figure 7-18 Radio-Direction Finder

(2) Thunderstorm Locating Devices. Various devices[16]/ have been designed to detect the locations of thunderstorms between 0 and 1000 km from the observer by the storm's electromagnetic radiation produced by lightning discharges. The signals (atmospherics or sferics) may last only one or two microseconds, and amplitudes may vary in sign and amount.

There are two main types of instruments for locating thunderstorms: the CRT (cathode ray tube) type and the narrow-sector recorder. With the former, the signals are received by two similar CRT's; the latter is a radio direction finder in which atmospherics from a limited sector are recorded.

The British Meteorological Office has established four sferic stations in the British Isles and three stations in the Mediterranean region for determining the place of origin of a group of atmospherics (foyer). Each station is equipped with two directional aerials which receive and transmit the signals to a CRT for display. A frequency of about 9 kHz is employed and the range of this system is about 2000 km. The following is a description of the British atmospherics direction-finder shown in Figure 7-18, appearing in the Handbook of Meteorological Instruments (see Footnote 16).

---

[16] A detailed description of the various devices may be found in Chapter 14 of: WMO. 1954. Guide to International Meteorological Instrument and Observing Practices. WMO, Geneva, Switzerland.

"The Atmospherics Direction-Finder, Mark 3, consists of two enclosed loop aerials and a console housing the receiver and associated units. The loops, which are wound in two layers, are about 1 m square and are oriented in the north-south and east-west directions. They are connected by coaxial cable to twin radio-frequency amplifiers which are normally tuned to a frequency of 10 kHz. Around this frequency the energy of the spectrum of an atmospheric is relatively high and there is very little interference from manmade signals or other sources. Provision is made, however, for tuning in other predetermined frequencies by means of switched capacitors.

Each amplifier consists of a cathode-follower input stage, two resistance-capacitance coupled amplifier stages, a phase splitter, and a push-pull output stage. Only the aerial circuits and the output transformers are tuned, the variable capacitors in these circuits serving mainly to adjust the phase response. Adjustment of the amplifier gains is made by means of a pair of matched attenuators.

The signals from the amplifiers are applied to the deflection plates of a cathode ray tube, the screen of which is displayed in the middle section of the console. They produce on the screen a diametral trace representing the resultant of the signal components in the planes of the aerial loops and therefore with a direction corresponding to the direction of arrival of the atmospheric (but with an ambiguity of 180 deg). A perspex cursor round the circumference of the screen enables the bearing to be read accurately. There is also a brilliance-modulator unit for momentarily increasing the brilliance of the trace; it is actuated by atmospherics with amplitudes exceeding a predetermined level. It is essential that the components of the atmospheric signals should reach the cathode ray tube in exactly the same phase. This is achieved by careful design of the receiver, which provides linear amplification over a very wide range of signal strengths, and by regularly matching each stage of amplification of the receiver. For this purpose at test signal generator is provided; the test signals are injected into the aerials and corresponding stages in the two amplifiers are adjusted to have the same gain and phase-shift characteristics over a small band of frequencies. These adjustments are carried out stage by stage, switches between corresponding stages of the two amplifiers being provided to enable the grids of the valves to be connected or either one earthed.

The console of the direction finder incorporates built-in telephone equipment and an automatic selector for use in the coordination of observations at the stations in the network. At any one station each atmospheric of sufficient intensity to trigger the brilliance modulator produces an audible pulse in the telephone system which serves as a signal for observers at the other stations to read the bearing. The pulse is followed by a quiescent period long enough for the bearings to be recorded. A sense channel, used with a short vertical aerial, is available for resolving the 180 deg ambiguity in the bearings but it is not normally required when observations from two or more stations are plotted."

7.2.3. <u>Meteorological Radar</u>. RADAR, coined from the expression <u>RA</u>dio <u>D</u>etec-tion <u>A</u>nd <u>R</u>anging, is an electronic device for detecting the slant range of airborne targets, atmospheric or non-atmospheric in origin, by means of micro-waves. In meteorological usage, atmospheric targets refer to clouds, hydro-meteors, lithometeors, tornadoes, etc.; non-atmospheric targets include bal-loons, corner reflectors, and transponders. Weather radar is used to measure atmospheric targets, while tracking radar is designed to detect non-atmospheric objects. The basic principle and mechanism of the two, however, are the same.

(1) <u>Tracking Radar</u>. The basic mechanism employed in tracking radar is to trace the trajectory of a balloon-borne target with a ground radar set. The radar antenna emits a directional beam in the form of pulses about a micro-second (μs) in duration. The target, which intercepts the energy of a radar pulse, returns a small portion of this energy to the antenna. The returned signal is detected and amplified in the radar receiver, and then displayed on various radar scopes called cathode ray tubes (CRT).[17] The time duration between the emitted and received signal is a measure of the slant range of the target. This slant range and the azimuth and elevation angles obtained from the radar scope determine the trajectory of the airborne target.

(a) <u>Airborne Unit</u>. The airborne targets used for tracking upper wind velo-cities are of two distinct types: the passive target and the active target. The passive target, a mechanical device, reflects or scatters back energy to the radar, and it is known as a radar reflector. The active target, an elec-tronic device, amplifies the input signal at a different frequency and sends it back to the radar. Also known as a responder or transponder, the active target is equipped with a receiver and transmitter and is capable of operating over long ranges with a low expenditure of power. A transponder also can be designed as a telemetering system to give additional information for the deter-mination of its location. In other words, it does not require as much power for a transponder to produce a large response than for an efficient reflector to generate an echo at the same range. Any radar using reflectors is called a primary radar and one using transponders, a secondary radar. A brief discus-sion of the two airborne targets follows:

Primary radar requires the following characteristics for the target:

  i. It must be an efficient reflector or reradiator of the incident beam;

 ii. It reflects or scatters approximately isotropically (or that the echo is independent of the angle of incidence);

iii. It must be light and compact in order to reduce weight and thus the aerodynamic drag of the airborne assembly; and

 iv. It should be a good conductor of electricity.

---

[17] The CRT is a vacuum tube consisting of an electron gun that produces a con-centrated electron beam (or cathode ray). The cathode ray impinges on a phosphorescent coating on the back of a viewing face (or screen). The ex-citation of the phosphor produces light, the intensity of which is controlled by regulating the flow of electrons. Deflection of the beam is achieved either electromagnetically by the current in coils wound around the tube, or electrostatically by voltages on internal deflection plates.

Two kinds of passive targets are available. One is a resonator, usually in the form of a half-wave dipole, which <u>absorbs</u> and reradiates energy from an incident beam. A dipole target for upper wind measurement usually consists of three conductors, each a half wavelength long, fixed mutually at right angles. The conductors are made of aluminum foil or other metal strips. Other passive targets are untuned (non-resonating) reflectors of various shapes and designs which <u>reflect</u> or <u>scatter</u> energy from the incident beam with minimal absorption. The intensity of echo, I_e, from a passive target is a function of its effective echoing cross-sectional area ($\sigma = \pi r^2$ for an isotropic reflector with radius r), the intensity of incident beam, I_i, on the target, and the distance of the echo, R, or the slant range. Applying the inverse square law, then

$$I_e = \frac{I_i \, \sigma}{4\pi R^2} \quad \ldots \ldots \ldots \ldots \ldots \ldots \ldots \ldots \quad (46)$$

The effective echoing area of a dipole, however, is computed as

$$\sigma = 0.86 \, \lambda^2 \cos^2\gamma \quad \ldots \ldots \ldots \ldots \ldots \ldots \quad (47)$$

where λ is the wavelength of the radar and γ is the angle formed by the inclination of the dipole and the direction of the electric field. When a large number of dipoles are tied together and randomly oriented in all directions, $\sigma = 0.29N\lambda^2$, where N is the number of dipoles. For any reflector, σ is the projected area, S, of the reflector perpendicular to the incident beam. Therefore, it varies with the shape and orientation of the reflector, with the exception of the spherical reflector. The corner reflector shown in Figure 7-19 is the most commonly used target. The flat plate reflector is a very efficient reflecting target when fixed in a position normal to the direction of the incident beam. A spherical reflector is weaker in reflection,[18] but it is isotropic. The theoretical formulae for calculating effective echoing areas of several reflectors are given in Table 7-17. Calculations of effective echoing cross-sections and relative maximum ranges are given in Table 7-18 for several different wavelengths and reflectors.

Table 7-17 The Effective Echoing Cross-Sectional Area

Target	Size	Orientation	Projected Area	Echoing Cross-Section
Sphere	Radius, r	Independent	πr^2	πr^2
Corner reflector	Hypotenuse, ℓ	Corner symmetrical to beam	$\ell^2/2\sqrt{3}$	$\pi\ell^4/3\lambda^2$
Flat plate	Area, S	Normal to beam	S	$4\pi S^2/\lambda^2$

[18] The maximum range value relative to those of spheres with a diameter ten times the wavelength.

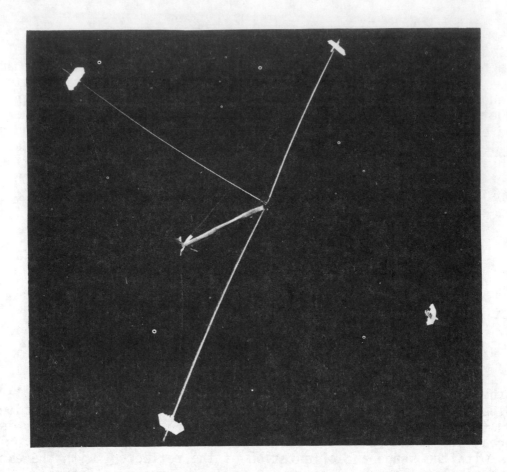

Figure 7-19   Corner Reflector

Secondary radar employs a transponder which consists of a receiver, triggering circuit, and transmitter, together with an antenna. A fraction of a micro-second delay always exists between the receiving and transmitting of the tran-sponder. The delay increases the indicated range on the CRT. The automatic computations of the tangential and radial components of wind in terms of their slant range, R, the azimuth angle, α, and the elevation angle, β, as well as their rate of change with time, dR/dt, dα/dt, and dβ/dt, are expressed by

$$V_T = R\frac{d\alpha}{dt} \cos \beta \ldots \ldots \ldots \ldots \ldots \ldots \ldots \ldots \ldots (48)$$

$$V_R = \frac{dR}{dt} \cos \beta - R\frac{d\beta}{dt} \sin \beta \ldots \ldots \ldots \ldots \ldots (49)$$

where V_T is the tangential wind component and V_R is the radial wind component. The height of the transponder is computed by

$$H = R \sin\beta + \frac{R^2 \cos^2\beta}{D} \quad \ldots \ldots \ldots \ldots \ldots \ldots \quad (50)$$

where H is the height of the transponder and D is the diameter of the earth.

The minimum power of the transponder, P_{oa}, required to produce a signal can be obtained from

$$P_{oa} = \frac{P_{tg} G_{tg} G_{ra} \lambda_t^2}{P_{ta} G_{ta} G_{rg} \lambda_a^2} P_{og} \quad \ldots \ldots \ldots \ldots \ldots \ldots \ldots \quad (51)$$

where the symbols, P, G, and λ are the power, gain, and wavelength, respectively. Subscripts g, a, t, r, and o refer to the ground, transponder, transmitter, receiver, and minimum value, respectively. For example, P_{tg} stands for the power of the ground transmitter; λ_a, the wavelength of the transponder transmitter; and λ_t, the wavelength of the ground transmitter.

(b) The Ground Unit. Radar systems have been briefly discussed previously. As shown in Figure 7-20, the basic mechanism of a tracking radar consists of a transmitting system and a receiving system with power supply and antenna in common.[19] The radio frequency signal from the antenna, in the form of pulses,

Table 7-18 The Effective Echoing Cross-Sections
and Relative Maximum Range

Wavelength	Reflectors	Size	Echoing Cross-Section	Relative Maximum Range*
10 cm	Sphere	100 cm dia.	0.79	1.0
3 cm	Sphere	30 cm dia.	0.071	1.0
10 cm	Flat plate	100 cm² area	1250	6.3
3 cm	Corner Decca 1	92 cm hypotenuse	725	10.2
3 cm	Corner Decca 2	65 cm hypotenuse	175	7.1
10 cm	Balloon cover	150 cm dia.	---	2.8

* The relative maximum range refers to the gain. Values for 10 cm wavelength are not comparable with those for 3 cm.

---

[19] Some radars use separate antennas for transmitting and receiving and in so doing eliminate the need for a duplexer.

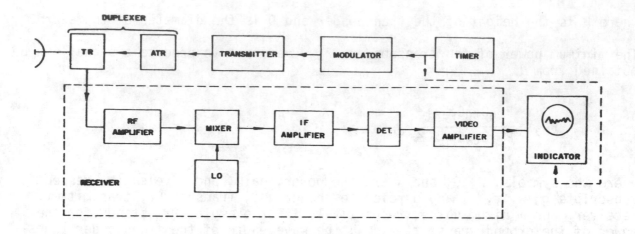

Figure 7-20 Block Diagram of a Pulse Radar System

is generated by a timer which triggers the modulator to send high voltage pulses to the transmitter. The pulses pass through the duplexer (which includes the anti-transmit-receiver and transmit-receive tubes) to the antenna for transmission.

A pulse travels a slant range, R (km), at the speed of light, c (299,792 km s^{-1}), to a target; it is reflected back to the radar in a time interval, Δt (s). Then, the slant range is

$$R = \frac{1}{2} c \, \Delta t \quad \dots \dots \dots \dots \dots \dots \dots \dots \dots \quad (52)$$

The number of pulses transmitted in a unit time is known as the pulse repetition frequency or PRF. The duration of a pulse is in the order of a microsecond (μs) whereas the time interval between two successive pulses (or the silent period) is about one millisecond. From Eq. (52), it is obvious that the minimum time interval of a silent period should be $\Delta t = 2R/c$, where R is now the maximum range of detection. The silent period allows sufficient time for a pulse to travel a maximum range to a target and return to the same antenna before the next pulse is emitted by the radar. This suggests that a low PRF or a prolonged silent period is preferable. But at the same time, the PRF should be high enough so that no moving target positions are missed by a revolving radar beam. Setting F equal to PRF

$$F = \frac{c}{2R} \quad \dots \dots \dots \dots \dots \dots \dots \dots \dots \dots \dots \dots \quad (53)$$

The length of a radar pulse, h (cm), is given by the product of the pulse duration, τ (s), and the speed of light, c (cm s^{-1}), or

$$h = \tau c \ \dotfill \ (54)$$

Thus, a one microsecond pulse is about 300 m or 1000 ft in length. The greater the length, the larger the amount of energy transmitted. For example, a five microsecond pulse has about five times as much energy as a one microsecond pulse. But the target resolution is greatly reduced.[20] The pulsed energy emitted from a paraboloid antenna has a conical radiation pattern[21] as shown in Figure 7-21.

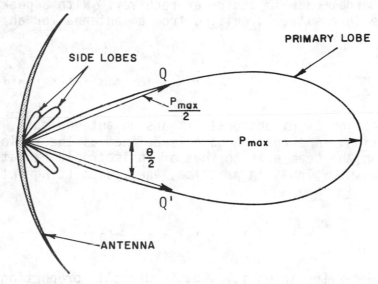

Figure 7-21   Cross Section of a Radar Beam from
a Paraboloid Reflector

The beam consists of a primary lobe and some side lobes. Since radar pulses are highly directional, their peak power, P_t, is directed along the axis of the primary lobe. One half of this peak power is directed toward the half power points, Q and Q', on the cross-section of the primary lobe. The beam width, θ, is defined as the angular distance between the half power points, and is a function of the wavelength, λ, and antenna diameter, d.

An approximation for the beam width, θ, from Wexler and Swingle[22] is

$$\theta = \frac{k\lambda}{d} \ \dotfill \ (55)$$

When λ and d are expressed by the same units and θ is expressed in deg, k is about 85. The value, k, however, may vary from 70 to 85 depending upon the

---

[20] At high resolution the radarscope is capable of differentiating one target from another; at low resolution it cannot.

[21] The radiation pattern, which differs according to the types of radar, is the radar beam.

[22] Wexler, R. and D.M. Swingle. 1947. Radar storm detection. Bull. Amer. Meteorol. Soc. 28: 159-167.

types of radar. For example, for the AN/FPS-77 λ = 5.5 cm, d = 244 cm (8 ft), k = 70, and the beam width is 1.6 deg (see Table 7-19). The narrower the beam width, the stronger the echo but the smaller the areal coverage. The sum of energy contained in the side lobes is quite small, and normally they present no problem when targets at a range of more than a mile or two are being considered. However, when targets are located at short ranges, the side lobes may return enough energy to produce spurious echoes or to distort actual echoes.

The peak power of transmission is on the order of a thousand kilowatts but only a very small portion of this energy, approximately 10^{-12} to 10^{-13} watts, is reflected by the target back to the radar receiver. With a peak power of transmission, P_t, the intensity, I, emitted from an antenna for an isotropic radiator[23] is

$$I = \frac{P_t}{4\pi R^2} \quad \cdots \cdots \cdots \cdots \cdots \cdots \cdots \cdots \cdots \cdots (56)$$

As radar transmission is directional, it has an antenna gain, G_t, in the most favorable direction. The antenna gain is defined as the ratio of the power transmitted along the beam axis to that of an isotropic radiator transmitting the same total power. Thus, in practice, the transmitting intensity, I_t, of a radar is

$$I_t = \frac{P_t G_t}{4\pi R^2} \quad \cdots \cdots \cdots \cdots \cdots \cdots \cdots \cdots \cdots (57)$$

Obviously, the receiving intensity, I_r, is directly proportional to σ and I_t; thus

$$I_r = \frac{\sigma I_t}{4\pi R^2} \quad \cdots \cdots \cdots \cdots \cdots \cdots \cdots \cdots \cdots (58)$$

The echo power received at the radar, P_r, on the other hand, depends upon the effective aperture of the antenna, A_e. Thus Eq. (58) may be written as

$$P_r = \frac{\sigma A_e I_t}{4\pi R^2} \quad \cdots \cdots \cdots \cdots \cdots \cdots \cdots \cdots \cdots (59)$$

Substituting Eq. (57) into Eq. (59)

$$P_r = \frac{\sigma A_e P_t G_t}{16\pi^2 R^4} \quad \cdots \cdots \cdots \cdots \cdots \cdots \cdots \cdots (60)$$

But the receiver gain can be expressed as

---

[23] An isotropic radiator emits exactly the same intensity in all directions.

$$G_r = \frac{4\pi A_e}{\lambda^2} \quad \ldots \ldots \ldots \ldots \ldots \ldots \ldots \ldots \quad (61)$$

where λ is the wavelength of radar. Solving for A_e in Eq. (61) and substituting it into Eq. (60)

$$P_r = \frac{\sigma G_r G_t P_t \lambda^2}{64\pi^3 R^4} \quad \ldots \ldots \ldots \ldots \ldots \ldots \ldots \quad (62)$$

When the antennas for the transmitter and the receiver are of the same dimension, then $G_t = G_r = G$, and Eq. (62) becomes

$$P_r = \frac{\sigma G^2 P_t \lambda^2}{64\pi^3 R^4} \quad \ldots \ldots \ldots \ldots \ldots \ldots \ldots \quad (63)$$

The receiving system operates as follows. A fast-acting switch called the transmit-receive (TR) switch (see Figure 7-20) disconnects the receiver during transmission. After passage of the transmitted signal, the TR switch reconnects the receiver to the antenna. The anti-transmit-receive (ATR) switch, which has no effect during the transmission portion of the cycle, acts to channel the received signal into the receiver. The TR and ATR together are called the duplexer. The radio frequency pre-amplifier (RF pre-amplifier) is needed to amplify the received signal because the latter is very weak. However, it amplifies the noise[24] as well. The mixer and the local oscillator (LO) convert the radio frequency (RF) to an intermediate frequency (IF). After amplification by the IF amplifier, the IF pulse signal is converted to video by the detector and further amplified by the video amplifier for presentation on an oscilloscope (CRT) or oscillograph. Different types of CRT provide various combinations of displays on an oscilloscope: the azimuth angle (RHI) versus the horizontal range (PPI),[25] and the vertical range (A-R scope) versus the horizontal range (PPI), and so on.

Decca Radar, Limited (London) has manufactured several types of tracking radars such as WF1, WF2, and WF44X. The WF1 radar shown in Figure 7-22 operates with an RF of 9375 ± 30 MHz, a peak power of 20 kW, a pulse duration of 0.5 or 0.25 µs, and a PRF of 1000 pulses s^{-1}. The antenna is a paraboloid reflector 2.5 m in diameter and with a beam width of 1 deg. The movement of the antenna system in azimuth and elevation is manually controlled through direct gearing by a pair of hand wheels. The angles are indicated on dials which can be read to 0.1 deg.

The received signal, upon being converted into an IF of 30 MHz, is displayed on a single CRT with an I-scope indication. A complete circle is shown on the

---

[24] The noise comes from the RF amplifier tube and other tubes.

[25] PPI stands for plan position indicator and RHI stands for range height indicator.

308

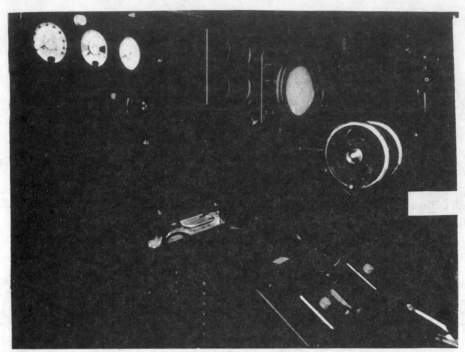

By courtesy of Decca Radar Ltd.
INTERIOR OF CABIN OF DECCA WIND-FINDING RADAR, TYPE W.F.1

By courtesy of Decca Radar Ltd.
EXTERIOR VIEW OF DECCA WIND-FINDING RADAR, TYPE W.F.1

Figure 7-22　The WF1 Decca Radar

oscillograph when the target is within all positions of the conical scan of the radar beam. This is on target. If it were off target there would be a gap in the circle or it might even appear as an arc. The radius of the circle is proportional to the slant range. A coarse-time-delay will help the operator to bring the target on the screen. To take readings, a fine-time-delay is switched into the circuit. The error in short range measurement does not exceed 1 per cent of the 10 km increment. The error in angle measurement (azimuth and elevation) does not exceed 0.13 deg.

(2) <u>Weather Radar</u>. Weather radar is designed to detect the physical characteristics of echoes as well as their distribution and movement. An echo, the target of weather radar, can be one of many weather conditions: rain, snow, cloud, lightning, tornado, and the like. They are presented in the form of solid (ice crystals, hailstones, and snowflakes), liquid (droplets of rain, cloud, and fog), and gaseous (water vapor and oxygen) particles. The detectable characteristics of these particles are described by such properties as their state, size, shape, albedo, structure, boundary, motion, and location.

Tracking radar and weather radar have many aspects in common. Characteristics and specifications, such as the energy distribution pattern of a paraboloid reflector (Figure 7-21), the empirical equation of beam width (Eq. 55), and the computation of pulse repetition frequency (Eq. 53) are the same. Also, similarities can be found in Figure 7-20 and Eq. (58); but Eq. (56) requires some modifications when applied to weather radar.

Dissimilarity between the two radar systems arises from the differences in physical aspects of the targets. Targets employed in tracking radar are well defined, small in size, and their dimensions and physical properties can be precisely predetermined. Weather radar targets are ill-defined and differ in size and physical properties. Weather radar involves the choice of operating wavelength, depending upon the types of target (weather conditions) to be measured. The desirable wavelengths for some common targets are tabulated in Table 7-19. While the wavelength used in primary radar for upper wind measurements varies from 6 to 50 cm, and has an order of magnitude of 100 to 200 cm for secondary radar, wavelengths for weather radar seldom exceed 20 cm. Drop size determination, for example, requires a short wavelength for high received power (see Eq. 68). The shorter the wavelength the smaller the drops that can be detected. The sizes of drops have an important bearing on the echoes; a large drop is exaggerated while a small drop is neglected. For example, if a large raindrop is 10 times the size of a small raindrop, then the radar echo is a million times stronger for the large one than the small one, if all other factors are equal (see Eq. 65).

A weather radar with high transmission energy requires a long pulse duration of at least 2 μs, and a high peak power. Useful ranges of 200 to 300 n. miles are obtained with a power of about 500 kW at 10 cm wavelength, and 100 kW at 3 cm. A high gain antenna, with a beam width of 2 deg or less, is desirable for best concentration of energy in the path of transmission. If feasible, separate antennas for transmitting and receiving are recommended. Different scanning arrangements and types of display are necessary depending upon the type of target. In cloud area determinations, weather radar often detects a small section of the cloud, depending upon the type of radar and the size and range of the cloud. In precipitation determination, a portion of the precipitation area is

Table 7-19 Selection of Wavelength for Various
Applications of Weather Radar

Wavelength, cm (frequency, MHz)[1]	Band Designator[2]	Detectable Target	Representative Radar Set
0.86 (35,000)	K_A	Clouds and cloud droplets	AN/TPQ-11
3 (10,000)	X	Light rain or snow	AN/CPS-9
5 (6,000)	C	Moderate rain or snow	AN/FPS-77
10 (3,000)	S	Heavy rain and severe storms	WSR-57
20 (1,500)	L	General weather surveillance	AN/FPS-7

[1] Computation is made by $\nu = c/\lambda$.

[2] Band designators are military usuages of the wavelength used in radar, and it is convenient to use bands instead of wavelength because the former gives a rounded-off number for the wavelength. For example, for the AN/FPS-77, $\lambda = 5.5$ cm and d = 224 cm.

generally intercepted by radar. Thus, the types of radar are classified by the size, intensity, and types of precipitation they detect. Furthermore, interpretation of radar data partially relies on the types of radarscopes employed.

The three most common types of radarscopes are: the plan position indicator (PPI), the range height indicator (RHI), and the A or R scope.

The antenna of a PPI radar rotates in a horizontal plane around a vertical axis. As shown in Figure 7-23, the traces on a radarscope form as spots of light which move along an axis from the center of the scope to its edge and return very rapidly back to the center. The procedure is repeated as the trace changes its bearing by scanning around the scope and is synchronized with the rotation of the antenna. As the antenna beam sweeps over a target, an echo appears as a bright spot and the scope retains the fluorescence long enough for the antenna to complete a full rotation. The stronger the return signal, the brighter the echo or the higher the signal level up to the point of saturation. Thus, the range and azimuth of an echo are indicated on a polar coordinates scope as the radius vector, R, and azimuthal angle, β. The radar antenna is located at the origin of the polar coordinates.

The antenna of the RHI (see Figure 7-24) scans a vertical cross-sectional development of the target along any azimuth. An echo is shown in a rectangular coordinate scope with the radar antenna at the lower end. The height, H, of the target is represented by the y-axis (i.e., elevation angle) and the range, R, representing the distance to the target is displayed on the x-axis.

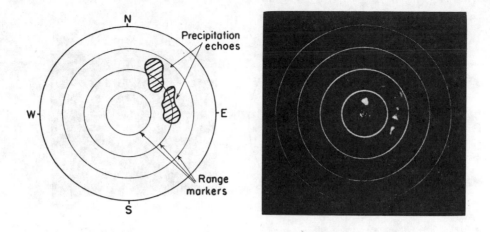

Figure 7-23 An Illustration of PPI Display

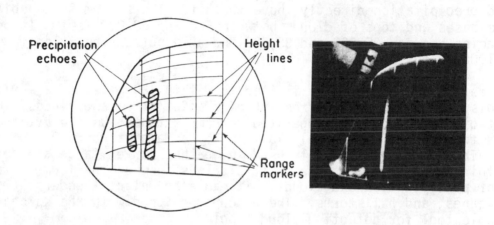

Figure 7-24 An Illustration of RHI Display

The A scope or R scope scans horizontally or vertically across the scope in rectangular coordinates with range versus relative signal intensity (or amplitude) received from the target. The echo is indicated as a pip or deflection on a horizontal illustrated by the A scope in Figure 7-25. The distance of the deflection from the origin of the trace is proportional to the range of the target, and the amplitude of the deflection indicates the intensity of the echo. A switch is sometimes provided to convert the A scope into an R scope. The latter is magnified portions of a selected range interval of the former and is useful in magnifying distant echoes of relatively weak intensity. The C or S band set can only be used for the detection of rain drops of diameters greater than 100 μm. Cloud and fog droplets with diameters of 2 to 65 μm are detected by the K_A band.

A typical radar cloud-detecting set, AN/TPQ-11 (manufactured by Lear-Siegler), is shown in Figure 7-26. This is a vertically pointing, two antenna, K_A band

312

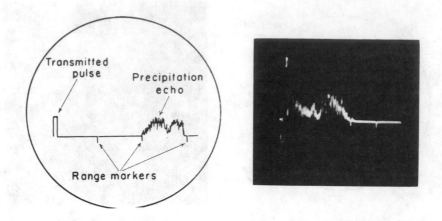

Figure 7-25   An Illustration of A Scope Display

radar system for detecting, displaying, and recording the density and height of clouds and precipitation directly above the set. The system is capable of measuring the bases and tops of clouds between 500 and 60,000 feet. It operates at frequencies between 35,200 and 35,500 MHz at a pulse repetition frequency (PRF) of 1000 pulses per second.

RHI scope detection is generally limited to within 100 miles of the radar and is good for display of the vertical extent of precipitation and clouds. However, the height of cloud tops is exaggerated to some extent. On the average, the radar echo tops exceed the actual cloud top by an amount up to 0.5 of the beam width.[26] The PPI scope is widely used in weather surveillance and detects echoes within a range of 350 miles. It displays a smooth uniform texture and pattern for precipitation and clouds. It can also detect thunderstorms, lightning, hurricanes, and hailstorms. The A scope equipped with the K_A band is the most reliable tool for detecting cloud droplets.

In weather radar, it is important to know the size, composition, shape, and orientation of hydrometeors in the atmosphere. Also, equations should be provided to calculate the power returned (echoing power) to a radar set by these hydrometeors. Thus, a brief discussion of these follows.

For either a duplexer or a radar with its receiving and transmitting antennas of the same size, $G_t = G_r$. Therefore, Eq. (61) can be rewritten as $G_t = 4\pi Ae/\lambda^2$, which is then substituted into Eq. (60). Hence

$$P_r = \frac{\sigma A_e^2 P_t}{4\pi\lambda^2 R^4} \quad \ldots\ldots\ldots\ldots\ldots\ldots\ldots\ldots \quad (64)$$

[26] Saunders, P.M. and F.C. Ronne. 1962. A comparison between the height of cumulus clouds and the height of radar echoes received from them. J. Appl. Meteorol. 1: 296-302.

When Eq. (64) is applied to the study of cloud droplets, the effective echoing area, σ, depends upon the intensity of scattering by nearly spherical droplets. Ryde,[27] utilizing the Rayleigh Law of Scattering, derived the scattering function, S, for a spherical particle whose diameter, D, is smaller than the wavelength, λ, in the direction of the source of energy. His theoretical expression for the scattering function is

Figure 7-26 A typical radar cloud-detecting set, AN/TPQ-11

[27] Ryde, J.W. 1946. The attenuation and radar echoes produced at centimeter wavelengths by various meteorological phenomena. In: Meteorological Factors in Radio-Wave Propagation, Conf. Rept. The Phys. Soc. and the Royal Meteorol. Soc., London. 169 p.

$$S = \frac{\pi^5 D^6}{\lambda^4} \left[\frac{\varepsilon - 1}{\varepsilon + 2} \right]^2$$

where ε is the permittivity, a complex dielectric constant of the droplets. The expression $(\varepsilon - 1)/(\varepsilon + 2)$ can be represented by a consant value, K. As cloud droplets are not uniform in size, the echoing power, P_r, is proportional to $\Sigma N_D S$, where N_D is the number of droplets of diameter, D, per unit volume, or

$$P_r = \Sigma N_D S = \frac{\pi^5 K^2}{\lambda^4} \Sigma N_D D^6 \quad \dots \dots \dots \dots \dots \dots \dots \quad (65)$$

If θ is the beam width in radians, h is the pulse length, and R is the slant range, the volume, V, of the cloud effectively illuminated by a pulse is

$$V = \frac{\pi R^2 \theta^2 h}{8} \quad \dots \dots \dots \dots \dots \dots \dots \dots \dots \dots \dots \dots \quad (66)$$

The effective echoing area, σ, equals $V \Sigma N_D S$, or from Eq. (66)

$$\sigma = \frac{\pi R^2 \theta^2 h \; \Sigma N_D S}{8} \quad \dots \dots \dots \dots \dots \dots \dots \dots \dots \quad (67)$$

Substituting Eqs. (65) and (67) into Eq. (64)

$$P_r = \frac{\pi^5 A_e^2 \theta^2 K^2 h P_t \Sigma N_D \; D^6}{32 \; \lambda^6 \; R^2} \quad \dots \dots \dots \dots \dots \dots \dots \quad (68)$$

But θ^2 is proportional to λ^2/A_e. Thus

$$P_r \propto \frac{A_e h \; P_t \Sigma N_D \; D^6}{\lambda^4 \; R^2} \quad \dots \dots \dots \dots \dots \dots \dots \dots \dots \quad (69)$$

The last expression implies that echo intensity varies directly with the droplet size, D, and its density, N_D, but inversely with range, R, for a given radar. For an improvement of the echo intensity for a radar, it is necessary to have a high peak power, P_t, a short wavelength, λ, a wide aperture antenna, A_e, a long pulse length, h, and a large beam width, θ. However, it is not always possible to have all the above in combination. For example, at wavelengths of less than 3 cm, the attenuation of an echo increases rapidly. Thus, a long pulse length would sacrifice the resolution of targets.

Doppler radar[28] is employed when the target to be detected is moving. It may

---

[28] Radar named after the Austrian physicist, J.C. Doppler, who first noted the effect of a moving sound source on the frequency of sound. (See Section 5.2.)

be used as tracking radar for upper wind measurement and as a weather radar for the measurement of moving cloud elements, tornadoes, and precipitation. When a target is in motion it causes a shift of frequency in the reflected pulse compared to the pulse transmitted, so that a Doppler shift[29] results.

If the original frequency and wavelength of a radar are ν and λ, respectively, and c is the speed of light, then a target at rest has a frequency, ν, expressed by $\nu = c/\lambda$. When a target is in motion at a speed of v, the new frequency, ν', can be expressed as

$$\nu' = \frac{c \pm 2v}{\lambda} \quad \dots \dots \dots \dots \dots \dots \dots \quad (70)$$

When the target is moving toward the radar, v is positive and it is negative for a target moving away from the radar.

The difference in frequency due to the Doppler effect is given by an amount, ν_D, where

$$\nu_D = \nu' - \nu = \pm\frac{2v}{\lambda} \quad \dots \dots \dots \dots \dots \dots \quad (71)$$

If v is expressed in knots, λ in cm, and ν_D in Hz

$$\nu_D = 103 \frac{v}{\lambda} \quad \dots \dots \dots \dots \dots \dots \dots \quad (72)$$

In practice, the radial velocity, v, for precipitation is given by

$$v = u \cos \alpha \cos \beta \pm w \sin \alpha \quad \dots \dots \dots \dots \dots \quad (73)$$

where u is the wind velocity; w, the falling velocity of the precipitation; α, the angle of elevation; and β, the azimuth angle with respect to wind direction. As v can be obtained from Eq. (71) and u, α, and β can be measured by radar, the falling rate of precipitation, w, can be determined.

Doppler radars are classified into pulsed and continuous wave types. The pulsed radar emits a signal of fixed duration at a single frequency. A moving particle of diameter, D, and radial velocity, v, would produce a single Doppler shift frequency and a signal intensity directly proportional to D^6. The difference from pulse to pulse provides a phase shift with time between transmitted and returned signal and the Doppler shift frequency. Limitations in measurement with pulsed Doppler radar occur if particles cover distances greater than one wavelength of the radar. Provided particle distances are less than the critical wavelength, this type of radar provides simultaneous range and velocity information.[30] However, information derived from pulsed Doppler radar becomes ambiguous

[29] Also known as the Doppler effect.

[30] Battan, L.J. 1973. Radar Observations of the Atmosphere. Univ. of Chicago Press, Chicago, Illinois. 324 p.

if the particles being tracked cover distances greater than one wavelength of radar.[31]/

The maximum Doppler shift frequency which can be detected for a radar having a pulse repeat frequency (PRF) of some fixed value is

$$\nu_{max} = \frac{PRF}{2} \quad\ldots\ldots\ldots\ldots\ldots\ldots\ldots\ldots\ldots\ldots\ldots\ldots (74)$$

And the maximum Doppler velocity detectable corresponds to

$$v_{max} = (PRF)\,\frac{\lambda}{4} \quad\ldots\ldots\ldots\ldots\ldots\ldots\ldots\ldots\ldots\ldots (75)$$

where λ is the wavelength. Thus, when measuring higher velocities, long wavelengths and high PRF's must be generated by the radar. The maximum range, R_{max}, is related to both ν_{max} and v_{max} by the following relationships.

$$\nu_{max} = \frac{1}{4}\,\frac{c}{R_{max}} \quad and$$

$$v_{max} = \frac{\lambda}{8}\,\frac{c}{R_{max}} \quad\ldots\ldots\ldots\ldots\ldots\ldots\ldots\ldots\ldots (76)$$

where c is the speed of light and λ, the wavelength.[32]/

Continuous wave Doppler emits microwave energy continuously. This type of radar measures all Doppler shift frequencies and velocities unambiguously, but does not provide range information between target and radar. Single Doppler radar in continuous mode (CW), operating in microwave frequencies, has been employed in severe storm and tornado studies, to name a few of its applications. More recent developments in improved frequency modulated, continuous wave (FM-CW), single Doppler radar now allow this system to be used for probing clear air motions between heights ranging from the near surface layer to the stratosphere.[33]/

Both pulsed and continuous wave Doppler radars have benefitted from the development of high speed data acquisition and processing systems. For instance, these innovations now permit real-time processing of radar signals and color-enhanced displays that are feasible for not only research purposes but for operational

---

[31] Doviak, R.J., D. Sirmans, D. Zrnić, and G.B. Walker. 1978. Considerations for pulse-Doppler radar observations of severe thunderstorms. J. Appl. Meteor. 17(2): 189-205.

[32] See Footnotes 30 and 31. Both provide in depth discussions of Doppler radar principles, limitations, and applications.

[33] Lemon, L.R., R.J. Donaldson, Jr., D.W. Burgess, and R.A. Brown. 1977. Doppler radar application to severe thunderstorm study and potential real-time warning. Bull. Am. Meteorol. Soc. 58(11): 1187-1193.
Gage, K.S. and B.B. Balsley. 1978. Doppler radar probing of the clear atmosphere. Bull. Am. Meteorol. Soc. 59(9): 1074-1093.

use.[34]

Dual Doppler radar is increasingly being used in meteorological probing of the atmosphere. Pulsed Doppler radars are operated, usually in pairs, with each transmitting on the same or on different frequencies. Both radars scan a set of common, tilted planes which intersect the baseline between the two units. The radars are each observing radial velocities of targets with the same motion at any point where targets are detected. The two dimensional velocity of a target as projected onto a plane is calculated as it is scanned nearly simultaneously by the two radars. The third velocity component is derived by calculations based on assuming that mass continuity applies. Data acquisition and real-time display require the use of high speed magnetic tape drives or other data storage devices, and a digital computing unit in order to calculate and display the motion fields. Thus, detailed analyses of convective clouds, tornadic and hailstorms, and so on are possible.[35]

7.2.4 Lidar and Laser. Although cloud heights can be measured by radar,[36] it has never become a routine approach for such measurements. Ceilometers presently are the only instruments serving this purpose. Two types of ceilometers are used: the triangulation ceilometer and the ranging ceilometer. The former is represented by the fixed beam ceilometer or search light device and the rotating beam ceilometer; the latter, by the Swedish ruby laser and the French lidar. Triangulation methods using the fixed and rotating beam ceilometers have been described in Section 6.7.1. The method of ceiling determination using the ranging ceilometer is described here.

The general principles of the light-pulsed ranging ceilometer are similar to those employed by radar: the time interval is measured between emission of a light pulse from a projector and its detection after reflection from a cloud base. The Swedish ruby laser ceilometer generates a 30 nanosecond pulse of approximately monochromatic light centered at $\sim$ 6943 Å (694.3 nm). Part of the pulse reflected from the cloud base is detected by a ground receiver equipped with a photomultiplier component as part of its solid state receiver system. The ruby laser system is capable of measuring cloud bases to a maximum range of 16,000 ft (4877 m). The compact laser system is provided with a special shield to protect the user from inadvertant exposure to the intense laser beam.

---

[34] Gray, G.R., R.J. Serafin, D. Atlas, R.E. Rinehart, and J.J. Boyajian. 1975. Real-time color Doppler radar display. Bull. Am. Meteorol. Soc. 56(6): 580-588.

[35] Lhermitte, R. 1975. Real-time monitoring of convective storm processes by dual-Doppler radar. Atmos. Techn. 7: 26-33.
Eccles, P.J. 1975. Developments in radar meteorology in the National Hail Research Experiment to 1973. Atmos. Techn. 7: 34-45.
Ray, P.S., R.J. Doviak, G.B. Walker, D. Sirmans, J. Carter, and B. Bumgarner. 1975. Dual-Doppler observation of tornadic storm. J. Appl. Meteorol. 14(8): 1521-1530.

[36] Wilk, K. 1958. Evaluation of the AN/APQ-39(XA-3) cloud detector radar. Ill. State Water Survey, Urbana, Ill. 34 p.
United Aircraft Corp. 1964. Preliminary operational application techniques for AN/TPQ-11, USAF Air Weather Service Techn. Rept 180. 83 p.

The French lidar has three main subsystems: the projector, the receiver, and the cloud height conversion unit. The projector and its housing consist of two tungsten electrodes, a 20 in. parabolic mirror with a 10 in. focal length, a tilted glass top for rain protection, two bubble levels, and a drying unit with air heater and ventilation system. The two electrodes are fixed at the focus of the parabolic mirror; they generate a peak power of 1 mW and produce blue to ultraviolet light pulses with a peak wavelength centered at blue. The pulse has a length of approximately 1,968 ft (600 m), a rate of 30 Hz, and a time interval between pulses of 2 ms. The ground receiver and its housing are located 25 ft from the projector. This subsystem contains a cesium-antimony photoelectric cell with maximum response to light in the blue range, a parabolic mirror identical to the projection mirror with a 1.92 deg field of view, a pair of bubble levels, and a drying unit. The cloud height conversion unit controls the entire ceilometric system. It consists of a controller, video amplifier, noise amplifier, traveling gate, and recorder. The controller performs the following functions simultaneously and continuously: regulates the spark of the tungsten arc, controls the shutter of the photoelectric cell (including initiation and synchronization of the recordings), and performs a calibration check, all by automatic control. The scan pulse, regulated by a traveling gate, permits cloud echo pulses for various stepped intervals to be transmitted to the recorder. Thus, at the first pulse the traveling gate scans the range of 100 to 166 ft within an 0.15 ms gate covering a 66 ft height range; the second pulse occurs 0.03 s later, scanning the range of 166 to 182 ft within 0.15 ms covering a 16 ft height range, and so forth in increments until the total 4,966 ft (1514 m) range of the instrument is covered in incremental steps over a 10 s time interval.

Table 7-20 compares the performance of several different ceilometer systems.

Table 7-20 Specifications of Several Ceilometers[37]

Ceilometer	Rotating Beam	Fixed Beam	French Lidar	Ruby Laser
Type	Triangulation	Triangulation	Ranging	Ranging
Light Discrimination Technique	Modulation	Pulsation	Pulsation	Pulsation
Baseline	400 ft	250 ft	None	None
Light Source	Incandescent iodine vapor	Xenon	Tungsten Arc	Ruby Laser
Measuring Range (approx.)	100-4000 ft	100-2500 ft	100-5000 ft	100-16,000 ft
Measuring Frequency	10 per minute	2 per minute	5 per two minutes	1 per minute
Light Beam Divergence	2 deg by 5 deg	12 ft cone	55 ft cone	5 ft cone

[37] Table based on data from George, D.H. 1972. A comparison of currently used cloud height sensors, In: Second Symposium on Meteorological Observations and Instruments. Amer. Meteorol. Soc.

7.2.5 <u>Wiresonde</u>. Although the routine use of kites and captive balloons for measuring the lower stratospheric air has been abandoned since 1900 (see Table 7-1), the kytoon (a combination of kite and balloon) has been in use as a meso-scale airborne platform for over three decades. A wiresonde is a kytoon equipped with temperature and humidity sensors, whose signals are telemetered through wire cables to a recorder at ground level.

The kytoon, also known as the tethered balloon, is fastened by a cable to the ground. It is designed to fly at heights of a few thousand feet. The kytoon is so named because it is streamlined and aerodynamically designed so that at low wind speeds it behaves like a captive balloon and at high wind speeds it lifts like a kite. The kytoon is unique in that it remains at an almost constant level, unless under strong gust conditions. The wiresonde airborne unit accurately measures the temperature and humidity in the lowest 2000 ft (610 m) of the atmospheric boundary layer. This is the region where radiosondes fail to give reliable measurements. For this reason kytoons with wiresondes are employed for the studies of heat and moisture fluxes of the lower boundary layer of the free atmosphere.

(1) <u>Wiresonde Airborne Unit</u>. The ML-441/UMQ-4 wiresonde airborne unit, shown in Figure 7-27, consists of mounting facilities for the meteorological sensors, a removable radiation shield, and a housing for the blower motor and the battery. The meteorological sensors are in spring clips attached to a lucite center post. The entire airborne unit is secured to the kytoon by means of a harness and is connected into the data cable which carries data to a ground recorder.

The kytoon used with the ML-441/UMQ-4 wiresonde is a barrage-type kite balloon with inflatable balloon bladders inserted into the casing. Wing struts, mounted between the tail plate and the tail fins, spread the tail fins and a tail post keeps them taut.

A pair of bead thermistors are enclosed in glass and extend slightly from the metal casing surrounding the sensors. The casing, which is impregnated with a solution, serves as one conductor and a single wire insulated by glass from the casing serves as the other conductor. One thermistor is used as the dry bulb and the other as the wet bulb. Distilled water is supplied to the wick through a small container built into the lucite post. For temperatures below 0C, a lithium chloride element takes the place of the wet bulb thermistor. The humidity range is 10% to 100% for temperatures above freezing, and 20% to 100% for temperatures below freezing. The lithium chloride element has an accuracy of ±10% for humidity measurements; that of the wet bulb thermistor for temperatures above freezing is ±3%. In this wiresonde system, a summation bridge containing four Wheatstone bridge circuits is used for temperature measurements and the other two for humidity measurements. One circuit in each pair is used for measurements above, and the other for below, freezing temperature conditions.

A wiresonde can be designed in many different ways with variations in the data transmission method, overall weight, and accuracy of height determinations.

(2) <u>Data Transmission</u>. Aside from wire communication, data can be transmitted to the ground by radio or recovered from the airborne recorder. The radio communication method requires the installation of a small radio transmitter in the airborne unit. However, the power supply has a life expectancy of only a few

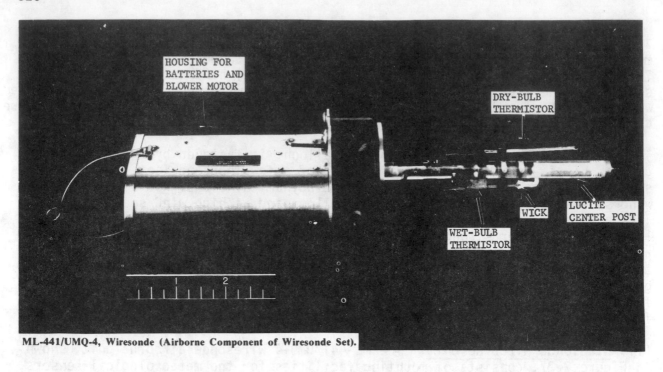

HOUSING FOR BATTERIES AND BLOWER MOTOR

DRY-BULB THERMISTOR

WET-BULB THERMISTOR

WICK

LUCITE CENTER POST

ML-441/UMQ-4, Wiresonde (Airborne Component of Wiresonde Set).

RL-156/UMQ-4, Hand-Cable-Reeling Machine.

TETHERED BALLOON WITH WIRESONDE PRIOR PRIOR TO ASCENT.

Figure 7-27 Wiresonde Set

days. Also, the radio communication method requires a special ground receiving and control unit similar to that of a radiosonde system. The airborne recorder can also store the data which are then recovered directly from the recorder once it is winched to the ground by means of a pulley. The same pulley can be used to lift the instrument and recorder unit to the airborne kytoon. The pulley is operated either by a hand reel or a motor-driven reel. Either method of data transmission (radio communications or recorder recovery) adds extra weight as well as added cost to the kytoon.

(3) System Payload. The airborne component of a wiresonde set weighs less than the cables. An FAA regulation (USA) requires visible markers attached at 50 ft intervals on the cable starting between 500 and 1000 ft above the ground (marker height depends upon the urban region and the airport involved). These markers add appreciable weight to the kytoon, contributing to the momentum of the entire system, particularly on windy days. Thus, the operational efficiency of the system is affected, including the height achieved by the kytoon in flight.

(4) Height Determination. There are several methods for determining kytoon height. A counter indicates approximate height in terms of cable length letout. An elevation indicator gives an approximate zenith angle of the cable. A clino-meter can be mounted on a clamp and raised as desired. A sensitive altimeter can be installed in the airborne unit for height recording. A triangulation method, such as the double theodolite technique, has been used occasionally for height determination but the operation is rather laborious. Under certain con-ditions, a stadia technique can be used effectively (see Section 7.1.3).

As mentioned, the wiresonde technique with kytoon carrier is one of the methods employed in the measurement of boundary profiles. In general, the effective-ness of wiresonde measurements is dependent upon the wind factor. Wind is nec-essary for the kytoon flight; however, high winds affect the stability of the kytoon and often create difficulties in obtaining accurate data for altitudes above 1000 ft. Also, the operation of a large carrier requires experienced crews; the use of helium for safety purposes reduces the lifting power of ky-toons.

A modified version of the kytoon has been employed in the Boundary-Layer Instru-mentation System (BLIS) for the GARP Atlantic Tropical Experiment (GATE). BLIS has the capability of measuring temperature, humidity, pressure, altitude, and total wind vector (speed and direction of both horizontal and vertical compon-ents) in the lowest 1500 m of the surface layer.[38] The kytoon used to support the BLIS instruments is a US Navy modified Class C tethered balloon with a vol-ume of 97 m^3 (3500 ft^3), a length of 12 m, and a diameter of 4.36 m. The bal-loon, when filled with helium, can remain aloft for 48 hrs at 1500 m and lift 45 kg (static net lift) at sea level. It holds altitude in winds up to 30 kt and can survive winds to 50 kt. Multiple instrument packages are attached to the tether line using a ball-and-ring mounting attachment, allowing instrument placement at several lengths beneath the kytoon and at angles independent of the

---

[38] Burns, S.G. 1975. Boundary-layer instrumentation system. Atmos. Techn. 6: 123-128.

tether line angle. The Boundary-Layer Instrumentation Packages (BLIPs) contain a three-cup anemometer facing upwind, instruments with associated electronics (pressure aneroid capsule, wet and dry bulb thermistors, carbon hygristor, reference high and low temperature resistors, tilt angle indicator, and frequency-modulated transmitter) at mid-section, and a cylindrical tail fin. A shipboard Portable Data Acquisition System (PODAS) consists of several crystal-controlled, narrow band FM receivers, each centered at the frequency specified for individual BLIPs. Data are encoded and displayed either as binary coded decimal (BCD) numbers or converted by a polynomial fit into engineering units. Data are stored on nine track magnetic tape after processing through a shipboard PDP-11 (Digital Equipment Corp.) mini-computer, and can be read out on strip charts.

7.2.6. <u>Diffusion Measurement Instrumentation</u>. The objectives of diffusion measurements are to relate such atmospheric properties as momentum, heat, water vapor, and gaseous or particulate matter to the state of the atmosphere, as expressed by the microstructure or mesostructure variables of wind, temperature, and humidity. The diffusion characteristics of the atmosphere can be determined through experimental techniques that permit direct quantitative measurements. The technique involves a controlled emission of relatively small amounts of tracer materials to the atmosphere and subsequent sampling and analysis of the same, using both tracing and ground sampling methods.

As the tracer materials are present at such low concentrations in the sample, great care must be taken to guard against excessive background concentrations and to include the measurement and correction for the background in any program. Many types of materials have been used in tracer techniques,[39] and those that are discussed here include:

      (1) Oil fog tracer technique
      (2) Fluorescent particle tracer technique
      (3) Other aerosol tracer techniques
      (4) Gaseous tracer technique
      (5) Visible plume tracer technique

(1) <u>Oil Fog Tracer Technique</u>. Smoke plumes produced by oil fog generators have been proven to be useful in atmospheric diffusion studies (e.g., Church[40] and Lowry, Mazzarilla, and Smith.[41] Usually military oil fog generators, such as the M2 or M3A2, which produce large volumes of smoke over extended periods of time, have been used in those studies. The M2 generator produces smoke by forcing a mixture of oil and water into a coil which is heated to vaporize the two liquids. When the vapor escapes into the air, it chills rapidly and a fog of submicron size droplets is formed. In the M3A2 generator, the smoke is produced when oil is injected into the engine tube of a pulse jet engine where it is vaporized. The desired particle size ranges from 0.5 to 1.0 μm. The average

39 Slade, D.H., ed. 1968. Meteorology and Atomic Energy. USAEC
    T10-24190: 293-298.

40 Church, R.D.M. 1949. Dilution of waste stack gases in the atmosphere.
    Ind. Eng. Chem. 41: 2493-2493c.

41 Lowry, P.H., D.A. Mazzarilla, and R.E. Smith. 1951. Ground-level measurements of oil-fog emitted from a hundred meter chimney. Meteorol. Monogr. 1: 30-35.

droplet size is controlled by the concentration of the condensing vapor and rate of cooling. The life of a properly generated smoke cloud is determined almost solely by meteorological conditions.

Methods of analysis include photometric, gravimetric, and fluorometric. In the photometric method, the concentration of smoke in small volumes of air is determined by the amount of light scattered by the aerosol particles. The photometer may be mounted on a moving platform such as a truck, or carried aloft in the smoke by means of a balloon. This method permits the use of continuous recording devices. In the other two methods, the oil is collected on a filter and analyzed either gravimetrically or fluorometrically. The latter method is based on the fluorescent property of the oil used to produce the fog.

(2) Fluorescent Particle Tracer Technique. The use of insoluble fluorescent particles (FP) as tracer material for diffusion studies was initiated in the late 1940's (see Leighton et al.[42]/ for the development of the technique and recent improvements). In this technique, the tracer particles are dispersed to the atmosphere by means of a blower generator. Figure 7-28 shows the blower generator, Model 5, manufactured by Metronic Associates, Inc. This system permits aerosolization over periods of two minutes to eight hours or more, and has a dispersion range of 20 to 450 g of tracer particles per min. The system can be mounted on a tower, moving ground vehicle or aircraft, in addition to stationary ground operation. The fluorescent pigments used are mostly zinc silicates, zinc sulfides, and zinc-cadmium sulfides, with various activator elements (pigments) added in small amounts. The pigments fluoresce when irradiated with ultraviolet light. The best fluorescent colors are yellow and orange, although green and red are considered acceptable. It is important to avoid colors which would make it difficult to distinguish natural fluorescent particles already present in the environment. The optimum particle size is between 1 and 3 μm. A full knowledge of the tracer source strength is required to obtain quantitative results with the FP tracer. Normally, the source strength, Q, is expressed in terms of the number of particles released. In this case $Q = WF_S$, where W is the weight of FP fed through the generator, and F_S is the observed number of airborne particles per unit weight. The main requirements for a sampling device are that it must be able to collect the particles with high diffusion rates in the air and deposit them individually on a smooth surface which is suitable for the excitation of the fluorescence and for counting. The three most common sampling devices are the (a) membrane filter, (b) drum impactor, and (c) Rotorod.

(a) Membrane filter. Of the various membrane filter materials, porous cellulose acetate-nitrate membranes are advantageous in that the collection efficiency is virtually 100%, most of the particles are deposited on the upstream face rather than in pores, they may be dyed to give a background favorable for counting, and they may be stored as a permanent record. The main disadvantages are that the membranes may easily be damaged and the flow rates obtainable are limited.

Nucleopore filters (General Electric Company), shaped as fitted disks with

---

[42] Leighton, P.A., W.A. Perkins, S.W. Grinnell, and F.X. Webster. 1965. The fluorescent particle atmospheric tracer. Journ. Appl. Meteorol. 4: 334-348.

known pore geometry,[43/] have been found by Twomey[44/] to measure different aerosol sizes accurately. Different degrees of filtration are provided by varying the flow rate and/or the length of tube, duct, or other filtering element, so that aerosol size separation occurs between 0.001 and 1 μm radii. In this process air is pushed through sets of four or five Nucleopore filters at varying rates of flow; a Pollak photoelectric nucleus counter measures the emerging particle concentrations. The size distribution captured by each set of filters is estimated indirectly by a count of the numbers of particles surviving varying degrees of filtration. A mathematical inversion procedure[45/] can be used to calculate the filter transmission, $K_i(x)$, as a function of the size, x (log radius), of the aerosol particles, followed by an iterative calculation of its size distribution. Twomey[46/] derived a numerical estimate of the solution to

$$\int K_i(x) \, f(x) \, dx = g_i + \varepsilon_i \quad \ldots \ldots \ldots \ldots \ldots \ldots \quad (77)$$

Where $K_i(x)$ is filter transmission; x, size of aerosol particle (log radius); $f(x)$, size distribution; g_i, measured particle concentrations; ε_i, error term (due to g_i and left side of equation); and i = 1, 2,...N.

The size distribution, after corrections have been made for diffusion and electrostatic losses to the walls of the container (a problem with aluminized Mylar collection bags)[47/] yields

$$g_i = \int_0^\infty K_i(x) \, \exp\left[-\eta t_i \, D(x)\right] f(x) \, dx \quad \ldots \ldots \ldots \ldots \quad (78)$$

Where η is the geometry and boundary layer coefficient of the container; D, diffusion rate; and t_i, time interval. This second equation is solved by an iterative procedure on a computer to derive particle sizes at various stages of filtration.[48/]

(b) Drum impactor. As shown in Figure 7-29, a strip of aluminum tape, coated

---

[43] Spurny, K.R., J.P. Lodge, E.R. Frank, and D.C. Sheesley. 1969. Aerosol filtration by means of Nucleopore filters: structural and filtration properties. Environ. Sci. Techn. 3: 453-464.

[44] Twomey, S. 1975. Comparison of constrained linear inversion and an iterative nonlinear algorithm applied to the indirect estimation of particle size distribution. J. Comput. Phys. 18: 188-200.

[45] See Footnote 44.

[46] Twomey, S. 1976. Aerosol size distributions by multiple filter measurements. Journ. Atmos. Sci 33(6): 1073-1079.

[47] Cooper, G., G. Langer, and J. Rosinski. 1979. Submicron losses in aluminized mylar bags. Journ. Appl. Meteorol. 18(1): 57-68.
Stein, R.L., W.M. Ryback, and N.W. Sparks. 1973. Deposition of aerosol in a plastic chamber. Journ. Colloid Interface Sci. 42: 441-447.

[48] See Footnote 46.

on one side with silicone grease or rubber cement, is mounted on a drum which is enclosed in a housing and can be rotated by increments beneath the rectangular orifice of a conical nozzle. The collection efficiency of the drum impactor depends on the jet velocity, the orifice size, the clearance between the orifice and the collecting surface, the nature of the collecting surface, and the materials used in the construction of the nozzle and collecting drum. The collection efficiency of the current models is 95% or better.

(c) <u>Rotorod sampler</u>. In this system (see Figure 7-30), the sampling is made by moving the collector through the air instead of pumping the air through the collector. Two thin metal bars coated with silicone grease are attached by a cross-arm to the shaft of a small battery-driven motor so that they are parallel to the axis of rotation. A small, lightweight Rotorod can be mounted on a tethered balloon cable. A series of as many as 20 units, mounted at intervals along the cable, have been used for sampling up to 750 ft aloft.

Figure 7-28 Blower Generator

Figure 7-29 Drum Impactor

Figure 7-30   Rotorod Sampler

The collection rate of the Rotorod is little affected by wind speed, v, as long as it is smaller than the rotation speed, w, of the collector arm. When v exceeds w, the collection rate increases. For v/w = 1.5, the increase is about 10%, and for v/w = 2, it is about 22%. High winds increase the drag on the rod and, hence, the load on the motor. The use of multiple arms appears to have little effect on the collection efficiency. With collecting surfaces of 0.38 x 60 mm, a rotation radius of 60 mm, and a rotation speed of 2400 rpm, the effective sampling rate of this device is approximately 40 l min^{-1}, or 90 l min^{-1} cm^{-2} of surface. The power consumption is 0.7 W. The chief disadvantage of the Rotorod is that its collection efficiency is low, since it is dependent on particle size.

Either visual or photoelectric counting can be employed to determine particle concentration. An automatic fluorescent particle counter manufactured by Mee Industries automatically samples, counts, and determines concentration in real time. An intense ultraviolet source emits radiation which closely matches the excitation spectrum (the design of the instrument provides maximum sensitivity to the zinc sulfide particles). The air sample containing tracer particles passes through the excitation chamber at a rate designed to guarantee that the particles reach maximum fluorescence. A sensitive photo-multiplier detector system then scans and counts each zinc sulfide particle contained in the continuous flow of sample air. The detector output is displayed digitally, showing either counts per unit volume or total counts, and can be recorded in either digital or analog form. The instruments can be operated unattended at remote sites.

Leighton et al. (see Footnote 42) found that for ground releases of tracer, the losses by fallout and impaction can range from 1 to 10% during the first few miles of travel, depending on the rate of rise of the fluorescent particle clouds due to turbulent mixing and terrain capture. Releases from aircraft have no losses until the cloud reaches the ground.

A typical application of this technique is a study of pollutant transport in the San Francisco Bay area by Sandberg et al.[49] To simulate the transport and diffusion of airborne contaminants, point source releases of fluorescent tracer materials were made near various urban centers and some 50 samplers were arrayed in expected downwind directions. As the tracer could be assessed with high sensitivity over great distances, the study provided a quantitative indication of pollutant dispersion across an extensive metropolitan area.

Studies have also been made on the use of water soluble fluorescent uranin dye as the tracer material, such as those conducted by Robinson et al.[50] and Dumbauld.[50] The concentrated liquids, when dispersed into the atmosphere by a high pressure, air-aspirated, spray nozzle, dry and form aerosol particles of 1 to 10 μm. In this technique, the sample particles collected on the filters are dissolved with distilled water, and the fluorescence of the solution is measured by a fluorometer.

(3) Other Aerosol Tracer Techniques. Schaefer[52] made a study of atmospheric

---

[49] Sandberg, J.S., W.J. Walker, and R.H. Thuillier. 1970. JAPCA 20: 593-598.

[50] Robinson, E., J.A. MacLeod, and C.E. Lapple. 1960. Atmospheric tracer technique using aerosols. Instrument Society Conference, San Francisco.

[51] Dumbauld, R.K. 1962. Meteorological tracer technique for atmospheric diffusion studies. Journ. Appl. Meteorol. 1: 437-443.

[52] Schaefer, V.J. 1958. The use of silver iodide as an air tracer. Journ. Meteorol. 15: 121-122.

diffusion with the use of silver iodide as the tracer material. In his study, the tracer plume was measured with a portable cloud chamber in which the concentration of ice crystals formed by seeding with silver iodide was visually estimated. Temperatures at formation and crystal appearance made the differentiation of silver iodide nuclei and natural ice nuclei relatively uncomplicated. The major limitation of this technique is the cumbersome sampling procedure.

Hay and Pasquill[53], in England, used Lycopodium spp. spores in diffusion tracer studies. The spores were released from a point source and collected downwind by natural impaction on exposed cylinders. Sample catches were evaluated through microscopic identification of the distinctive spore particles.

Diffusion in the boundary layer also is studied using chaff, an artificial radar target, in conjunction with Doppler radar to track the dispersing needles.[54] The chaff is dispersed as point or line sources by airplane, helicopter, or balloon. Serafin and Strauch[55] estimate the diffusion of chaff needles into 1 km^3 of convective boundary layer air in approximately 15 min. Under relatively still air conditions, however, the chaff needles settle quite rapidly to the surface.

(4) Gaseous Tracer Technique. Several gases including Freon, sulfur dioxide, and sulfur hexafluoride have been used successfully as tracers. Freon 12 (dichloro difluoromethane) was first used by Schultz[56] and further improvements in the technique were made by Collins et al.[57] The gas was released from a cylinder through appropriate flow meters and control valves. Evacuated flasks were used to collect the samples which were then analyzed for Freon by such techniques as gas chromatography (see Section 8.4.5). Concentrations as low as 0.05 ppm have been detected.

---

[53] Hay, J. and F. Pasquill. 1959. Diffusion from a continuous source in relation to the spectrum and scale of turbulence, In: Advances in Geophysics, Vol. 6 Academic Press, New York: 345-365.

[54] Kropfli, R.A. and N.M. Kohn. 1976. Dual-Doppler radar observations of the convective mixing layer at St. Louis, In: Preprints, 17th Radar Meteorological Conference, American Meteorological Society, Boston, Mass.: 321-325. Kropfli, R.A. and N.M. Kohn. 1977. Persistent horizontal rolls in the urban mixing layer as revealed by Dual-Doppler radar. Presentation, 6th American Meteorological Society Conference on Inadvertant and Planned Weather Modification, Champaign, Illinois.

[55] Serafin, R.J. and R. Strauch. 1978. Meteorological radar signal processing, In: Air Quality Meteorology and Atmospheric Ozone, ASTM Spec. Techn. Publ. 653, American Society for Testing and Materials: 159-182.

[56] Schultz, H.A. 1957. Measurement of concentrations of gaseous halide tracers in air by positive ion emission techniques. Anal. Chem. 29: 1840-1842.

[57] Collins, G.F., F.E. Bartlett, A. Turk, S.M. Edmonds, and H.L. Mark. 1965. A preliminary evaluation of gas air tracers. Journ. Air Pollution Control Assoc. 15: 109-112.

Sulfur dioxide is a common air pollutant and many fine methods have been developed for measuring its concentration in the atmosphere, either instrumentally or chemically (see details in Section 8.4.4). The use of sulfur hexafluoride (SF_6) is also mentioned by Collins, et al. Gas chromatographic techniques are used to analyze SF_6. A problem in the use of SF_6 is that it is commonly used as an insulating gas in electric power installations and, hence, may already be present in the sampling area.

(5) <u>Visible Plume Tracer Technique</u>. Measuring the density of visible plumes on a photograph offers a possibility of directly estimating atmospheric diffusion.[58] Stereoscopic techniques can be used to size individual smoke parcels in space.[59] For visual studies, smoke is generated for short periods of time by smoke pots or flares, and for extended periods by an oil aerosol generator. Smoke releases can be made either on the ground or aloft from tethered balloons or towers. Several smoke sources ignited simultaneously at various heights can be used to estimate the variations in diffusion with altitude.

The tracer techniques so far described are essentially for determining diffusion, turbulence, and air trajectories near the ground. When applied to the upper air and upper atmospheric region, the sampling techniques are different due to the nature of the information sought. With a few exceptions, tracer techniques of the upper atmosphere as compared to those of the near surface layer differ in the following respects:

1. Uncontrolled tracer sources of natural and artificial origin exist, as opposed to controlled emissions from blowers;
2. Broader measurements are involved that include the dynamic, chemical, and radiative states of the atmosphere, as opposed to simply air trajectories;
3. A great variety of platforms are utilized, including aircraft, balloons, sounding rockets, satellites, and ground-based equipment, as opposed to ground-based platforms only; and
4. Many types of samplers, instruments, and techniques for both <u>in situ</u> and remote sensing are employed, as opposed to capture techniques.

The exceptions, however, include the use of aircraft to trace low-level plume propagation, the release of chaff for tracking with high precision radar, and the employment of satellite sensing for tracing materials carried by stratospheric motions.

In this section only a brief description of sampling techniques appropriate to the upper air and upper atmosphere is given, with elaboration to be found in later chapters. Two types of sampling used in upper atmospheric research are (a) capture and (b) instrument techniques.[60]

---

58 Högström, V. 1964. An experimental study on atmospheric diffusion. Tellus 6: 205-251.

59 Clark, R.D.M. 1956. Photographic Technique for Measuring Diffusion Parameters. USAEC Report TID-7513, Pt. 1, Division of Reactor Development: 186-199.

60 NCAR. 1978. Instruments and techniques for stratospheric research. Atmos. Techn. 9. 103 p.

(a) <u>Capture Techniques</u>. Material is collected or captured for later return to the surface and subsequently analyzed in the laboratory. Some of the techniques employed include the following.

Whole air or grab samples of unfractionated ambient air are collected, usually by means of opening an evacuated vessel to the surrounding environment and waiting until an equilibrium state is attained before sealing the vessel. The quantity of ambient air sampled can be enhanced by collecting the air in vessels cooled with liquid neon, hydrogen, or helium. This method of cryogenic sampling increases the amounts and reduces the reactivity of the gases collected in this manner. Particles and gases can be collected by reacting them with chemically impregnated filters. The ambient substances are either captured on, dissolved in, or reacted with the selected reagents and/or solvents on the filters. The particular substances most effectively sampled by this method are those unaffected by the low pressures and temperatures of the ambient air. Impactors have been adapted for upper atmospheric sampling of particles. These include fibers, threads, or rods moved through or upon which is impinged the aerosol to be sampled, and the movement of the aerosol through jets or apertures towards a collecting surface. Filters are used to collect particles, with the fiber types rather than membrane types used predominately. The fiber filter (either made of cellulose or polystyrene) has a very low pressure drop across its face, increasing its efficiency in collecting particles as small as Aitken size and organic polymers. All of the above techniques require that the materials sampled must be returned to the surface for the follow-up analysis.

(b) <u>Instrument Techniques</u>. Several types of instruments accomplish the necessary analysis of upper-air constituents <u>in situ</u>. A mass spectrometer, fitted to a rocket, balloon, or aircraft platform, can give the mass distribution of the ambient air inducted into the device. Photomultiplier tubes are used as detectors of the chemiluminescence given off when ambient substances react with a suitable substrate. Balloons or aircraft usually are employed to carry this apparatus to its sampling levels. Lidar (light detecting and ranging or optical radar) directs light energy towards aerosol layers; the amount of energy backscattered by the aerosol provides information on the aerosol properties. This method usually employs a ground-based lidar and detection system. Optical systems on orbiting satellites or rockets are used to detect the attenuation of solar radiation as it passes through an upper atmospheric layer. The photoelectric particle counting device (also known as a dustsonde or aerosolsonde) is carried aloft by balloon, where it measures the scattering properties of particles drawn through the instrument. Likewise, a laser polar nephelometer detects the light energy scattered by aerosols drawn into the instrument. This device commonly is carried by aircraft. A resonance lamp with a photomultiplier tube is used to detect energy absorbed and emitted by ambient air which fluoresces at certain wavelengths. This system can be mounted on either a balloon or aircraft. Lastly, an infrared spectrometer, carried aloft by aircraft or balloon, can be used to determine the emission spectra of atmospheric gases.

7.3 <u>Aircraft Techniques</u>. In Germany the first aerological work with airplanes was undertaken in 1912 at Frankfurt-am-Main by Lt. von Hiddessen, but it was during the First World War that regular flights for weather observations were initiated. Since then, aerological observations have been incorporated into three groups of aircraft flights. They are:

1. Commercial and Military Flights. Since neither commercial nor military flights are flown specifically for aerological purposes, the variety of observations is restricted and the quality of data are generally low. Only a few meteorological instruments for temperature, humidity, and pressure measurements are installed for these flights.

2. Atmospheric Research Flights. Most research flights are flown for a certain specific investigation. Here again, observational data are limited. The total aerial coverage and the duration of flight are restricted. Moreover, research experiments are usually conducted for a short period of time. Lastly, findings are not made available for immediate use.

3. Weather Reconnaissance Flights. While research flights are more or less made on an unscheduled basis, most weather reconnaissance aircraft are flown at scheduled intervals and according to predesignated flight patterns to provide data for basic weather analysis and forecasting. These aircraft are generally well equipped with aerological instruments, special communication systems, and trained observers. In addition, dropsondes are frequently provided at selected locations for vertical aerological soundings.

Of the above three groups, the weather reconnaissance flights are the best organized and provide the greatest contributions to aerological research. In 1974, the Air Weather Service of the US Air Force began operational testing of the Airborne Weather Reconnaissance System (AWRS). The AWRS is the best planned and equipped weather reconnaissance system established to date. The AWRS is briefly described here.

The AWRS program is operated on board a USAF Lockheed WC-130B turboprop aircraft equipped to measure 32 meteorological and aeronautical sensing (instrumental) and observing (visual) variables in addition to the variables of the atmospheric vertical profile measured by the dropsonde. Three flight patterns for the aircraft have been designed for obtaining a close view of the fine structure of storms, including penetration and periphery. Data are gathered, processed, displayed, and recorded at a rate of one input per second with four forms of system output: digital onboard displays, magnetic tape recordings, printout data, and graphs in real time, as well as radio meteorological messages. With the use of the onboard computer and by means of a high frequency, single side band communications link, the radio message is transmitted directly from the aircraft to a ground station.

The range, accuracy, and resolution of the aerological output data for AWRS are summarized in Table 7-21, in which all accuracies are specified by three standard deviations (3σ).

Aside from the specifications listed in Table 7-21, information gathered includes air density (g m^{-3}), relative humidity (%), D-value (ft), and equivalent potential temperature (deg K) of the aerological variables, in addition to such aircraft flight variables as true airspeed (kts), position (nm), drift angle (deg), ground speed (kts), and roll and pitch (deg).

Table 7-21a The AWRS Specifications for Horizontal
Flight Data Output

Flight Level Variable	Range	Steady Rate Accuracy	Resolution
Wind Speed	3 to 350 kts	±4.25 kts	0.1 kt
Wind Direction	0 to 360 deg	±25.0 (10 kts)	0.1 deg
		±10.0 (25 kts)	0.1 deg
		± 5.0 (50 kts)	0.1 deg
		± 2.5 (100 kts)	0.1 deg
Temperature	-90C to +50C	±0.49C	0.1C
Dewpoint	-20C to +40C	±0.75C	0.1C
Temperature	-40C to -20C	±2.7C to 0.75C	0.1C
	-60C to -40C	±4.7C to 2.7C	0.1C
	-73C to -60C	±8.5C to 4.7C	0.1C
Sea Surface Temperature	0C to +35C	±0.6C	0.1C
Static Pressure	1060 mb to 300 mb	±0.325 mb	0.1 mb
Radar Altitude	500 ft to 35,000 ft	±20 ft	1.0 ft

Table 7-21b The AWRS Specifications for Vertical
Profile Data Output

Vertical Profile Variable	Range	Steady Rate Accuracy	Resolution
Temperature	-85C to +50C	±0.46C	0.1C
Dewpoint	-60C to +40C	±0.5C	0.1C
Temperature	-73C to -60C	±1.1C to 0.5C	0.1C
Pressure	1060 mb to 150 mb	±2.0 mb	0.1 mb

As mentioned earlier, in research investigations the choice of aerological variables for study depends upon the nature of the research. The variables may
range from atmospheric pollutants to the characteristics of clouds. Taking
cloud physics investigations with aircraft as an example, the chief measurements
may be the liquid water content, the types of nuclei, the particle size and
distribution of cloud elements, the rate and amount of precipitation in and

334

below clouds, as well as the meteorological variables listed in Table 7-21.

7.3.1 <u>Characteristics of Aircraft Weather Observations</u>. In view of the aerodynamics of aircraft, there are certain general performance characteristics
which influence weather observations. There are also certain special characteristics pertaining to a specific type of aircraft. Because many different
types of aircraft have been used for weather observations, it would be imprac-
tical to specify them all. However, three major classes are readily recognized.
They are:

(1) <u>Helicopters</u>. Helicopters are relatively low speed and low altitude vehicles
and, therefore, they provide a better platform for aerological observations than
airplanes. For this reason, helicopters have been used increasingly for the exploration of the lower tropospheric air in recent years. In addition, helicopters have flexibility in both vertical and horizontal flight patterns and do
not require a runway for takeoff and landing. But they are usually subject to
greater vibration than airplanes and, therefore, aerological instruments are
susceptible to damage, particularly during landing. Also, helicopters collect
more dirt and dust than airplanes so that protective caps are needed for all
equipment. Additional drawbacks are the limited payload and cabin floorspace.

(2) <u>Airplanes</u>. Airplanes have a greater carrying capacity (payload and cabin
space) than helicopters, in addition to having less vibration and fewer dust
problems. But airplanes usually are flown at higher altitudes and at faster
speeds than helicopters. This creates undesirable impacts to the aerological
instruments onboard, such as aerodynamic heating, icing, and so forth. These
impacts, which are elaborated in the remaining section, are important criteria
in instrument design.

A subcategory of the airplane is the unpowered glider which has been adapted
for meteorological research. The glider must be towed into the air where, once
it releases the towline, the machine depends on its high aspect ratio wings
to provide efficient lift in updrafts or very gradual sink otherwise. The
glider trades speed, altitude, and payload for slow vertical and forward motion
with less aerodynamic problems to instruments than with powered aircraft. A
brief summary of a instrument-equipped glider for meteorological research appears
at the end of this section.

(3) <u>High Altitude and High Speed Aircraft</u>. Space vehicles, such as the U-2 aircraft and the supersonic airplane, reach the central and upper stratosphere
while traveling at extremely high speeds. The U-2 has a ceiling of about 30 km with
features specifically designed into this small plane, along with high maneuver-
ability and low operating costs. It has been utilized to measure such atmospheric
variables as turbulence, jetstreams, the Sierra wave, hurricanes, tornadoes,
atomic radiation, ozone, humidity, solar radiation, sky brightness, atmospheric
optics, cosmic rays, dust concentration, and clouds. All these aircraft types,
at the high altitudes and supersonic speeds at which they operate, have environmental effects that differ from regular aircraft because of the high ultraviolet
radiation, low temperatures and pressures, and high wind velocities they en-
counter.

Of the three classes above, the airplane used for aerological soundings still is the preferred method. For urban meteorological studies, the helicopter is increasingly being utilized. The supersonic airplane and the U-2 aircraft, however, are presently important only to upper air atmospheric research. Therefore, the performance characteristics of conventional airplanes are stressed here.

An airplane, being a fast moving, elevated platform, generally suffers in performance characteristics due to sudden changes in position, vibration, and acceleration. In addition, constraints are imposed on the use of instruments onboard as related to payload and space occupancy, particularly in compliance with FAA safety regulations. Aerological instruments for airplanes must be designed so as to respond to the rapid changes in altitude and the ambient environment. When the airplane ascends, there is an abrupt increase in solar heating during the day and cooling at night. There is a sudden increase in moisture as the airplane enters clouds and a decrease in moisture as it passes through an inversion layer. There are numerous environmental impacts on aerological instruments. Examples are the occurences of low pressure and low temperature at high altitudes; the dynamic heating caused by the rapid passage of the aircraft; and the frost and icing due to the presence of freezing moisture. In addition, precipitation, gusts, clear air turbulence, and many other unfavorable weather conditions affect the flight. All of these factors must be taken into consideration in the design of aerological instruments.

7.3.2 <u>Aircraft Instruments for Weather Reconnaisance</u>. Based upon the above criteria, the principles and mechanisms of the five major conventional instruments for aircraft are discussed here. These instruments measure wind, pressure, temperature, moisture, and altitude. The instruments used to measure other variables are discussed later in the following chapters. Emphasis has been placed upon the instruments for conventional airplanes rather than for helicopters, the U-2, and supersonic aircraft.

(1) <u>Wind Measurement</u>. Both wind speed and direction, based upon air navigation procedures,[61] can be determined by the following four quantities without direct measurements with an anemometer. They are: heading, θ; true airspeed, v; ground speed, g; and drift angle, α. As shown in Figure 7-31, let w be the wind speed, and ϕ be the wind direction; then the wind speed can be determined by use of the cosine law as follows

$$w = [v^2 + g^2 + 2vg \cos \alpha]^{1/2} \quad \ldots \ldots \ldots \ldots \ldots \ldots \quad (79)$$

---

[61] To navigate the course of an airplane, every pilot uses air navigation procedures. These involve terminology such as heading, track, ground speed, drift angle, and true airspeed. The direction in which the plane is pointing as it flies through the air is called the heading. The path and speed over the ground are the track and ground speed, respectively. The angle between the heading and the track is known as drift angle. The speed with which the airplane travels is the true airspeed. For additional information, see Jeppesen Sanderson, Inc. 1978. Aviation Fundamentals. Sanderson, Inc., Denver, Colorado.

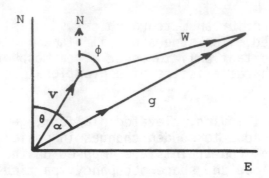

Figure 7-31 Vector Diagram Illustrating
Aircraft Wind Determination

The wind direction can be determined by one of the following equations

$$\cos \phi = \frac{1}{w} [g \cos(\theta + \alpha) - v \cos \theta]$$

$$\sin \phi = \frac{1}{w} [g \sin(\theta + \alpha) - v \sin \theta]$$

. . . . (80)

where θ is measured by a compass; g and α are determined by navigational aids; and v is derived by standard methods from the indicated airspeed, the atmospheric pressure, and temperature. Here the main problem is how to obtain the ground speed, or the g term, accurately. This can be accomplished by one of four methods: (a) pilotage, (b) dead reckoning, (c) celestial navigation, and (d) use of optical or electronic aids. In the first method, the ground speed is obtained by flying the airplane a known distance from one visible landmark to another. The second method utilizes the computation of direction and distance from known position. The third method includes reference to celestial bodies such as the sun, moon, stars, and planets. The fourth method uses navigational beams from airports or by use of radar techniques, either pulse or radio frequency. The latter is the most reliable. The Doppler radar designed for the National Center for Atmospheric Research (NCAR) combines a Doppler radar with an onboard computer. The radar sends down beams in four directions, 90 deg apart. The computer keeps two beams aligned along the track. This system has the advantage of not only giving instantaneous readings, but also can record the variance of the wind. As indicated in Table 7-21a, with a Doppler radar system the accuracy of wind speed can reach ±4.25 kts with a resolution of 0.1 kt. Wind direction measurement accuracy is determined by the wind speed. For example, with a low wind speed of about 10 kts, the error can be as large as ±25 deg, whereas at a high wind speed of 100 kts, the error is only ±2.5 deg.

(2) Pressure and Altitude Measurements. Design criteria for atmospheric pressure instruments include: (a) a wide pressure range between sea level and the ceiling of the flight; (b) the capability to withstand strong vibrations; (c) the minimization of the dynamic pressure impact; and (d) the reduction of thermal effects. Based upon the above criteria, barometers have been modified to suit aircraft usage. In fact, only aneroid capsules have been widely employed.

(a) <u>Altitude Effects</u>. The usual pressure measuring range by an aneroid barometer covers 1060 mb to 150 mb. For high altitude aircraft, this has to be extended at least to 10 mb or less. Here, perhaps, the hypsometric technique is suitable. The overall accuracy of the aneroid barometer can be around ±1.0 mb. However, accuracies between horizontal and vertical flight are appreciable. In the AWRS (see Table 7-21) the accuracy of the former is ±0.325 mb and that of the latter, ±2.0 mb. Another altitude effect on the accuracy of barometric readings is the decrease of accuracy with the increase of altitude. The root mean square (rms) error of six British aircraft aneroid barometers (Mark 2B models), for example, is 1.0 mb for the 500 mb level and 2.8 mb for the 150 mb level. Also, a slight departure from linearity in barometric readings occurs at high altitudes.

(b) <u>Vibration Controls</u>. Vibration is a common problem to all moving platforms in environmental monitoring. The degree of vibration of an aircraft varies directly with the intensity of atmospheric turbulence, acceleration, flight pattern, and make of airplane. The instrument panel is usually supported on anti-vibration mountings. In the assembly of an aneroid barometer, various mechanical devices are designed for further reduction of the oscillation of the recording pen or pointer and the reduction of vibration of the capsule.

(c) <u>Dynamic Impacts</u>. Static instead of dynamic pressure should be measured by the aircraft barometer. The former is the true ambient atmospheric pressure, whereas the latter is the differential pressure produced by movement of the aircraft. In order to obtain the true pressure, the aneroid capsule is first placed in an airtight case. Then, the nipple of this airtight case is connected through a pipe to the static vent or the static head in the pitot static system. It must be noted that much care must be taken to achieve the proper exposure of the barometer since the air rushing past the fuselage causes pressure differences around the aircraft. This is affected by the speed and altitude of the craft. A static vent must be so arranged that the pressure is equal to the true free atmospheric pressure. Even by doing this with care there still remains a small error in pressure, dp, and a correction for airspeed, dv_i, must be made. Let v_i be the indicated airspeed and ρ_0, the density at which the airspeed indicator is calibrated; then

$$dp = -\rho_0 v_i \, dv_i \quad \ldots \ldots \ldots \ldots \ldots \ldots \quad (81)$$

Another dynamic impact on barometric readings is the sudden ascent and descent of the aircraft. When the craft is climbing, the indicated pressure is higher than the true atmospheric pressure and vice versa when descending. As a result, there is a cyclic maximum aneroid deflection of more than 5 mb. This cyclic process is known as creep and this phenomenon is generally known as hysteresis. The magnitude of aneroid deflection depends upon the rate and amount of the ambient pressure changes. But when the cyclic process is repeated several times, the aneroid deflection is gradually reduced.

(d) <u>Temperature Effects</u>. As shown in Table 7-21b, an aircraft aneroid is frequently exposed to a wide range of ambient temperature varibility on the order of -85C to +50C within a very short period of time. Of course, this does not occur in surface temperature measurements. The expansion and contraction of such components of an aneroid as the chamber and the spring depend on the value of the coefficient of thermal expansion for each different material and, hence, are a function of temperature. The choice of material for the construction of these components is important. An alloy, Nispan, is used for the diaphragm of the aneroid capsule since it results in the least thermal stress with the greatest elasticity. In practice, however, more emphasis is placed upon the structure of bimetallic strips for temperature compensation. As an illustration of an aircraft aneroid barometer, the British Meteorological Office Model Mark 2B,[62] is described below.

"It is designed to cover the range 1050 to 150 mb. As shown (Figure 7-32) the dial, which is about 7.5 cm in diameter, is divided into units from 0 to 100, the units being millibars for the long pointer and tens of millibars for the short pointer; the pointers, which are concentric, move counterclockwise with decreasing pressure. The case of the instrument is made of moulded bakelite and is airtight, except for the connecting nipple at the back, the toughened glass front being sealed with a rubber ring.

A double capsule is used as the sensitive element and the mechanism linking it to the indicator is illustrated (in Figure 7-32). The deflection of the capsule C is transmitted through the jointed link LA to a shaft S, to which a toothed sector H is fixed. This sector is geared through an intermediate pinion I to another pinion J, which moves the long pointer P. Gears driving the short pointer are housed in a recess behind the dial mounting and a screw for adjusting the pointers is provided; this screw, which is covered when it is not in use, meshes with gear teeth round the circumference of the dial mounting. A hairspring attached to the staff of the intermediate pinion removes backlash from the mechanism."

Two types of instruments generally are available for altitude determination on board aircraft: the barometric and the radar altimeters. The former is con-

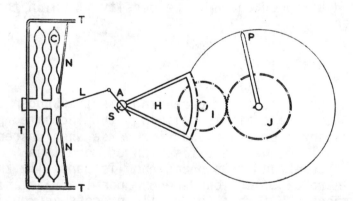

Figure 7-32 Aircraft Aneroid Barometer

62 See Footnote 11.

structed following the aneroid principle; the latter adopts the radar system. Radar altimeters can be further differentiated into radio frequency modulation and pulse altimeters.

When a barometric altimeter is employed, the magnification of the aneroid capsule deflection is provided by a mechanism similar to that in the sensitive aircraft aneroid shown in Figure 7-32. With the provision of levers, dials, and gears, the scales of the barometric altimeter can have a resolution of about 25 ft in height and its zero-setting can be adjusted readily.

Similar to the aircraft aneroid, the barometric altimeter is normally connected to the static pressure vent. It is also subject to position error, because of the variation in airspeed. Hence, Eq. (81) can be rewritten as

$$dH_i = + \frac{v_i dv_i}{11.0 \, \sigma} \quad \ldots \ldots \ldots \ldots \ldots \ldots \ldots \ldots \quad (82)$$

where dH_i and σ are the height correction (ft) and the relative air density at the ICAO height, H_i, [63] respectively.

In the radio frequency modulation altimeter, a continuous carrier radio wave, modulated at about 450 MHz within a certain frequency, is transmitted from the aircraft to the ground surface below. The reflected signal from the ground differs from the transmitted signal of the aircraft in phase. This difference is a measure of the altitude of the aircraft. In the pulse radar altimeter, the difference in time (or the transit time) between the transmitted pulse of the aircraft and the echo pulse from the reflected surface is a measure of the height of the aircraft.

Of the three altimeters, the radio frequency modulation type offers the highest accuracy, whereas the pulse radar and the barometric type are useful at high altitudes. A summary of the three altimeters is given in Table 7-22.

Table 7-22   A Comparison of Aircraft Altimeters

Type	Approximate Altitude (ft)	Accuracy (ft)	Resolution (ft)
Barometric Altimeter	Sea level to 50,000	70	25
	at 25,000	450	25
	at 50,000	750	25
Radar Altimeter (frequency)	Up to 4,000 max. height	10	1
Radar Altimeter (pulse)	Up to 40,000 max. height	50	1

[63] By using the International Civil Aviation Organization (ICAO) Standard Atmosphere, the relative air density, σ, is determined.

The discussions so far have been on instrumentation for wind, pressure, and altitude measurements onboard aircraft. The common denominator for accuracy determination of all these measurements is the exact time and location of the aircraft in space. A more accurate method for establishing the aircraft position in real time and space is by the use of a ground radar network equipped with a computer program. Although this method is expensive, it is useful for aircraft weather reconnaissance. Further improvements can result from the use of electronic in place of mechanical transducers. This increases precision and reduces the time response required for measurements.

(3) Temperature Measurements. The measurement of atmospheric temperature with electric sensors from aircraft is complicated by such impacts as (a) dynamic heating, (b) radiative heating or cooling, (c) evaporative cooling with the presence of liquid water, and (d) icing or frost effects. Should remote-sensing techniques be applied, these impacts would be minimized. But they create different types of problems. These problems and their solutions are discussed in Chapters 8, 15, and 16. The discussion here is confined to the contact type thermometer.

(a) Dynamic Heating. Dynamic heating is caused by two factors: frictional and adiabatic heating resulting from the flow and compression of the airstream passing through the sensor. At the nose of the craft or inside a pitot head, the relative velocity is zero. Therefore, there will be no frictional heating, and the adiabatic heating can be given by

$$\Delta T = \frac{v^2}{2Jc_p} \quad \dots \dots \dots \dots \dots \dots \dots \dots \dots \dots \dots \dots \dots \quad (83)$$

where v is the true airspeed (cm s^{-1}); J is the mechanical equivalent of heat (4.2×10^7 erg cal^{-1}); and c_p is the specific heat of dry air at constant pressure (0.240 cal g^{-1} $^\circ$C^{-1}). It is clear that adiabatic heating would be a serious problem only when the airspeed is 194 kts (i.e., 100 m s^{-1}) or more. This is common to modern aircraft. For $v = 100$ m s^{-1}, $\Delta T \simeq 5$C as computed from Eq. (83).

Terada and Yamamoto devised a method which has become a standard technique for the computation of frictional heating on a flat plate thermal sensor located at the center of a Venturi tube.[64] They made use of the adiabatic cooling effect in the throat of a Venturi tube where there is an increase in the speed of airflow accompanied by a pressure drop. Similar to Eq. (83), the amount of cooling is given by

$$T_a - T_t = \frac{v_t^2 - v_a^2}{2Jc_p} \quad \dots \dots \dots \dots \dots \dots \dots \dots \dots \dots \quad (84)$$

where T_a and v_a are the ambient air temperature and airspeed at the Venturi inlet, respectively; and T_t and v_t, the values at the throat, respectively. As

---

[64] Terada, K. and G. Yamamoto. 1947. Methods for measuring air temperature on a high-speed airplane. J. Amer. Meteorol. Soc. 4: 201.

shown in Figure 7-33, a flat-plate nickel wire sensor is placed in the Venturi throat and its temperature, T_p, will exceed that of the air passing over it by

$$T_p - T_t = \frac{\frac{1}{2}\sigma v_t^2}{2Jc_p} \quad \dots \dots \dots \dots \dots \dots \dots \dots (85)$$

where σ is the Prandtl number; or $\sigma = \mu c_p/K$ (μ is the dynamic viscosity and K, the thermal conductivity of the air). For complete compensation, $T_p = T_a$, and from Eqs. (84) and (85), the following relations exist

$$1 - \frac{v_a^2}{v_t^2} = 0.85 \text{ [for laminar flow]} \quad \dots \dots \dots \dots \dots (86)$$

Let r_i and r_t be the radii of the inlet and the throat of the Venturi tube, respectively. Assume the air density remains constant and then $r_i^2 v_a = r_t^2 v_t$. Replacing the ratio, v_a^2/v_t^2 of Eq. (86) by r_t^4/r_i^4 gives

$$\left(\frac{r_t}{r_i}\right)^4 = 0.15 \quad \dots \dots \dots \dots \dots \dots \dots \dots \dots (87)$$

In Figure 7-33, r_t and r_i are 1.2 and 2.0 cm, respectively. A flight test at a height of 4 km indicated that a temperature rise of 0.7C was obtained when the airspeed was increased from 140 to 220 kts. This shows that the Venturi tube thermometer does not offer complete compensation.

In place of a Venturi tube thermometer, Vonnegut[65] used a somewhat similar temperature compensation principle by constructing a vortex thermometer. The air that is forced by the aircraft speed into the chamber of a vortex thermometer whirls down a circular vortex, resulting in a reduction of pressure and, hence, adiabatic cooling.

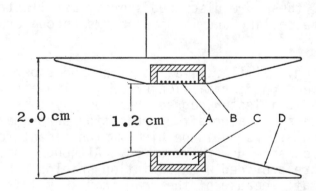

A, wire resistance element; B, ebonite; C, glass-wool; D, steel.

Figure 7-33 Venturi Aircraft Thermometer

[65] Vonnegut, B. 1950. A vortex thermometer for measuring true air temperature and true air speeds in flight. Rev. Sci. Instrum. 21: T36.

As shown in Figure 7-34a, the airstream enters a hollow cylinder tangentially and produces vortices along its axis where a thermistor is located. Thus a reduction in pressure is accompanied by adiabatic cooling. The amount of adiabatic cooling can be regulated by an inlet valve which controls the volume of air intake. From numerous flight tests, it has been found that the vortex thermometer showed satisfactory performance in clear air up to 8 km, over a speed range from 120 to 180 kts. It also has been found that the thermometer is useful in measuring cloud temperatures since the whirling action is strong enough to prevent cloud liquid or solid particles from depositing on the sensor. With the presence of supercooled cloud droplets, however, there is a possibility of ice or frost forming at the inlet of the cylinder which would hinder the passage of the airstream. Should such a condition exist, the usefulness of the vortex thermometer is greatly reduced. An improved version of Vonnegut's vortex thermometer was developed by the introduction of an axial inlet as shown in Figure 7-34b. There the axis of the thermometer is parallel to, instead of tangent to, the direction of flight. The vortex of airflow is produced by a spiral vane located inside the hollow cylinder.

The vortex thermometer (AN/AMQ-13), manufactured by the Bendix Aviation Corporation, is mounted outside the aircraft in a position at which the airflow is not greatly disturbed by the motion of the aircraft. It consists of a cylindrical radiation shield containing a fixed spiral vane and a resistance wire thermometer. Similar to all vortex thermometers, the air flows axially into the housing and is forced by the vane to form a vortex around the sensor. The cooling effect at the center of the vortex compensates for dynamic heating in the airspeed range of 130 to 400 kts. The manufacturer's performance specifications indicate that (i) the probe is accurate within 0.5C with a response time of about 10 s; (ii) rain, fog, and snow introduce no detectable errors; and (iii) the effect of icing conditions is negligible until 3/16 in. of ice has built up on the probe.

Several other types of aircraft thermometers are noted: the flush bulb, the reverse flow, the total head, and the stagnation types. Since these are not based on the principle of adiabatic cooling for the compensation of dynamic heating, only two of these are discussed shortly as illustrations in conjunction with evaporative cooling and icing problems encountered in aircraft measurements.

(b) _Radiative Heating_. Similar to all surface and upper-air (e.g., radiosonde) temperature probes mentioned, radiation error is also a problem for aircraft thermometers, though to a lesser degree than most systems. Ventilation is always adequate in high speed aircraft so radiation errors are minimal. On the other hand, radiation errors could be highly significant for a slow flying aircraft, where radiation shields are required. Although most lower speed aircraft thermometers are provided with radiation shields, they are not suitable for high speed aircraft because of the drag that they introduce. In this case, a highly polished surface or the application of a radiation coating to the sensor is advisable, instead of a cumbersome radiation shield. The conical head thermometer of the British Air Ministry, for example, uses this technique. In-flight tests of their thermometers at an airspeed of 500 kts resulted in errors of 0.2C at the 900 mb level and 1.0C at the 100 mb level due to radiative heating. As to radiative cooling at night, its effect is negligible and can be ignored.

343

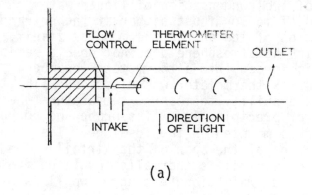

(a)

(b)

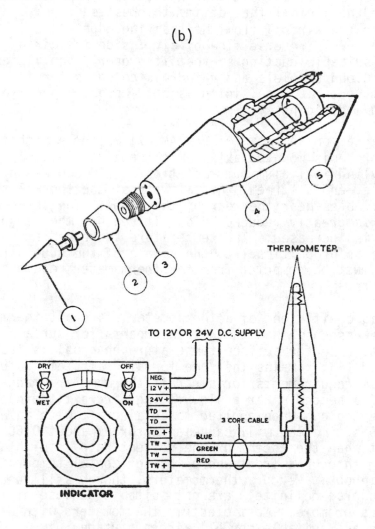

Figure 7-34 Vortex Aircraft Thermometer

(c) <u>Evaporative Cooling and Icing</u>. Both evaporative cooling and icing result from the presence of water substances in the atmosphere. While evaporative cooling is a problem of the condensation of liquid water on a thermal sensor, icing is the formation of ice or frost either at the entrance of the thermometer housing or the deposition of ice on the thermometer itself. The dry snow and liquid water content in the atmosphere does not pose a serious problem when a high speed airstream is forced into a thermometer chamber. This was explained previously when the vortex thermometer was discussed. The Cook reverse-flow thermometer also has the ability to measure temperature when the aircraft is flying through clouds or precipitation. Its probe and supporting strut are airfoil-shaped structures mounted on the fuselage. The supporting strut comprises about three-fourths of the span of the airfoil. The outer temperature-measuring portion of the airfoil is thermally insulated from the strut portion and chrome plated to reduce radiation heating. Two lines of slots, cut in both the top and bottom surfaces of the sensor in its thickest section, extend along the span of the airfoil. A third line of slots is cut in the trailing edge of the airfoil. When air flows past the probe, the air pressure is greater at the trailing edge of the airfoil than at its thickest section, and an air current is created inside the airfoil from the trailing edge slots to the slots at the thickest section. This reverse flow of air inside the airfoil is converged and and directed by suitable ducting so that it properly ventilates the sensor contained inside. Cloud droplets and raindrops do not enter the trailing edge because of their momentum. Also, icing is not a problem because this instrument is equipped with a de-icing system.

The Kollsman stagnation thermometer (AN/AMQ-7) measures both temperature and humidity. The probe is mounted axially in the airstream. It consists of a double-walled cylindrical aluminum radiation shield containing de-icing heater, an airflow baffle, and a screen. Air flows past the thermal sensor, and out through 12 small holes near the rear of the probe. The speed of air relative to the aircraft is greatly reduced before it reaches the sensor but the pressure of the air is increased. The temperature increase caused by dynamic heating in the probe is proportional to the square of the true airspeed. The true air temperature must be computed from the thermometer readings and the true airspeed measurements.

Regarding the lag coefficient of a thermometer, aircraft thermometers require a much faster time response than their counterparts for surface measurement. The step function response for surface temperature changes, as described in Chapter 4 by Eqs. (3) and (4), remains the same for aircraft temperature measurement. The time response requirements for aircraft thermometers range from a few seconds to a fraction of a second. For a slow-flying aircraft, perhaps a time lag of 10 s may be acceptable. When applied to an aircraft with an ascent rate of 6 m s^{-1}, this 10 s time lag will cause an error of about 0.6C. Errors become more significant when the ascent or the descent rate is high and when the aircraft is passing through a front where the horizontal temperature gradient density is discontinuous. Mercury thermometers, though still used as aircraft psychrometer sensors, definitely are <u>not</u> suitable because of their large time lag of about 30 s or more. Most electric thermometers with a lag coefficient of less than 2 s are suitable for all aircraft usage.

(4) <u>Humidity Measurement</u>. Aircraft humidity measurements become a very complicated matter. The complexity of surface humidity measurement was described in

Section 6.6. For aircraft, a wider operating range, a greater accuracy, and a faster time response are required. At high altitudes, the relative humidity may be as low as 1% to 2%, whereas near the ground, it is rarely lower than 30%. In order to measure the small amount of water vapor that may occur at high altitudes, a large volume of air must be sampled. As an aircraft penetrates into a super-saturated cloud deck and reaches the sunny (warm and dry) inversion layer above the cloud, it measures a wide range of humidities and temperatures. This abrupt change can occur within minutes. Since no single humidity instrument meets the requirements, several instruments may need to be installed in the aircraft. For example, one instrument can be employed for measuring humidity at temperatures of -30C and above while the other, for temperatures of -60C and below.

In the following sections, the mixing ratio indicator (MRI), the dewpoint hygro-meter of the Bendix Corporation, the vortex dewpoint hygrometer, and the heated air psychrometer of NRL[66] are discussed.

(a) <u>The Mixing Ratio Indicator</u>. As shown in Figure 7-35, the MRI Model 901 mixing ratio indicator is a ventilated, battery operated instrument utilizing the principle of an electrolyzing water vapor sensor cell.[67] In this case, phosphorous pentoxide is used as the moisture sensor. The electrolytic cell consists of two platinum electrodes wound as a tight double helix on the in-side of a small glass tube. At a flow rate of 4.5 cc min^{-1} of STP air through the moisture cell, 1 ppm of water vapor (a mixing ratio of 0.001 g kg^{-1}) cor-responds to a current of 1 μA. The output signal can then be transmitted and registered. The cell calibration depends on the fact that 93.5 μg of water ab-sorbed by the cell corresponds to one coulomb of charge passed by the cell. A simplification of the chemical process of this system can be expressed as

$$P_2O_5 + xH_2O \xrightarrow{\text{Hydration Process}} P_2O_5 \cdot xH_2O$$

[10% solution in acetone]

$$P_2O_5 \cdot xH_2O \xrightarrow{\text{Current}} P_2O_5 + H_2O$$

[Hydrate] [Anhydride] [Vapor]

---

[66] The following three references all come from Wexler, A., ed. 1965. Humidity and Moisture. Reinhold Publishing Corp., New York.
  (1) "Mixing Ratio Indicator" by Paul B. MacCready, Jr. and John A. K. Lake, pp. 512-521.
  (2) "Design and Applications of High-Performance Dewpoint Hygrometers" by Louis C. Paine and H. R. Farrah, pp. 174-188.
  (3) "Hygrometer Developments at the U.S. Naval Research Laboratory" by R. E. Ruskin, pp. 643-650.

[67] Keidel, F.A. 1959. Determination of water by direct amperometric measure-ment. J. Anal. Chem. 31: 2043-2048.

$$P_2O_5 + H_2O \xrightarrow{\text{Hydration}} P_2O_5 \cdot xH_2O \xrightarrow{\text{Electrolysis}}$$

[Anhydride]   [from atmosphere]   [Hydrate]

$$P_2O_5 + H_2 + 1/2\ O_2 + \text{Current}$$

[Output signal]

The process is continuous since all the moisture trapped is immediately electrolyzed and the drying agent is always kept dry.

As shown in Figure 7-35, the mass flow sensor monitors the mass flow rate and, through the flow servo, causes the pump to keep the mass flow constant in spite of variations in ambient conditions. The temperature servo controls the temperature to maintain the calibration of the mass flow sensor. Several flight tests were conducted during the prototypic stage of development of this indicator. It was found that the indicator can measure a wide range of mixing ratios under a broad range of ambient temperature and pressure conditions ranging from the surface to the stratosphere. The mixing ratio measured ranges from 0.03 to 30 g kg^{-1} with a time lag as small as 1 s. Its fast response, high sensitivity, and high accuracy qualify this indicator to measure the ambient humidity with the aircraft in flight.

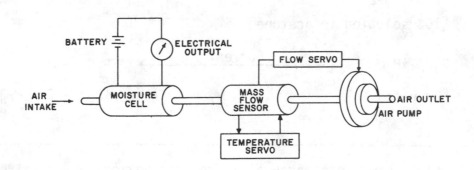

Figure 7-35 Mixing Ratio Indicator

(b) <u>The Dewpoint Hygrometer</u>. The principles and mechanisms of the Bendix and Cambridge dewpoint hygrometers were described in detail in Section 6.6.4. When applied to aircraft usage, some modifications which have to be made include: (i) the air sample pressure control; (ii) protection against vibration and other unusual flight conditions; and (iii) automatic maintenance of the instrument system. Only the Bendix aircraft hygrometer system is described here.

As shown in Figure 7-36, the ambient air sample regulated by the pressure transducer and the pressure regulator valve is controlled by the pressure ratio servo whereby a constant ratio of mirror chamber pressure (PM) to the static pressure (PS) is maintained at PM/PS = 1.7. Thus, this air pressure control system establishes the mirror chamber pressure at a fixed ratio above the static pressure over the normal flight profile of the aircraft. The partial pressure of the water vapor is changed according to the same ratio. While the pressure regulator valve keeps the excess air from entering into the sampling chamber, the pressure transducer with the pressure ratio servo system adds a fixed amount of air into the mirror chamber. To minimize shock and vibration on the components of the aircraft hygrometer, a completely transistorized system is used. In addition, the mirror heater protection together with the pressure limiter are used to protect the hygrometer against unusual operating conditions. The reference junction temperature of the thermopile regulated by boiling liquid oxygen is controlled by a pressure relief valve. An automatic mirror cleaning and photocell balancing system are provided. A provision is also made for a digitized dewpoint computer and its output is corrected to static pressure. The measuring range for both dewpoint and frostpoint temperatures is between +30C and -75C. As an example of its quick recovery time, in a flight test of the Bendix dewpoint hygrometer the frost reformed on the mirror within 10 s after the mirror had been completely cleared by heating to +70C.

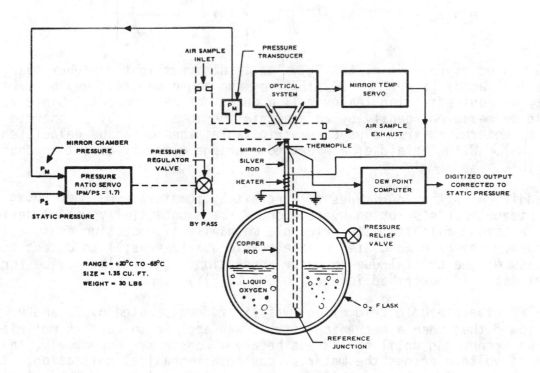

Figure 7-36 Aircraft Hygrometer

Two types of humidity instruments have been developed by the US Naval Research Laboratory (NRL) for aircraft usage: the vortex dewpoint hygrometer and the heated air psychrometer.

(c) The Vortex Dewpoint Hygrometer. The vortex dewpoint hygrometer consists of a 1/2 hp air compressor with 35 psi inlet pressure, a vortex cooling tube, a heat exchanger, a small metallic mirror, and a servo valve. The compressor pumps the sample air into a vortex cooling tube through a small heat exchanger which removes some of the heat of compression. As the air enters into the vortex tube tangentially, it expands and cools to a temperature of 45C below the inlet ambient air temperature. While the hot air produced by compression is discharged from one end of the vortex tube, the cold air passes through a small hole at the other end. The stem of a small metallic mirror is inserted into the cold air hole to facilitate the formation of dew on the mirror. The presence of dew is detected by balanced photodetectors. The servo valve controls simultaneously both the air pressure at the inlet of, and the hot air discharge from, the vortex tube. A fine resistance wire thermometer or a thermocouple is used for measuring the dewpoint temperature of the mirror at the instant when a film of dew is formed.

(d) The Heated Air Psychrometer. This device is designed in accordance with the Ferrel Equation (see Section 6.6.2, Eq. (38) for the explanations of symbols)

$$e = e_w - 0.000367 \, P(T - T_w) \left[1 + \left(\frac{T_w - 32}{1571} \right) \right] \quad \ldots \ldots \ldots \text{(88)}$$

In a heated psychrometer, the T_w value is always slightly higher than 32F, thus the factor $[1 + (T_w - 32)/1571]$ of the above equation may be taken as unity without affecting the overall accuracy of measurement. Since T_w and T could be measured readily by an electric thermometer, P by an aneroid capsule, and e_w determined from a psychrometric table, then e can be calculated by Eq. (88). With the aid of a Wheatstone bridge circuit, direct readout of e is possible.

In addition to the four types of aircraft hygrometers mentioned above, the piezoelectric sorption hygrometer may be a potentially useful device for high altitude flights. Its principle mechanism of operation is to utilize the frequency change of a coated piezoelectric quartz crystal to detect the change in mass of the crystal due to water absorption. (A similar device for ground level usage is described in Section 6.4.2.(5).)

In 1880, the piezoelectric effect was first demonstrated by J. and P. Curie, who found that when a mechanical stress was applied to certain materials a voltage was generated until the stress became a constant. Conversely, the application of voltage across the material caused a mechanical deflection. It was also found that an AC voltage applied across the material caused the material to vibrate mechanically at the frequency of the applied voltage. Again, the converse is true, i.e., an AC voltage is developed when the material is made to vibrate mechanically.

The piezoelectric effect does not occur in a material having a center of symmetry. It occurs in 20 out of the 32 crystal classes. Two of the most commonly used are quartz and Rochelle salt crystals. Tourmaline is another commonly used material.

Piezoelectric crystals are used in many applications. The crystal microphone and crystal phonograph pick-up are examples, usually using Rochelle salt. Quartz crystals are commonly used to control the frequencies of radio transmitters and receivers, and as frequency selective filters in amplifiers. From these applications, it can be seen that crystals can be used as tuned resonant circuits; that is, they can be made to resonate at a desired frequency.

Electrical characteristics of quartz crystals are determined by several factors. The resonant frequency is an inverse function of the crystal thickness. Temperature stability is dependent on the orientation of the crystal slice. The AT cut[68] is practically insensitive to temperature fluctuations.

Examination of many crystals of the AT cut show that a linear relationship exists between a mass added to the crystal surface and a frequency change. That is, the resonant frequency of the crystal is a direct and linear function of the mass added to the crystal surface. Let Δf be the frequency change in Hertz and Δm, the total mass change in g, then the constant of proportionality, k, would be a function of the material, cut, and size of the crystal. Thus

$$\Delta f = k \Delta m \quad \dots \dots \dots \dots \dots \dots \dots \dots \dots \dots \dots \dots (89)$$

Therefore, by coating the crystal with a hygroscopic material, the crystal can be used as a humidity detector. Placed in an oscillator circuit, such as that shown in Figure 7-37, the frequency of oscillation will be a function of the mass of the crystal. The crystal is provided with gas conduits to hold the water vapor to be analyzed.

The signal from the detector oscillator is a radio frequency which can be readily measured in a number of ways. The most rapid and precise method utilizes a digital frequency counter. Accuracies of ±0.01 Hz can be obtained by this method. Other measurement methods include heterodyning a reference oscillator with the detector oscillator to develop an audio frequency which is then read out on an analog audio frequency meter.

The relationship existing between the absorption of the hygroscopic coating and the detector signal can be expressed as follows. Let Δf_0 be the frequency change due to the dry coating; Δm_0, the mass change of the dry coating; Δf, the frequency change due to moisture absorbed; and Δm, the mass increase due to moisture absorption. Substituting the above expressions into Eq. (89)

$$\Delta f_0 = k \Delta m_0 \quad \dots \dots \dots \dots \dots \dots \dots \dots \dots \dots \dots \dots (90)$$

---

[68] There are four different ways to cut a crystal. Each cut has different piezoelectric properties. The AT cut provides the greatest temperature stability. Other cuts are called X cut, Y cut, and XY cut.

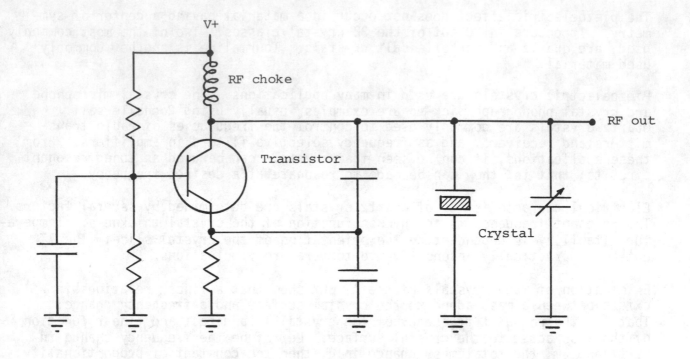

Figure 7-37   Crystal-controlled Oscillator

Dividing Eq. (89) by Eq. (90) and rearranging

$$\Delta f = \frac{\Delta f_0}{m_0} \Delta m \quad \cdots \cdots \cdots \cdots \cdots \cdots \cdots \cdots \cdots \quad (91)$$

in which the constant of proportionality, $\Delta f_0/\Delta m_0$, is a function of the types of hygroscopic coating. This, in turn, determines the sensitivity and range of the measurement.

A number of hygroscopic materials are used as absorbents on the crystals. Molecular sieves made of zeolite-type material are the most sensitive moisture detectors in the low partial pressure range. Polyethylene glycol and other similar polar liquids make rapid linear detectors. Since these are not as permanent or selective as solid absorbents, a compromise on linearity, operating range, and sensitivity by using hygroscopic polymers and natural resins is achieved.

Hygroscopic polymers enable a single detector to cover a wide range of humidities. On one unit that was tested, water concentration of 0.1 ppm gave a signal change of 0.5 Hz; 100 ppm gave a signal change of 310 Hz; and 30,000 ppm gave a signal change of 3900 Hz. The noise level was about 0.5 Hz, which corresponds to a minimum detectable amount of 0.1 ppm.

Response time, although better than that of many other hygrometers, is not very fast. The above-mentioned unit requires 1.5 min to measure a 98 per cent change. The main disadvantage is the amount of time required to dry out the crystal after making a measurement.

Therefore, a piezoelectric sorption hygrometer is suitable for high altitude flights with respect to its accuracy and range, but not to its response time. More research is needed to reduce the time requirement for the drying out process.

Another type of aircraft hygrometer is the Dunmore lithium chloride electrolytic device. As a hygroscopic substance, a saturated lithium chloride salt solution absorbs atmospheric moisture upon exposure to moist air. When the atmospheric vapor pressure is higher than that of the salt solution, an equilibrium in vapor pressure can be established by raising the temperature of the salt solution. This temperature then determines the atmospheric vapor pressure at no expense to the atmospheric moisture. Dunmore at the US National Bureau of Standards developed the first lithium chloride radiosonde hygrometer in 1938.[69] Until 1963, when the carbon strip was adopted by the US Weather Bureau, the lithium chloride flat-strip was the only humidity sensor for all US radiosondes. Even with the adoption of the carbon strip, lithium is still in use for both surface and some other types of air measurements. For a detailed description of lithium chloride hygrometers, the reader may refer to any text on meteorological instruments, particularly Humidity and Moisture, edited by Wexler.[70]

So far we have discussed aerological soundings with the airplane as the platform. When helicopters are employed, different operating requirements must be reconciled for mesoscale measurements. Helicopters usually operate at low altitudes up to a few thousand feet and fly at speeds between 100 and 200 mph. The rotor blades of a helicopter produce considerable turbulence due to strong downwash, in addition to lifting of dust upon landing. The vibrations of a helicopter are generally greater than those of a regular airplane.

One method of instrument exposure is to install a frame outside the copter, several feet away from the window of the pilot. When using this method, protective shields against radiation, turbulence, rain, and dust can be provided for the sensors. Another method is to suspend an aspirated housing 40 ft below the helicopter. As shown in Figure 7-38, this housing is called a suspended sonde, in which sensors are installed.

Figure 7-38 Suspended Sonde

[69] Dunmore, F.W. 1938. An electric hygrometer and its application to radio meteorology. J. Res. Natl. Bur. Std. 20: 723.

[70] Vol. 1, Section III, "Electric Hygrometry." See Footnote 66.

It is controlled by a cable reel and a balancing device. Signals which are transmitted by a data cable are usually recorded by a digital recorder.

High speed and high altitude aircraft, such as the U-2 and supersonic airplanes,
have operational requirements similar to those of the airplane. Exceptions are
the exposure of sensors to high ultraviolet levels and dynamic heating. Once
the high altitude aircraft climbs above thunderstorms or cumulonimbus clouds,
the liquid water content problems become insignificant.

The glider, an unpowered sailplane, has been used as a vehicle for atmospheric
measurements. The high aspect ratio of the glider wings results in a very efficient lifting surface. As a result, once towed aloft and released, the craft
either slowly loses altitude or efficiently gains lift in updrafts.

Instrumented gliders have been used for measurements of lee wave motions and
cumulus cloud, thunderstorm, and hailstorm structure and behavior, to name a few.
Instrument packages carried by the craft are limited by the size and space capacity, since excessive weight results in decreased lift and gliding efficiency
of the glider. Typical instrumentation carried by a glider used for cloud
studies (such as a two place Schweizer sailplane) might include a pressure altitude transducer, reverse flow temperature probe, liquid water content probe (such
as a hot-wire device), particle measuring probe (such as a forward scattering
spectrometer), and cloud particle camera (for in situ photography of particles).
Other instrumentation, described previously in the powered aircraft section, is
used depending on the particular application. Radio telemetry of data to the
ground is one method employed in data transmission, thereby eliminating the
weight of onboard chart recorders and other data logging devices.

The slow forward speed of the glider generates considerably less turbulence and
dynamic heating compared to powered aircraft and helicopters. Some of the instrument problems associated with powered flight are minimized in the glider. However, such things as rapid changes in altitude and bank angles introduce errors
in measurements such as altitude, pressure, atmospheric humidity, and liquid
water content, particularly when the glider moves rapidly in and out of clouds.

7.3.3 Aircraft Instruments for Research Studies. Significant advancements in
using aircraft for probing the atmosphere and its boundaries have been achieved
in a variety of fields of investigation during the last decade. Two major types
of techniques are noted: in situ and remote-sensing.

(1) In Situ Aircraft Instruments. Since no single aircraft can fulfill all requirements, different types of aircraft are utilized for special research needs.
The Beechcraft Queen Air of the National Center for Atmospheric Research (NCAR)
in the USA is small enough for low-level flights in populated or agricultural
areas. But it has insufficient electrical power, size, or weight-carrying capacity to accomodate all the desirable instruments and equipment. Also, it cannot
be employed for long range or long duration flights. The Lockheed Electra of
NCAR is large enough to house a complete array of instrumentation. In addition,
its nose boom is massive enough to accomodate an Inertial Navigation System (INS),
which can then be rigidly coupled to the air velocity sensors at the tip of the
nose boom (see Figure 7-39). For thunderstorm studies, neither the Queen Air
nor the Electra are suitable. Examples of aircraft that can be used for this

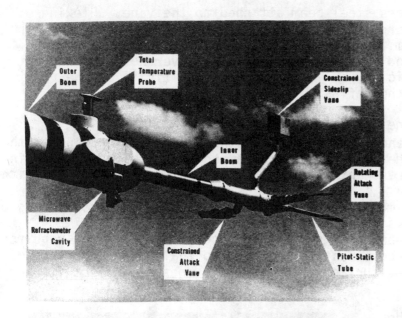

Figure 7-39 Aircraft Nose Boom

purpose are the T-28 operated by the South Dakota School of Mines & Technology and the F-101 operated by Colorado State University.

Much effort is being devoted to three-dimensional wind measurements, since the windfield cannot be measured with the desired accuracy utilizing classical approaches. Therefore, it receives foremost coverage in this section. Another very important topic is updated research in improving atmospheric humidity measurements. Finally, this section is concluded with brief descriptions of cloud physics, atmospheric electricity, and aerosol measurements.

(a) Research in Wind Measurements. As mentioned earlier, wind speed is determined from two large vector quantities, the ground speed vector and the airspeed vector. Since typical aircraft fly at speeds about one order of magnitude faster than the wind speed, and since the natural wind is always fluctuating and three-dimensional, great precision is required for the measurements of the two incidence angles of an airstream relative to the aircraft (to an accuracy of 0.001 rad). This degree of precision applies to attitude angle accuracy as well.

One research study by NCAR developed the fixed-vane sensor system. As shown in Figure 7-40, this system consists of a pair of fixed-vane sensors, A (sideslip) and B (attack), a rotating attack vane C, and a pitot-static tube. The two fixed-vane sensors mounted on the nose boom are used to measure the angle of attack and the angle of sideslip, respectively.[71] These two sensors have a resolution of 0.01 deg. For example, at an aircraft speed of 100 m s⁻¹, a lateral or vertical gust of 0.1 m s⁻¹ is equivalent to an airflow angle variation of 0.6 deg. The rotating vane aligns itself with the airstream and a differential manometer senses the pressure difference across a symmetrical set of ports at varying flow angles. The pitot-static tube measures the longitudinal component of the air velocity.

The principle of measurement of the fixed-vane sensors is based upon the change of electric resistance in each of the fan strain gages, which are mounted on

---

[71] The angle of the airstream with respect to the aircraft in the vertical plane of the aircraft is the angle of attack, and that in the horizontal plane is the angle of sideslip. While one fixed-vane sensor measures the incidence angle of attack, the other measures sideslip.

the opposite sides of a stainless steel beam attached to the vane. Flexing of the beam as a result of wind force is responsible for resistance changes in the strain gages. The sensing beam is inside a polyester-fiberglass vane housing. A stainless steel front edge and side fins for the vane cover not only prevent erosion by sand and water particles, but increase the sensitivity of the vane by about 30% over the cover without fins.

The advantages of the fixed vane for use on aircraft are:

(i) It has a higher frequency response than either the rotating vane or the differential manometer. The sensing beam is damped with a silicon fluid in order to reduce the natural resonant frequency. The fluid changes little with temperature.

(ii) Its structure has no moving parts and no water accumulation. It has been reported that a de-icing system is in the making.

(iii) Its mass is counter-balanced so that an accelerometer is not required.

(iv) It has a temperature-compensating resistor so that the large range of temperature changes between the surfaces and upper air does not affect its accuracy.

Because of these advantages, the fixed vane is applicable to studies of clear air turbulence and cloud physics.

(b) <u>Research in Humidity Measurements</u>. One promising new humidity device is the radiation attenuation hygrometer. It is based on the absorption of Lyman-alpha radiation by water vapor present in the air.[72]

Collimated monochromatic radiation is emitted by a DC-excited cold-cathode glow, discharge tube filled with hydrogen gas and a uranium-uranium hydride mixture (U and UH_3). The latter mixture is added to replenish hydrogen lost to the walls and electrodes of the sensing head; it permits the sources of Lyman-

Figure 7-40 Aircraft Gust Probe

[72] Buck, Arden L. 1976. The variable-path Lyman-alpha hygrometer and its operating characteristics. Bull. Amer. Meteor. Soc. 57(9): 1113-1118; _____. 1978. Note on the fabrication of a fixed-path Lyman-alpha hygrometer. NCAR RSF #040-032-004. Boulder, Colo., 27 p.; _____. 1971. Models for absorption hygrometer. Proc. of the Techn. Program, Electro-Optical Systems Design Conference-1971 West. Chicago Industrial and Sci. Conf. Managemt, Inc., Chicago.; Tillman, J.E. 1965. Water vapor measurements utilizing the absorption of vacuum ultraviolet and infrared radiation, In: Humidity and Moisture Measurement and Control in Science and Industry, Vol. 1, Principles and Methods of Measuring Humidity in Gases. Reinhold, New York.

alpha radiation to be operated at very low pressures (0.0013 kPa with neon added at 1.3 kPa to act as a buffer and improve the firing characteristics of the source) needed for spectral purity. A heater at a temperature of 130 ±0.5C for firing surrounds the source. Magnesium fluoride windows and the ion chamber have wavelength cutoffs of 132 nm and 115 nm, respectively, which bracket the Lyman-alpha line. Figure 7-41 shows the spectral output of a conventional 1 kPa hydrogen lamp versus the 0.0013 kPa uranium hydride source. Detection of transmitted radiation is accomplished by a nitric oxide chamber. Figure 7-42 shows the control chassis which powers the other two units and contains the path length and heater control circuitry, along with the sensing head and humidity display.

The emitted radiation is attenuated as a function of the intensity of the transmitted wavelengths of radiation, I_{0i}; path length, x; absorption coefficients of absorbing gases, k_j; concentration of the absorber in the sensing region, ρ_i; and concentration of the absorber, ρ_0, at standard temperature and pressure (STP: OC and 1013 mb). If the absorbing gases are water vapor, oxygen, and ozone, and if the intensity of the transmitted radiation is known, then the received intensity provides a direct measure of absolute humidity provided extraneous spectral lines and effects of other absorbers are accounted for or eliminated. Beer's law, suitably modified for more than one absorbing material, is written as

$$I = I_{01} \exp\left[-(k_{11}\rho_1/\rho_{01} + k_{12}\rho_2/\rho_{02} + \dots)x\right]$$
$$+ I_{02} \exp\left[-(k_{21}\rho_1/\rho_{02} + k_{22}\rho_2/\rho_{02} + \dots)s\right]$$
$$+ \dots$$
$$= \sum_{i=1}^{m} I_{0i} \exp\left(-\sum_{j=1}^{n} k_{ij}\rho_j \, x/\rho_{0j}\right) \quad \dots \dots \dots \dots \quad (92)$$

where I is the total received intensity and the subscripts refer to the ith wavelength and the jth absorbing gas. If a continuum of wavelengths is involved, Eq. (92) is replaced by

$$I = \int_{\lambda_1}^{\lambda_2} I_0(\lambda)$$

$$\exp\left[-\sum_{j=1}^{n} k_j(\lambda)\rho_{0j}\right] d\lambda$$

$$\dots \dots \dots \dots \quad (93)$$

The Lyman-alpha line, an emission line in the far ultraviolet (UV) at 121.56 nm for atomic hydrogen, is strongly absorbed and attenuated in a few mm by

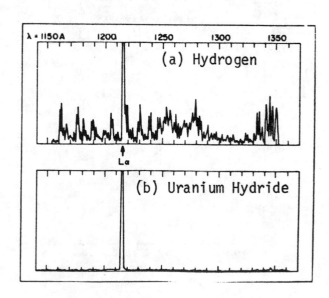

Figure 7-41 Spectral Purity

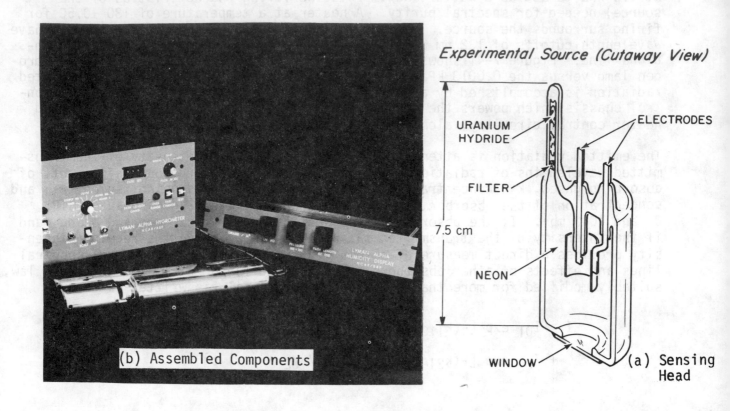

Figure 7-42   Lyman-Alpha Hygrometer

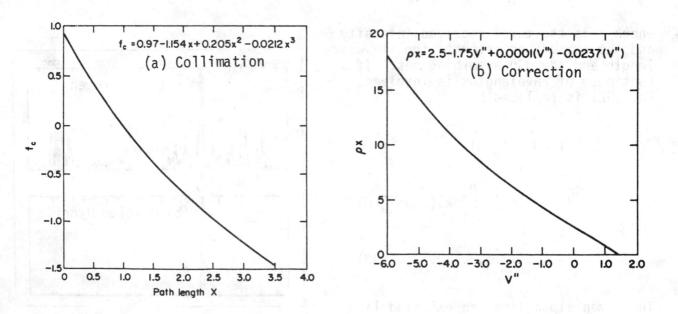

$$f_c = 0.97 - 1.154x + 0.205x^2 - 0.0212x^3$$

(a) Collimation

$$\rho x = 2.5 - 1.75V'' + 0.0001(V'') - 0.0237(V'')$$

(b) Correction

Figure 7-43   Collimation and Correction Curves

water vapor, and less so by oxygen and ozone. The three absorption coefficients at STP are

$$H_2O: \quad k_1 = 387 \text{ cm}^{-1}$$

$$O_2: \quad k_2 = 0.35 \text{ cm}^{-1} \text{ (slightly pressure sensitive)}$$

$$O_3: \quad k_3 = 640 \text{ cm}^{-1} \text{ (significant absorption in stratosphere)}$$

The sensed voltage equivalents of Eq. (93) are

$$V = \int_{\lambda_1}^{\lambda_2} V_0(\lambda) \exp\left[-\sum_{j=1}^{3} k_j(\lambda) \rho_j x / \rho_{0j}\right] dx \quad \ldots \ldots \ldots \ldots \quad (94)$$

where V_0 is the system gain, proportional to I_0, window transmission, and detector quantum efficiency. The absolute humidity is empirically determined by a simplified expression for Eq. (94). The absolute humidity, ρ, is

$$\rho x = f_V(V'') \quad \ldots \ldots \ldots \ldots \ldots \ldots \ldots \ldots \ldots \ldots \ldots \quad (95)$$

V'' is ln V corrected for the effects of V_0, nonplanar propagation, and the other absorbers. The interaction of the $\rho_j x$ terms is negligible for the normal range of water vapor concentrations.

Self-calibration of the Lyman-alpha hygrometer (see Figure 7-42) is possible if a reference dewpoint hygrometer is not available. This is accomplished by varying the path length between x_1 and x_2 and solving for ρ

$$V_1 = V_0 \exp\left[-k\rho x_1 / \rho_0\right] \ldots \ldots \ldots \ldots \ldots \ldots \ldots \ldots \quad (96)$$

$$V_2 = V_0 \exp\left[-k\rho x_2 / \rho_0\right] \ldots \ldots \ldots \ldots \ldots \ldots \ldots \ldots \quad (97)$$

Dividing V_2 by V_1 yields

$$V = V_2/V_1 = \exp\left[-k\rho(x_2 - x_1)/\rho_0\right] \quad \ldots \ldots \ldots \ldots \ldots \quad (98)$$

Output is independent of variations in system gain as reflected by changes in V_0. In practice, the more complicated forms of Eq. (95) are used, and V_1 and V_2 are measured sequentially.

Digital processing, taking into account gain changes, absorption by oxygen, and imperfect optical collimation, and data inputs for atmospheric pressure and temperature, path length, and Lyman-alpha detector output, are used to derive the corrected voltage

$$V'' = \ln V - \ln V_0 - f_c(x) - f_{21}(\rho x/T) \quad \ldots \ldots \ldots \ldots \ldots \quad (99)$$

where f_c is the collimation correction and f_{21} is the correction for oxygen absorption. Both f_c and f_{21} are empirically determined (see Figure 7-43). The data reduction algorithm follows Eq. (95).

The operating characteristics of the Lyman-alpha hygrometer are summarized, as follows:

Time response, ms	<15
Accuracy, deg C dewpoint	1
Stability (up to 500 h), deg C dewpoint	1
Resolution, deg C dewpoint	<0.03
Operating range, deg C dewpoint	-40 to +30
Noise, mV, peak-to-peak	10
Gain change, per cent h^{-1}	0.8 typical
Environmental range	
Temperature, deg C	-60 to +40
Pressure, mb	200 to 1100
Useful path length, cm	0.3 to 5
Path control accuracy, cm	±0.005
Effect of extraneous light, cosmic rays	negligible
Source lifetime, h	>2000
Detector lifetime, h	>400

Self-calibration is done a few times per day by changing the path length; the latter, however, must be known to 0.5%. Several readings are averaged to prevent errors due to humidity changes during the brief self-calibration period. If frequent baseline calibrations are made, two methods are suggested. Hourly checks are done using a fixed path hygrometer with a conventional hydrogen source. Calibrations, made at one to eight hours of on-time, are done using a fixed path hygrometer with a uranium hydride source. Window degradation is reversible by washing the glass with alcohol or acetone, followed by light polishing with a fine-grained abrasive. The hygrometer is unaffected by fog or cloud droplets (estimated error of 0.004% for measurements made in heavy fog) provided no condensation occurs on the windows. A film of condensate prevents the hygrometer from functioning. A dewpoint temperature error of <0.1C is caused by air density errors to 10% of actual value; however, a path length error of only 0.005 cm causes a dewpoint temperature error of 0.1C. Thus, Eq. (99) is highly sensitive to the accuracy of the path length measurement.

(c) <u>Research in Water Content Measurements</u>. For measurements of total water content, two research instrument systems have been developed: the PMS sensors, manufactured by the Particle Measuring System, Inc., and the TWC instrument of Meteorology Research, Inc. (MRI). The former is a remote-sensing system, whereas the latter is an <u>in situ</u> instrument. In this section, only the latter device is described.

The MRI continuous total water content instrument, TWCI,[73] was designed on the
principle that the ratio of water, R, in a water/fluid mixture is directly related to the dielectric constant of the mixture. Since the dielectric constant
is also directly related to capacitance, a resistor-capacitance (R-C) circuit
was designed so that the outputs from the reference, A, and sensor, B, capacitors controlled oscillators in such a way that the output frequencies, F_1 and
F_2, were also inversely proportional to the square root of the water ratio, R.
The fluid employed for the water/fluid mixture is butyl carbitol. The functional
relationship between the above factors is expressed empirically in the form of

$$R = A \frac{(1 - B) \Delta T}{F_1{}^2} - \frac{1}{F_2{}^2} \quad \ldots \ldots \ldots \ldots \ldots \ldots \ldots (100)$$

Here A and B are reference and sensor capacitances, respectively; ΔT, the differences in fluid temperature of the reference and sensor capacitors, respectively;
and F_1 and F_2, the frequency outputs in the above two capacitors.

Let w be the collection rate of water from the atmosphere, in volume per unit
time, and θ be the volume fluid flow rate at sensor capacitor; hence

$$w = R\theta/(1 - R) \quad \ldots \ldots \ldots \ldots \ldots \ldots \ldots \ldots (101)$$

Then the TWC is obtained by

$$TWC = w\rho/V \quad \ldots \ldots \ldots \ldots \ldots \ldots \ldots \ldots \ldots (102)$$

where ρ is the density of water from sampled hydrometeors and V is the air volume
sampling rate (i.e., inlet area x true airspeed).

Figure 7-44 shows the schematic of the components of TWCI in which butyl carbitol
is pumped from the supply tank F to the reference capacitor A. Thus, the dielectric constant of the unmixed fluid is established. This fluid is then injected in the sampling inlet C through a shower head (12 holes of 0.012 in.
diameter). At this stage, both sampled hydrometeors and the butyl carbitol are
mixed, impacting the inner walls of cyclone separator D by inertia and by the
vortex flow pattern created by the cyclone separator. This vortex directs the
mixed fluid downward and outward in a spiral pattern along the inner walls. Then
a portion of the fluid moves upward into the air exhaust E. The extension of
the exhaust is shown in the lower right of Figure 7-44. The other portion of
the fluid is settled by gravitational force and by the downward dispersion of
the vortex, moving along the heated walls into the sensor capacitor B and
finally entering the holding tank G by a pump located near the sump of the
assembly.

[73] Takeuchi, D.M., L.J. Jahnsen, and S.M. Thomas. 1978. An evaluation of the
MRI Continuous Total Water Content Instrument (TWCI), In: Fourth Symposium
on Meteorological Observations and Instrumentation. Amer. Meteorol. Soc.,
Boston. 563 p.

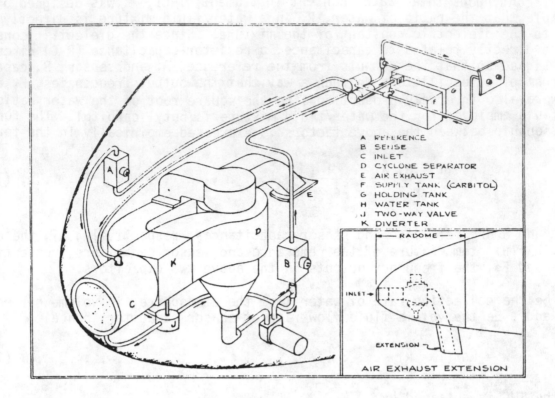

A REFERENCE
B SENSE
C INLET
D CYCLONE SEPARATOR
E AIR EXHAUST
F SUPPLY TANK (CARBITOL)
G HOLDING TANK
H WATER TANK
J TWO-WAY VALVE
K DIVERTER

Figure 7-44 Total Water Content Instrument (TWCI) Schematic

The TWCI is capable of measuring mass concentration of hydrometeors from 0.1 to 3.0 g m^{-3} in the atmosphere up to 14 km, at aircraft true airspeeds ranging between 50 to 200 m s^{-1}. The conclusions on the performance of the TWCI by Takeuchi, et al. (1978)[74]/ are cited below:

i. "In rain, TWCI results for average water content agree closely with PMS values--within 20 per cent. This result indicates a high degree of confidence in the TWCI measurements since the PMS is known to be accurate in rain.

ii. In ice conditions, possible limitations to the PMS 1-D measurements indicate that additional measurement such as radar or 2-D probes be used in the evaluation.

iii. The TWCI has an instrument delay of about 20 to 30 seconds and is related to the amount of water content being sampled.

iv. The variable delay of the instrument and the time constant of about 15 to 24 seconds indicate poor spatial resolutions in widely varying conditions of water content."

[74] See Footnote 73.

(d) Research in Cloud Physics. Although cloud physics is an important subject in upper-air studies, its measurements involve many basic techniques which are elaborated in the forthcoming chapters in Volume II. Therefore, only a brief discussion on aircraft in situ measurements in cloud physics is given here. For the sake of convenience, the NCAR's airplane, the Beechcraft Queen Air, used for environmental cloud physics programs,[75] has systems that illustrate research for ice nuclei (IN), cloud condensation nuclei (CCN), rainwater, and hail.

Two twin engine Queen Airs, 304 and 306 Delta, have been employed to assist in the following programs:

Florida Area Cumulus Experiment (FACE)	1975
National Hail Research Experiment (NHRE)	1976
The Western Piedmont Area of Virginia (Virginia)	1977

In FACE, the distribution of silver iodide released during seeding activities, as well as an overall air and water quality budget, were investigated. In the subsequent experiment program, NHRE, the spectrum of Aitken, CCN, and IN within the hail research area were measured. Finally, during the Virginia Experiment, the goal was to study the quality of air and rainwater, as well as the spectrum of CCN and IN. The basic meteorological measurements were the same for the three experiments. They are: stagnation temperature, moisture content, pressure, altitude, airspeed, magnetic heading, cloud liquid content, weather radar imagery, forward and side-looking cloud photography, drift angle, ground speed, vertical motion, and accelerations associated with updrafts.[76] The cloud physics instrumentation varied mostly with the types of experiments.

The two instruments used to collect Aitken nuclei during NHRE in 1976 were the Gardner counter and the E/1 CN counter. The former is a CN type, weighing 4.5 kg, made by Gardner Associates, Inc. This is a battery-operated electronics system with a hand-operated sample pump. It has a range of 200×10^7 nuclei cm^{-3} with an accuracy of $\pm20\%$. It counts particles from 0.01 to approximately 0.1 μm once a minute. The E/1 Counter, made by Environment/One Corporation, weighs 17.7 kg. Its linear range is from 200 to 300,000 nuclei cm^{-3}, becoming nonlinear for values up to 10^7.

Two different membrane filter samplers were used during the three programs for ice nuclei development and chemical analysis. They were the rotating sampler and the sequential membrane filter system. The former is a 10.2 cm diameter membrane filter and it rotates over a 20 deg wedge-shaped suction area of 220 mm^2 in an adjustable time period.[77] This suction provides a flow rate of 4.5

---

[75] Wisniewski, J. and G. Langer. 1978. The usage of NCAR's Queen Air for environmental cloud physics studies, In: Fourth Symposium on Meteorological Observations and Instrumentation. Amer. Meteorol. Soc, Boston. 563 p.

[76] NCAR. 1973. Atmos. Techn. This publication has details of these instruments.

[77] Dye, J.E., G. Langer, V. Toutenhoofd, T.W. Cannon, and C.A. Knight. 1976. The use of a sailplane to measure microphysical effects of silver iodide in cumulus clouds. J. Appl. Meteorol. 15(3): 264-274.

liters min^{-1}, using a 0.45 μm millipore filter at 120 mm Hg pressure drop. Twelve field monitors are phased in sequentially and automatically by a rotating valve.

A continuous ice nucleus counter was designed by NCAR, for use in all three of the programs. This counter detects the ice nuclei concentration at a fixed temperature of -20C. The counting rate maximum is 18,000 cpm. There is a 20 s delay before nuclei are detected after entering the system.

A rainwater collector scoop was used to collect rainwater for chemical and ice nuclei analyses during the FACE (1975) and Virginia (1977) programs. The scoop is made of a durable fiberglass frame but lined with polyethylene to prevent contamination. The scoop is mounted on top of the aircraft fuselage toward the front of the aircraft. In this way, it is capable of collecting between 250 to 1000 ml of rainwater on a traverse through moderate or heavy precipitation.

In addition to the NCAR Queen Air utilization as described above, other research using aircraft to study atmospheric electricity, aerosols, and turbulence has been conducted at NCAR and elsewhere. Since these subjects are elaborated upon in Volume II, the reader is referred to this source for additional information. Additional suggested references on these subjects can be found in the Fourth Symposium on Meteorological Observations and Instrumentation, 1978, of the American Meteorological Society and Atmospheric Technology, 1978-79, of NCAR.

(2) Remote-Sensing Aircraft Instruments. Three major categories of remote-sensing instruments are generally recognized: radar, radiometry, and photography.

One of NOAA's research aircraft, the WP-3D Orion (a four engine turboprop), is equipped with all of the three. Its C-band PPI radar is installed in the nose, with an areal coverage of ±120 deg in azimuth (i.e., 240 deg scan) and ±20 deg in elevation, referenced to the aircraft's longitudinal axis. Another C-band PPI, located in the lower fuselage, has an areal coverage of 360 deg in azimuth and ±10 deg in elevation. An X-band RHI radar is in the tail, covering 360 deg perpendicular to, and ±25 deg fore and aft of, the longitudinal axis of the aircraft. This WP-3D radar system features range coverage to 200 nm and accurate contoured display of precipitation to 50 nm (i.e., if 10 mm hr^{-1} precipitation is detected at 5 nm, the return will not grow in amplitude as the range to the return decreases). Thus, the moisture content and moisture gradient can be determined within 50 nm with great accuracy.

In the area of radiometry, the WP-3D aircraft utilizes a microwave radiometer to sense liquid water content, a modified Barnes PRT-5 infrared radiometer to measure sea surface temperature, and a CO_2 radiometer to measure air temperature. When the data measured by the infrared radiometers were compared with those of the Airborne Expendable Bathythermograph (AXBT)[78] for sea surface temperature, the mean difference was ±0.16C, with a standard deviation about the mean of ±0.12C. This result strengthened confidence in both remote and direct measurements.

---

[78] Black, P.C. and T. Schricker. 1979. Direct and remote sensing of ocean temperature from an aircraft, In: Fourth Symposium on Meteorological Observations and Instrumentation. Amer. Meteorol. Soc., Boston. 563 p.

In the photography field, infrared, ultraviolet, or visible light of natural origin is utilized for measuring cloud amounts, shapes, and distribution. This is another passive technique similar to radiometry.

In summary, the remote-sensing techniques, both active and passive, utilize microwave, visible, infrared, and to a lesser extent ultraviolet light spectra for the measurements of hydrometeors, including precipitation and clouds in the atmosphere. Descriptions of remote-sensing technology are elaborated in Volume II, including the use of such platforms as satellites, spacelabs, ships, and towers, as well as ground-based stations. The techniques involved are SODAR and satellite radiometry. In recent experiments, SODAR and radiometry, for example, were utilized to measure wind, temperature, and integrated water vapor. As explained by Little,[79] for wind measurements, the typical rms errors for reasonable signal-to-noise ratios and integration times were of the order 0.1 m s^{-1}. For atmospheric temperature profiles at a given height, Decker, et al.[80] found they were 1 to 2K, even under cloudy conditions. For total integrated water vapor at a given height, they were 2 g m^{-3}. It should be noted that most of these measurements were performed by ground-based stations, rather than aircraft.

7.3.4 <u>Dropsondes with Aircraft as Carriers</u>. In the past, dropsondes have been carried by aircraft, super-pressure balloons, and rockets. They functioned as radiosondes, measuring temperature, pressure, and humidity. A new dropsonde, designed by NCAR, is a windfinding dropsonde, measuring the wind field in addition to temperature, pressure, and humidity. A brief description of this NCAR dropwindsonde[81] is given.

The dropwindsonde contains sensors, batteries, electronics, and antennas. The position of the dropwindsonde is sensed by signals it receives and compares from the worldwide Omega navigation network.[82] The Omega receiver in the dropwindsonde compares the phase differences of the Omega signals, all of which transmit at 13.6 KHz but at different distances from the dropwindsonde so signal phase shifts determine the sonde's position location.

---

[79] Little, C.G. 1978. Remote sensing calibration and standards, In: Fourth Symposium on Meteorological Observations and Instrumentation. Amer. Meteorol. Soc., Boston. 563 p.

[80] Decker, M.T., E.R. Westwater, and F.O. Guirand. 1978. Experimental evaluation of ground-based micro-wave radiometric sensing of atmospheric temperature and water vapor profiles. J. Appl. Meteorol. 17(12): 1788-1795.
They used a 5-channel radiometer with three oxygen band channels, a 1.45 cm water vapor channel, and a window channel at 0.9 cm.

[81] Smalley, J.H. 1979. Aircraft dropwindsonde system. Atmos. Techn. 10 (Winter 1978-79): 24-28; Cole, H.L., S.A. Rossby, and P.K. Govind. 1973. The NCAR wind-finding dropsonde. Atmos. Technol. 2: 19-24; Govind, P.K. 1975. Dropwindsonde instrumentation for weather reconnaissance aircraft. J. Appl. Meteorol. 14(8): 1512-1520.

[82] Omega refers to the low frequency, aid-to-navigation system of eight ground stations located worldwide that provide signals for position-finding.

Sensors are carried in the sonde package which, with accompanying electronics and other equipment, weighs 2 kg. Temperature is sensed with a 95 mil bead thermistor located in the nosecone of the sonde. Airflow over the bead as the package drifts toward the surface is on the order of 5 m s^{-1}. The circuitry for the thermistor thermometer is separate from that used for the other instruments and includes a CMOS oscillator circuit which is capacitance controlled. Humidity is measured by a carbon hygristor. Its data are transmitted using a separate CMOS oscillator circuit. Pressure is measured using an improved aneroid capsule, each half of which is placed symmetrical to a ceramic plate to insure reduced drift and increased reliability over traditional aneroid capsules. The capacitance between the aneroid capsule and the fixed plate is proportional to barometric pressure. The Omega receiver is centered on the 13.6 KHz Omega frequency (VLF). Position of the dropsonde and windfinding data are based on phase measurements of Omega navigation signals received by the sonde and retransmitted to the monitoring station (usually an aircraft).

The three data oscillators and the Omega receiver have their signals multiplexed by an FM transmitter, then radioed to the launching aircraft for subsequent recording. Up to three sondes transmitting at frequencies close to 400 MHz can be received by the monitoring aircraft.

Table 7-23   Dropwindsonde Specifications

Parameter	Accuracy	Operating Range
Pressure	±2 mb	150 mb to 1050 mb
Temperature	±0.5C (0C<T<40C)	-55C to +40C
Humidity	±5% (0C$\leqslant$T$\leqslant$56C) ±8% (-20C$\leqslant$T<0C) ±13% (-40C$\leqslant$T$\leqslant$-20C)	-40C to +40C
Wind	1 m s^{-1} rms vector wind error with 4 min averaging and favorable Omega	
Size	9 cm diameter and 46 cm length	
Mass	8 kg	
Fall rate	5 m s^{-1} at sea level	

The aircraft data processor checks the sonde for proper operation and calibration prior to launch. The processor includes a built-in computer to handle the checks, and later, the data processing. After the launch of the sonde, the processor amplifies, demodulates, and demultiplexes the signals using a multiple filtering method in order to maximize data recovery. The Omega signal phase changes are determined and zonal winds computed for the sonde and for the aircraft. Frequencies derived from temperature, humidity, and pressure instruments are converted and calibration checks made against the sonde baseline calibration curve. Real time data in scientific units are provided inflight as hard copy to the scientist on board the aircraft. Magnetic tape records are also made of the data from the aircraft, Omega, and sonde (both raw and processed data). The NCAR drop-windsonde is employed in various atmospheric experiments over regions of the world's ocean where data are sparse or non-existent. Projects using the dropwind-sonde include such programs as GARP, GATE, MONEX, and the Global Weather Experiment.

7.4 Reliability and Representativeness of Aerology. Considering all the available platforms to date, balloons are still heavily used for upper-air soundings. Even in such operations as weather satellite observations, constant-level balloons and dropsondes serve as navigation aids. In addition, dropsondes are often employed for obtaining vertical profiles of the atmosphere for airplane reconnaissance. Since balloon techniques are the main source of information for upper-air synoptic analysis, and since the choice of platform is the first consideration of upper-air soundings, the reliability and representativeness of balloon soundings as related to methods of monitoring must be considered. Two major concerns are noted: locating the balloon in space and time and the performance of meteorological sensors carried by the balloon. The first problem can best be illustrated by errors in wind measurements. The second problem is conveniently represented by errors in temperature and humidity measurements.

7.4.1 Errors in Wind Measurements. Since the balloon has been used as a tracer of air movement, its trajectory, represented by various positions of the balloon in space and time, must be determined accurately. This includes the determination of the height, the horizontal and slant range, as well as the azimuth and elevation angles of the balloon in real time with the earth as the frame of reference. Methods of determining these parameters vary from the single theodolite to the tracking radar as described previously.

As shown in Figure 7-7a and Eq. (22), the horizontal range, R, of a balloon is equal to the product of its height, Z, and the cotangent of the elevation angle, ϕ, thus $R = Z \cot \phi$. The error in height, ΔZ, is caused by improper inflation of the balloon, presence of a strong vertical draft, or errors in timing. As explained previously, the constant ascent rate assumption is the main drawback for single theodolite determinations of balloon position. In addition, the error in elevation angles, $\Delta\phi$, can be caused by such weather conditions as light, variable winds throughout the flight or a strong, gusty wind at the early stage of the flight. When errors occur in elevation angle, $\Delta\phi$, alone, the resulting error in horizontal range, ΔR, is determined by first-order error analysis. Thus

$$\Delta R_\phi = (\partial R/\partial Z)_Z \, \Delta\phi \quad\dots\dots\dots\dots\dots\dots (103)$$

expresses ΔR as a function of $\Delta\phi$ alone with Z at a given level. Taking the derivative from Eq. (22) and substituting in Eq. (103) yields

$$(\partial R/\partial Z)_Z = -Z \csc^2\phi \quad\dots\dots\dots\dots\dots\dots (104)$$

and

$$\Delta R_\phi = -Z \csc^2\phi \, \Delta\phi \quad\dots\dots\dots\dots\dots\dots (105)$$

Inspection of Eq. (105) shows that ΔR increases linearly with Z, is an inverse function of ϕ, and rises very rapidly as ϕ approaches zero. For light winds, the wind direction is more or less uniform at all levels. Hence, ϕ decreases with increase of Z and this causes high-level wind measurements to be extremely inaccurate. As an extreme case, take Z = 10 km, ϕ = 5 deg, and $\Delta\phi$ = ±0.05 deg, then ΔR_ϕ = ±(10)(11.5)(0.05/57.3) = ±1.15 km.

When the error occurs in Z alone, the resulting error in R can be expressed as

$$\Delta R_Z = (\partial R/\partial Z)_\phi \Delta Z \quad\quad\quad\quad\quad\quad\quad\quad (106)$$

or from Eq. (22),

$$R_Z = \cot\phi\,\Delta Z \quad\quad\quad\quad\quad\quad\quad\quad\quad (107)$$

Thus, ΔR_Z is an inverse function of ϕ and so is again most inaccurate at low elevation angles. It should be borne in mind, however, that since the error in Z is usually due to either the density effect [see Eq. (9) and Table 7-5] or an incorrect free lift, the value of ΔZ tends to increase in magnitude with time, i.e., with increasing Z. This once again causes the greatest errors in R to occur at high altitudes. The overall error from both $\Delta\phi$ and ΔZ can be given by

$$\Delta R = [(\partial R/\partial\phi)_Z^2(\Delta\phi)^2 + (\partial R/\partial Z)_\phi^2(\Delta Z)^2]^{1/2} \quad\quad\quad (108)$$

which yields, on substitution of the derivatives

$$R = \csc\phi\,[Z^2\csc^2\phi(\Delta\phi)^2 + \cos^2\phi(\Delta Z)^2]^{1/2} \quad\quad\quad (109)$$

For an example, take Z = 3 km, ϕ = 30 deg, ΔZ = ±50 m, and $\Delta\phi$ = ±0.05 deg; then ΔR = ±87.2 m.

Since wind velocity is a vector quantity, the inaccuracy in measuring the azimuth angle, α, elevation angle, ϕ, slant range, r, horizontal range, R, and height, Z, on the vector wind error, $\Delta\vec{V}$, must be taken into consideration. The upper-air techniques, discussed so far are the single-theodolite, double-theodolite, rawinsonde, and tracking radar. Their methods of monitoring are different. The random errors of all these techniques are first examined in terms of the root mean square (rms) of the vector wind. The examples on the order of magnitude of these errors are given. It must be noted that a random error is far more important than a systematic error for determining the mean vector wind speed in a layer a few thousand feet in depth. In height measurements by either the single theodolite or tracking radar, however, systematic errors overcome the random errors. Symbols for the expressions of error equations other than those specified above are listed below:

δS and δD = random errors in wind speed, S, and direction D;

δR and δr = random errors in horizontal range, R, and slant range, r;

$\delta\phi$ and $\delta\alpha$ = random errors in elevation, ϕ, and azimuth, α;

T and t = the time to reach height, Z, and the time interval between consecutive observations, respectively;

δw = random variation in mean rate of ascent between consecutive soundings;

R_0 = the length of the baseline;

Q = the ratio of mean vector wind, $\vec{V}$, and the mean rate of ascent, $\bar{w}$;

δZ = random error in height.

The root mean square vector error in the computed wind may be expressed as

$$\delta\vec{V} = [(\delta s)^2 + \vec{V}^2(\delta D)^2]^{1/2} \quad\ldots\ldots\ldots\ldots\ldots (110)$$

When the error in height is taken into consideration, the random error in the mean height of a layer is $\delta Z/\sqrt{2}$ and the resulting error may be expressed as

$$\delta\vec{V}_Z = [1/2(\delta\vec{V}/\delta Z)^2 \, (\delta Z)^2]^{1/2} \quad\ldots\ldots\ldots\ldots\ldots (111)$$

Since $R = r\cos\phi$ and $\cot\phi = (r^2 - Z^2)^{1/2}/Z = \vec{V}/\bar{w} \equiv Q$, then from all the above formulae, the mean square error in the vector wind and the mean random error in height may be formulated as follows for:

A. Single Theodolite with Z Derived from the Assumed Rate of Ascent.

$$(\delta\vec{V})^2 = (\delta w)^2 Q^2 + \frac{2w^2T^2}{t^2}[(\delta\phi)^2(Q^2+1)^2 + (\delta\alpha)^2 Q^2] \quad\ldots\ldots (112)$$

$$(\delta z)^2 = (\delta w)^2 t^2/2 \quad\ldots\ldots\ldots\ldots\ldots\ldots\ldots (113)$$

B. Double Theodolite or Radiosonde Measuring α or θ, and α_1, with Respect to Baseline.

$$(\delta\vec{V})^2 = \frac{2(\delta\alpha)^2 (R_0^2 - 2R_0ZQ\cos\alpha_1 + 2Z^2Q^2)(R_0^2 - 2R_0ZQ\cos\alpha_1 + Z^2Q^2)}{R_0^2 t^2\sin^2\alpha_1}$$

$$\ldots\ldots (114)$$

$$(\delta z)^2 = \left[\frac{Z^4Q^4 + (R_0^2 + Z^2Q^2)^2}{R_0^2Q^2}\right](\delta\alpha)^2 + \frac{Z^2(Q^2+1)^2 (\delta\theta)^2}{Q^2} \quad\ldots (115)$$

C. <u>Tracking Radar Measuring r, α, and θ.</u>

$$(\delta\vec{V})^2 = \frac{2}{t^2}\left[\frac{(\delta r)^2\ Q^2}{Q^2+1} + (\delta\theta)^2\ Z^2 + (\delta\alpha)^2\ Z^2Q^2\right] \quad \ldots \ldots \ldots \quad (116)$$

$$(\delta Z)^2 = \frac{(\delta r)^2}{Q^2+1} + (\delta\theta)^2\ Z^2Q^2 \quad \ldots \ldots \ldots \ldots \ldots \quad (117)$$

Table 7-24 and 7-25 are extracted from the British Meteorological Handbook[83/] illustrating the relative accuracy of the above techniques. The idealized values for all the measuring components are:

$\delta r = 20$ m　　　　$\delta w = 1$ m s^{-1} (2 kts)　　$R_0 = 10$ km　　$\alpha_1 = 90$ deg

$t = 1$ min　　　　　$\delta\theta = 0.1$ deg　　　　$\delta\alpha = 0.1$ deg

With these values the vector error of wind as a function of height and of the ratio, Q, of the mean wind to the rate of ascent is given in Table 7-24. Using the same values of the random observational errors, the corresponding root mean square errors in height are given in Table 7-25.

Table 7-24 Vector Error of Wind as a Function of Height and of Ratio, Q, of Mean Wind to Rate of Ascent

| | Height (km) | | | | | | | | | | | |
| | 5 | | | 10 | | | 15 | | | 20 | | |
Q	a	b	c	a	b	c	a	b	c	a	b	c
						kts						
1	1	1	2	1	2	3	2	3	3	2	5	4
2	1	2	4	2	5	6	3	13	7	4	18	9
3	1	3	7	3	11	10	4	23	13	5	40	17
4	2	5	10	3	18	16	5	40	22	6	71	28
5	2	7	14	4	28	23	6	61	32	8	110	42

a, radar
b, double theodolite or direction finder
c, optical or radio-theodolite and assumed rate of ascent

---

[83] See Footnote 11.

Table 7-25 Root Mean Square Height Error as a Function
of Height and of Ratio, Q, of Mean Wind to
Rate of Ascent

	Height (km)											
	5			10			15			20		
Q	a	b	c	a	b	c	a	b	c	a	b	c
						meters						
1	17	27	43	22	51	43	29	85	43	37	130	43
2	19	29	43	35	69	43	52	136	43	69	216	43
3	26	34	43	51	92	43	77	187	43	102	314	43
4	34	45	43	68	123	43	102	244	43	136	415	43
5	43	55	43	85	163	43	128	310	43	170	515	43

a, radar
b, double theodolite or direction finder
c, optical or radio-theodolite and assumed rate of ascent

D. Double Theodolite Positional Errors. The double-theodolite method of
balloon tracking supposedly eliminates many of the inherent errors associated
with single-theodolite angular measurements and the constant rate-of-lift assumption for pilot balloons.[84] However, geometric considerations in the double-
theodolite method introduce errors, particularly those associated with deriving
the balloon's position from the three-angle method. Schaefer and Doswell[85]
addressed this source of error from the standpoint that the quality of the angular measurements is highly dependent on the geometric design of the system, a
problem inherent to triangulation methods applied not only to double-theodolite
systems but also to stereo-photogrammetry and dual-Doppler radar systems.

The three-angle method is used to derive the balloon's position on the horizontal
plane and its elevation. As shown in Section 7.1.3 (b), the theodolite angles,
ϕ_1 and ϕ_2, determine the length of sides of a triangle constructed by 2 theodolite
positions, separated by baseline, R_0, to the projection of the balloon onto the

---

[84] See the extended discussion of the ascent rate question by: Boatman, J.F.
1974. The effect of tropospheric temperature lapse rates on the ascent rates
of pilot balloons. J. Appl. Meteorol. 13: 955-961; and the extensive replies
to the problem in J. Appl. Meteorol. for 1975 and 1976.

[85] Schaefer, J.T. and C.A. Doswell III. 1978. The inherent position errors in
double-theodolite Pibal measurements. J. Appl. Meteor. 17(6): 911-915; range
of possible height errors is also discussed in: Arnold, A. 1948. On the
accuracy of winds aloft at low altitudes. Bull. Amer. Meteorol. Soc. 29:
140-141.

horizontal plane, point C. The errors in azimuth angles, designated $\delta\theta_1$ and $\delta\theta_2$, respectively, produce positional errors in an x-y coordinate system for the horizontal projection of balloon position. This is given as

$$\delta x = x \{[(-\tan \theta_1 - \cot (\theta_1 + \theta_2)] \delta\theta_1$$

$$+ [\cot \theta_2 - \cot (\theta_1 + \theta_2)] \delta y\} \quad \ldots \ldots \ldots \ldots \quad (118)$$

$$\delta y = y \{[\cot \theta_1 - \cot (\theta_1 + \theta_2)] \delta\theta_1$$

$$+ [\cot \theta_2 - \cot (\theta_1 + \theta_2)] \delta y\} \quad \ldots \ldots \ldots \ldots \quad (119)$$

Assuming, for simplicity, that the balloon projection position is perpendicular to the baseline, L; is equidistant from the observation points so that $\delta\theta_1 = \delta\theta_2$; and that the errors are due only to precision problems of the theodolite system. Then, the geometric location of the balloon in the horizontal plane is in error by

$$\delta L = (\delta x^2 + \delta y^2)^{1/2}$$

$$= \frac{R_0 \sin \theta_2}{\sin (\theta_1 + \theta_2)} \{1 - [\cot \theta_2 - 2 \cot (\theta_1 + \theta_2)]^2\}^{1/2} \delta\theta_1 \quad (120)$$

For example, a displacement error of 10% for a baseline 1 km in length, associated with azimuth accuracies to the nearest tenth of a degree, produces an uncertainty of the horizontally projected balloon position, L, of ±100 m for balloon distances greater than 5.4 km from the baseline.

Position errors directly relate to errors in derived wind speed. Schaefer and Doswell relate the speed error to sampling errors in theodolite measurements, δ_1 and δ_2, thus: $\Delta = (\delta_1^2 + \delta_2^2)$. Again, using previously stated assumptions, a balloon released midway along a 1 km baseline and blown perpendicular to it by a 10 m s⁻¹ wind and a theodolite accuracy of 0.1 deg, generate position and speed uncertainties as follows:

Uncertainties in:	Position	Speed
3.0 min	12 m	
3.5 min		Up to 0.4 m s⁻¹
4.0 min	21 m	
...		
8.5 min	92 m	
9.0 min		2.4 m s⁻¹
9.5 min	114 m	

The magnitude of these errors increases rapidly if the angular errors rise to 1 deg. Likewise, balloon height, Z, is affected by errors in either azimuth (θ_1 and θ_2) or elevation (ϕ_1 and ϕ_2) angles. Thus, errors in the computations of balloon vertical velocity can rapidly reach unacceptable levels. Using the same assumptions previously stated, the height error, δz, in this simplest case becomes

$$\delta z = \frac{1}{2} R_o \sec \theta_1 (\tan \phi_1 \tan \theta_1 \delta\theta_1 + \sec^2\phi_1 \delta\phi_1) \quad . . . . . (121)$$

where R_o is the baseline length.

Thus, the absolute error in the double-theodolite method, either δZ or δL, is proportional to R_o. The wind speeds derived from the balloon positions in the early part of the run have smaller errors if a short baseline is used, and the reverse is true for the case of the later part of the balloon tracking.

7.4.2 <u>Errors in Radiosonde Measurements</u>. It is difficult to accurately determine radiosonde observational errors. There are two standard procedures for such determinations: the simulation and the atmospheric laboratory procedures. In the simulation chamber, radiosonde sensors are calibrated against those of the standard instruments under different chamber environments. The drawback of this procedure is that the complexity of natural conditions cannot be reproduced in the simulation chamber. In other words, the chamber produces over-simplified environmental conditions by controlling one meteorological variable at one time. For example, the interactions of humidity, wind, and radiation on temperature measurements are highly significant, and yet it is impossible to simulate the interplay of all these variables in a real time basis under chamber conditions. In the atmospheric laboratory, the use of twin (or dual) soundings is a common practice. The drawback of this procedure is the lack of standard instruments for comparison. The ideal laboratory standard instruments are usually too heavy or bulky to be carried aloft. In addition, tedious laboratory procedures cannot be performed aloft. For this reason, only the reference radiosonde, which is not a standard instrument, was adopted by WMO in 1950 and has been in use since that time.

Three kinds of errors are recognized in radiosonde observations: the systematic, random, and sonde. The first two were explained previously in Section 7.2.1. The third or the sonde errors, refers to errors which are more or less systematic in the sense that they change only gradually during a flight for a given sonde, but the amount of error is different for different sondes of the same type. This type of error is responsible for the statistical scatter of upper-air measurements in space and time.[86]

Various statistical formulations are available for the evaluation of radiosonde errors from twin soundings. For a laboratory similitude test by NOAA, the indicated temperature and pressure must not deviate from the true values as obtained from the standard instrument by more than 1.0C and ±4 mb, respectively. The root mean square (rms) of the temperature error for the ten consecutive tests must not exceed a value of 0.45C. The formula for the computation of root mean square error (rmse) is

$$rmse = \left[\frac{\Sigma(Q_i - Q_t)^2}{N} \right]^{1/2} \quad . . . . (122)$$

---

[86] The sonde error was first defined by Harrison, N.D. 1962. The Errors of the Meteorological Office Radiosonde, Mark 2B. M.O. 724 (Scientific Paper No. 15). HMSO, London: 40 p.

where Q_i and Q_t are the indicated and true values of a meteorological quantity, respectively, and N is the total number of tests.

For twin soundings, the equivalent pressure error, which is the systematic error in the height of the pressure surface between two stations with corrections for gravity, humidity, and temperature, is expressed as

$$\Delta P = \frac{P}{K\,T}\,A \quad \ldots \ldots \ldots \ldots \ldots \ldots \ldots \ldots \ldots \ldots \ldots \ldots \quad (123)$$

where P is the standard pressure level (mb); K = R/g (R being the gas constant for dry air and g, acceleration of gravity); T, air temperature (deg C); and $A = \Delta Z_1 - \Delta Z_2$ (where ΔZ_1 and ΔZ_2 are the observational errors in height above station 1 and station 2, respectively).

The formulae for the sonde error, S_1, and the random error, S_2, in the twin soundings are shown in Eqs. (124) and (125), respectively.[87]

$$S_1{}^2 = \frac{1}{2}\left[\frac{\sum\limits^{m} \overline{X}^2}{m} - \frac{4\,\sigma_2{}^2}{3\,\overline{n}}\right] \quad \ldots \ldots \ldots \ldots \ldots \ldots \ldots \ldots \quad (124)$$

$$S_2{}^2 = \frac{3}{4}\left[\frac{\dfrac{\sum\limits^{m}\sum (X - \overline{X})^2}{m}}{\sum (n - 1)}\right] \quad \ldots \ldots \ldots \ldots \ldots \ldots \ldots \quad (125)$$

where m is the number of soundings; $\overline{X}$, the arithmetic mean of X (the values of the meteorological quantity); σ_2, the standard deviation of random error; and $\overline{n} = \sum^{m} n/m$ (n being the number of observations). In his analysis of radiosonde observations in Payerne, Switzerland, Delver[88] compiled the differences in pressure readings between the British Mark 2B and the mean of all the fourteen types of radiosondes. His statistical analysis is given in Table 7-26. It is interesting to note that the differences become larger at higher pressures than at lower pressures, and larger at night than during the daytime. Also, the general positive tendency indicates that the Mark 2B usually gives a higher pressure reading than all the others.

Delver has also manipulted the statistics of the root mean square errors on temperature and pressure as shown in Table 7-27. This is an estimate of the scatter of individual radiosondes of all types together after elimination of the systematic type differences. These were based on the means of five consecutive readings at one minute intervals.

---

[87] These two equations were given by Harrison; see Footnote 86.

[88] Delver, A. 1956. A Statistical Treatment of the Data of the Second World Comparison of Radiosondes. Meteorological Institute, De Bilt, Netherlands.

Table 7-26 Differences in Pressure Readings Between Mark 2B
Sonde and the Mean of All 14 Types of Radiosondes

Pressure Level (mbs)	Pressure Difference Day (mbs)	Night	Pressure Level (mbs)	Pressure Difference Day (mbs)	Night
850	0	0	100	+3	+5
700	0	+2	70	+2	+5
500	-1	+3	50	-1	+7
300	+1	+5	30	+3	+6
200	+2	+6			

Table 7-27 Root Mean Square Errors of Pressure and Temperature
Found from the Payerne Radiosonce Comparisons

Pressure Level mbs	Root Mean Square Error Day P mbs	T °C	Night P mbs	T °C
850	4.4	0.54	4.7	0.53
300	5.1	0.76	4.7	0.70
100	5.5	1.27	4.9	0.80

The sonde and random errors of the Mark 2B radiosonde in the seven twin flights have been compiled by Harrison[89] for pressure (P), temperature (T), and humidity (U). This is given in Table 7-28 in which the random error is, in general, far smaller than the sonde error.

[89] See Footnote 86.

Table 7-28 Errors in Mark 2B Sondes During Seven
Twin Soundings

	Low Levels			200 mbs	
	P mbs	T °C	U %	P mbs	T °C
Sonde error	2	0.4	3	7	0.7
Random error	2	0.3	2	1	0.3

SUBJECT INDEX